# Cybermedia

New Approaches to Sound, Music, and Media

Series Editors: Carol Vernallis, Holly Rogers, and Lisa Perrott

Forthcoming Titles:
*Biophilia* by Nicola Dibben
*David Bowie in Music Video* by Lisa Perrott
*Popular Music, Race, and Media since 9/11* by Nabeel Zuberi
*Popular Music and Narrativity* by Alex Jeffery

Published Titles:
*Transmedia Directors* edited by Carol Vernallis, Holly Rogers, and Lisa Perrott
*Dangerous Mediations* by Áine Mangaoang
*Resonant Matter* by Lutz Koepnick

# Cybermedia

## Explorations in Science, Sound, and Vision

Edited by Carol Vernallis, Holly Rogers, Selmin Kara
and Jonathan Leal

BLOOMSBURY ACADEMIC
NEW YORK • LONDON • OXFORD • NEW DELHI • SYDNEY

BLOOMSBURY ACADEMIC
Bloomsbury Publishing Inc
1385 Broadway, New York, NY 10018, USA
50 Bedford Square, London, WC1B 3DP, UK
29 Earlsfort Terrace, Dublin 2, Ireland

BLOOMSBURY, BLOOMSBURY ACADEMIC and the Diana logo are trademarks of Bloomsbury Publishing Plc

First published in the United States of America 2022

Volume Editors' Part of the Work © Carol Vernallis, Holly Rogers, Selmin Kara, and Jonathan Leal 2022

Each chapter © by the contributor

For legal purposes the Acknowledgments on p. x constitute an extension of this copyright page.

Cover design: Carol Vernallis, Simona Fitcal, Paul Fox and Hannah Ueno
Hannah Ueno created the duotone drawing.
Cover image: Alex Garland's *Ex Machina*, 2015 / Alamy.

All rights reserved. No part of this publication may be reproduced or transmitted in any form or by any means, electronic or mechanical, including photocopying, recording, or any information storage or retrieval system, without prior permission in writing from the publishers.

Bloomsbury Publishing Inc does not have any control over, or responsibility for, any third-party websites referred to or in this book. All internet addresses given in this book were correct at the time of going to press. The author and publisher regret any inconvenience caused if addresses have changed or sites have ceased to exist, but can accept no responsibility for any such changes.

A catalog record for this book is available from the Library of Congress.

ISBN: HB: 978-1-5013-5704-6
PB: 978-1-5013-5703-9
ePDF: 978-1-5013-5706-0
eBook: 978-1-5013-5705-3

Series: New Approaches to Sound, Music, and Media

Typeset by RefineCatch Limited, Bungay, Suffolk
Printed and bound in the United States of America

To find out more about our authors and books visit www.bloomsbury.com and sign up for our newsletters.

Carol Vernallis: *For Beatrice*

Holly Rogers: *For Daisy*

Jonathan Leal: *For my father, my first science teacher*

# Contents

| | |
|---|---|
| Acknowledgments | x |
| List of Editors and Contributors | xi |
| Introduction   *Jonathan Leal and Carol Vernallis* | 1 |

Part One  AI and Robotics

1  Could the AI of Our Dreams Ever Become Reality?
   *James L. McClelland*   15

2  Director Alex Garland Converses with *Cybermedia*'s Scientists and Media Scholars   *Jonathan Leal and Carol Vernallis*   31

3  (S)*Ex Machina* and the Cartesian Theater of the Absurd
   *Simon D. Levy and Charles W. Lowney*   45

4  Epiphany, Infinity and Transcendent AI   *Zachary Mason*   65

Part Two  Big Data, Sentience, and the Universe

5  A MASSIVE Swirl of Pixels: Radiohead's "Go to Sleep"
   *Steen Ledet Christiansen*   75

6  The Rise of the Machine: Body-Knowing, Neural Nets, and Emergent Freedom   *Charles W. Lowney*   97

7  The Quantum Computer: Sci-Fi Narrative's Favorite Character
   *Leonardo P. G. De Assis*   129

8  Composer Ben Salisbury Discusses Scoring Science for Alex Garland   *Ben Salisbury, Holly Rogers, John McGrath, Carol Vernallis, and Dale Chapman*   137

9  *Ex Machina* and the Question of Consciousness
   *Murray Shanahan*   143

Part Three  The Neuroscience of Affect and Event Perception

10  "A Solid Popularity Arc": Affective Economies in *Black Mirror*'s "Nosedive"  *Dale Chapman*  151

11  Cognitive Boundaries, "Nosedive," and *Under the Skin*: Interview with Jeffrey Zacks  *Carol Vernallis, Jonathan Leal, and Dale Chapman*  171

12  Toward an AI Future of Comics Study and Creation: A Cognitive-Affective Approach  *Frederick Luis Aldama and Laura Wagner*  187

Part Four  The Digital West

13  The Philosophy of *Westworld*  *Paul Skokowski*  207

14  New Visions of the Old West: AI, Self, and Other in *Westworld*  *Christopher Minz*  223

15  Scoring Music for *Westworld* Then and Now: A Cognitive Perspective  *Annabel J. Cohen*  237

Part Five  Interface, Desire, Collectivity

16  Director Terence Nance Discusses *Random Acts of Flyness*  *Carol Vernallis, Jonathan Leal, Holly Rogers, Elizabeth Reich and the contributors of* Cybermedia  273

17  The Gift of Black Sonics: Interface and Ontology in *Sorry to Bother You* and *Random Acts of Flyness*  *Elizabeth Reich*  283

18  Technology, Chaos, and the Nimble Subversion of *Random Acts of Flyness*  *Eric Lyon*  311

19  Expecting the Twist: How Media Navigate the Intersections Among Multiple Sources of Prior Knowledge  *Noah Fram*  329

20  Face Color  *Bevil R. Conway*  347

Part Six  Productive Neuropathologies

21  Digital Vitalism  *Marta Figlerowicz*  367

22  Neuroplasticity: From Experience to Healing   *Sara Ferrando Colomer*   389

23  Where is My Mind? *Mr. Robot* and the Digital Neuropolis
    *Patricia Pisters*   401

24  The Dopamine Circuits of Wanting, Liking, Habit and Goals:
    An Interview about *Mr. Robot* with Neuroscientist Talia Lerner
    *Jonathan Leal, Carol Vernallis and Patricia Pisters*   419

25  The Taste of Cybermedia: An Interview with Hojoon Lee,
    The Lee Lab at Northwestern University   *Julia Peres Guimarães,
    Selmin Kara, and Carol Vernallis*   431

Index   437

# Acknowledgments

Starlin Lemons, John and Polly Rogers, Daisy Rogers-McGrath, John McGrath, Raymond Sookram, Regan Bowering, Lanier Anderson, Michael Bratman, Jim Buhler, Rosa Cao, Mark Crimmins, Simona Fitcal, Sabrina Finke, Takako Fujioka, Kalanit Grill-Spector, Julia Peres Guimarães, Parker Hibbett, Jarek Kapusinski, Charles Kronengold, Anna-Sara Malmgren, Dani Oore, Randal Parker, Mark Reimers, Steven Shaviro, Alison Thrash, Hannah Ueno, Ann Vernallis, Margaret Vernallis, Jared Warren, Anthony Wagner, Sarah Wood, Dan Yamins, A24, Simona Fitcal, Beatrice Kronengold, Sabrina Finke, Helen Marshall and Paul Fox

# Editors and Contributors

## Editors

**Carol Vernallis**'s monograph *Experiencing Music Video: Aesthetics and Cultural Context* (CUP, 2004) is the first to articulate a theory of how music, lyrics and image can be placed in relation to, as well as provide detailed analyses of, individual videos. Her second, *Unruly Media: YouTube, Music Video, and the New Digital Cinema* (OUP, 2013), takes account of a new mediascape that is driven by intensified audiovisual relations. She is co-editor of *Transmedia Directors* and two Oxford Handbooks. Her *The Media Swirl: Politics, Audiovisuality, and Aesthetics* is near publication. She teaches at Stanford University, USA.

**Holly Rogers** is Reader in Music at Goldsmiths, University of London. Video installation art is one of Holly's main research areas and her monograph *Sounding the Gallery: Video and the Rise of Art-Music* (OUP, 2013) explores sound and the moving image in site-specific and immersive settings. She has also edited several books on audiovisual culture, including *Music and Sound in Documentary Film,* (Routledge, 2014), *The Music and Sound of Experimental Film* (OUP, 2017), *Transmedia Directors: Artistry, Industry and New Audiovisual Aesthetics* (Bloomsbury, 2019), *The Cambridge Companion to Music Video* (CUP, 2022) and *YouTube and Music* (Bloomsbury, 2022). Holly has also written a textbook on twentieth-century musics for Cambridge and is the founding editor of the MIT Press / Goldsmiths Press journal, *Sonic Scope: New Approaches to Audiovisual Culture*. Along with Carol Vernallis and Lisa Perrot, Holly edits this book series for Bloomsbury.

**Selmin Kara** is an associate professor of Film and New Media Studies at OCAD University in Toronto. She is the co-editor of *Contemporary Documentary* and her work has also appeared in *Post-Cinema: Theorizing 21st Century Film*, *The Oxford Handbook of Sound and Image in Digital Media*, *Screen*, *Sequence*, *Music and Sound in Nonfiction Film*, *The Philosophy of Documentary*, and *Studies in Documentary Film*.

**Jonathan Leal** is a postdoctoral researcher (and, beginning Fall 2022, Assistant Professor of English) at the University of Southern California. A native of the South Texas borderlands, he studies and creates music and narrative across sonic, visual, and textual media to unpack the legacies of colonialism in and beyond the U.S. His essays and criticism have appeared in the *Los Angeles Times*, *Boston Globe*, *Air/Light Magazine*, *Rumpus*, and elsewhere; and his scholarship has appeared in *ASAP/Journal*, *Journal for the Society of American Music*, *Critical Studies in Improvisation*, *Río Bravo*, *Quarterly Review of Film and Video*, and elsewhere. He is the co-creator, with Charlie Vela, of *Wild Tongue* (2018), a compilation album celebrating the Rio Grande Valley's musical geographies, as well as *Futuro Conjunto* (2020), a transmedia, Chicanx speculative fiction album named one of the best Latinx records of 2020 by *Pitchfork* and *Texas Highways* magazines.

# Contributors

**Frederick Luis Aldama** is Distinguished University Professor at the Ohio State University, USA. He is the award-winning author, co-author, and editor of 48 books. He is founder and director of the Obama White House award winning LASER: Latinx Space for Enrichment & Research as well as founder and co-director of the Humanities & Cognitive Sciences High School Summer Institute.

**Leonardo P. G. De Assis** is a theoretical physicist currently holding positions as lecturer at San Francisco State University, USA, and instructor at Stanford University, USA. With publications in different areas of physics he is also interested in Quantum Computing, Neurophysics, and Artificial Consciousness.

**Dale Chapman** is associate professor of music at Bates College, USA. His work has appeared in the *Journal of the Society for American Music*, *Popular Music*, and the *Oxford Handbook of Sound and Image in Digital Media*. He is the author of *The Jazz Bubble: Neoclassical Jazz in Neoliberal Culture* (2018).

**Steen Ledet Christiansen** is Professor of Popular Visual Culture at Aalborg University, Denmark. His books include *The New Cinematic Weird* (2021), *Drone Age Cinema* (2016), and *Post-Biological Science Fiction* (2019). He has also published broadly on visual popular culture, particularly in terms of affect, posthumanism, and post-cinema.

**Annabel J. Cohen** (PhD Queen's; Fellow, American Psychological Association), whose film-music research has been supported by the Social Sciences and Humanities Research Council of Canada, is Professor of Psychology (University of Prince Edward Island, Canada), past Editor of *Psychomusicology: Music, Mind, & Brain*, and co-editor of *Psychology of Music in Multimedia* (Oxford).

**Sara Ferrando Colomer** is a postdoctoral neuroscientist at Northwestern University, USA, where she studies the biological mechanisms of memory formation. After graduating from the University of Bonn, Germany, she aims to understand how experiences shape the brain and influence our mental state. With a professional ballet formation, she enjoys exploring the fine line between neuroscience and art.

**Bevil Conway** is a neuroscientist and an artist. He is a native of Zimbabwe, a transplant to Canada, and currently lives in Washington DC with his husband and their two children. Conway's artwork has been exhibited in solo and group shows, is in many private collections, and is in the permanent collection of the Boston Public Library and the Fogg Museum. Conway's neuroscientific discoveries have been reviewed widely, including in *The New York Times*, *The Atlantic*, *Discover*, *Scientific American*, and NPR. Conway's explanation for #thedress, reported in WIRED, is the standard account of the phenomenon. Conway has written and lectured extensively on the intersection between visual neuroscience and visual art, appearing on stage with Alan Alda, Mark Morris, and the Mill at SXSW. He received a BSc from McGill University, Canada, taught high school biology at Peterhouse in Zimbabwe, and then completed his academic training at Harvard University, USA, where he was a junior fellow in the Harvard Society of fellows. He held faculty positions at the University of Kathmandu Medical School, Nepal, M.I.T., USA, and Wellesley College, USA, where he was a founding member of the Neuroscience Department before moving his research laboratory to the National Institutes of Health.

**Marta Figlerowicz** is an Associate Professor of Comparative Literature, English, and Film and Media Studies at Yale University, USA. She is the author of two books, *Flat Protagonists* and *Spaces of Feeling*. This essay is excerpted from a new book project in progress.

**Noah Fram** is a PhD candidate in Computer-Based Music Theory and Acoustics at Stanford University, USA, where he works with Jonathan Berger. His research combines categorical perception, cultural psychology, auditory perception, cognitive modeling, and theories of prediction and expectation to probe the genesis, nature, and function of artistic genres. Along with his research, he is an active composer and playwright specializing in musical theatre, opera, and chamber music, as well as a lighting and projections designer. He earned an MPhil in Music Studies from the University of Cambridge, UK, in 2017, where he was advised by Ian Cross, and a BA in Theatre and Mathematics from Vanderbilt University, USA, in 2013.

**Julia Peres Guimarães** is a Screen Cultures PhD student at Northwestern University, USA. Julia holds two master's degrees, in Political Science (University of Hawaii, USA) and International Relations (PUC-Rio, Brazil), and a BSc in International Relations & History from the London School of Economics, UK. Julia researches science fiction, Afrofuturism, and feminist/queer theory in television and music videos.

**Talia Lerner** is an assistant professor of Physiology at Northwestern Feinberg School of Medicine, USA. Her lab studies the neural circuit basis of motivation, reward learning, and habit formation. She is particularly interested in how individual variations in dopamine circuit function relate to differences in behavior and neuropsychiatric disease risk. Dr. Lerner earned her BS in Molecular Biophysics & Biochemistry from Yale University, USA, her PhD in Neuroscience from UCSF, USA, and completed postdoctoral training at Stanford University, USA.

**Simon D. Levy** is a professor of Computer Science at Washington and Lee University, USA. He holds undergraduate and graduate degrees in linguistics (Yale University, USA, University of Connecticut, USA) and a PhD in Computer Science from Brandeis University, USA. His current research focus is neurorobotics, including biologically realistic control of micro aerial vehicles and other robots.

**Charles W. Lowney II** teaches at Hollins University, USA. He earned his doctorate in philosophy at Boston University, USA, with a dissertation on Gottlob Frege and Ludwig Wittgenstein. His research involves understanding tacit knowing (in epistemology) and emergent being (in metaphysics) and in developing what he calls "emergentist ethics" (in moral philosophy).

**Eric Lyon**'s work focuses on articulated noise, chaos, oracular processing, and spatial orchestration. Lyon's creative work has been recognized with a ZKM

Giga-Hertz prize, MUSLAB award, League ISCM World Music Days, and a Guggenheim Fellowship. Lyon teaches composition and music technology at Virginia Tech, USA, in the School of Performing Arts.

**Zachary Mason** is a novelist and a computer scientist specializing in artificial intelligence. He lives in California.

**James L. (Jay) McClelland** uses neural network models like those in contemporary AI systems to capture human mental abilities. He is the Lucie Stern Professor in the Social Sciences and the Director of the Center for Mind, Brain, Computation, and Technology at Stanford University, USA, and he is a consulting research scientist at DeepMind.

**John McGrath** is a lecturer in Music at University of Surrey, UK. John's writing has received positive reviews in *Music & Letters, Wire Magazine, Psychology of Music*, and the *Irish Studies Review*. McGrath is also an active guitarist and his compositions have been aired on various television programmes and international radio stations. His monograph *Samuel Beckett, Repetition and Modern Music* (Routledge, 2018) explores the interactions and cross-pollination of music and literature while more recent publications investigate the transmedial work of Laurie Anderson and David Lynch. He is co-editor of *Twenty-First Century Guitar* (Bloomsbury, 2022).

**Christopher Minz** is a PhD Candidate at Georgia State University, USA. He has spoken at conferences and published several essays on the American Western and psychoanalytic theory. He is currently working on a dissertation about The American Western as it pertains to aesthetics of myth, melancholy, and American Ideology in the 20th and 21st Century.

**Patricia Pisters** is professor of film at the Department of Media Studies of the University of Amsterdam, Netherlands. She is author of *The Neuro-Image: A Film-Philosophy of Digital Screen Culture* (2012) and *New Blood in Contemporary Cinema: Women Directors and the Poetics of Horror* (2020). See also www.patriciapisters.com.

**Elizabeth Reich** is Associate Professor of Film and Media Studies at the University of Pittsburgh, USA. She is author of Militant Visions: *Black Soldiers, Internationalism and the Transformation of American Cinema* and co-author of *Justice in Time: Critical Afrofuturism and the Struggle for Black Freedom*. She is published widely in field journals.

**Murray Shanahan** is a senior research scientist at DeepMind and Professor of Cognitive Robotics at Imperial College London, UK. His publications span artificial intelligence, robotics, machine learning, logic, dynamical systems, computational neuroscience, and philosophy of mind. He is active in public engagement, and was scientific advisor on the film *Ex Machina*. His books include *Embodiment and the Inner Life* (2010) and *The Technological Singularity* (2015).

**Paul Skokowski** is a fellow of St. Edmund Hall, Oxford University, UK, and Executive Director of the Center for the Explanation of Consciousness, Stanford University, USA, where he is a consulting professor in Symbolic Systems. Paul recently edited the book *Information and Mind* published by Stanford University's CSLI Press, 2020.

**Laura Wagner** has a PhD in Linguistics and is a professor in the Department of Psychology at the Ohio State University, USA. She is the director of the Language Sciences Research Lab, a working lab and visitor exhibit embedded inside a science museum. Her research focus is children's language development.

**Jeff Zacks** is Professor and Associate Chair of Psychological & Brain Sciences at Washington University in Saint Louis, USA. His writing includes more than 90 journal articles and 3 books, including *Flicker: Your Brain on Movies*, and pieces for *Salon*, *Aeon*, and *The New York Times*. His research is supported by NIH, NSF, DARPA, ONR, and the McDonnell Foundation.

# Introduction

Jonathan Leal and Carol Vernallis

*Cybermedia*'s project is simple. In this volume, scientists, industry practitioners, and humanists consider the same recent media objects with the aim of creating more public engagement with the latest advances in science. The collection builds on a distinction philosophers often make between knowing water as "$H_2O$" and as a substance one feels as it runs through one's fingers. *Cybermedia* proposes that when we link embodied experiences (including those obtained from viewing media) with scientific concepts, we create richer forms of knowledge.

We use the term "cybermedia" to reveal connections between science, information technologies, and popular media. In today's films and streaming shows, directors and producers commonly incorporate scientific devices, techniques, and concepts, then bring in scientists during scriptwriting or shooting to ensure details are accurate. Director Alex Garland, for example, recruited neuroscientist Murray Shanahan to ensure his depictions of artificial intelligence (AI) and the Turing Test in his hit film *Ex Machina* (2014) were sound; director Terence Nance worked with two neuroscientists as advisors for his Home Box Office (HBO) show *Random Acts of Flyness* (2018–). (Nance and Garland are interviewed in the collection, and Shanahan, too, shares his reflections.) Despite such fruitful collaborations, many films tend to direct viewers' attention to the relevant science with only a line or two of dialogue—there's little exegesis. Our collection provides that context.

*Cybermedia*'s chapters—written by computer scientists, analytic philosophers, media theorists, music scholars, neuroscientists, cognitive roboticists, creative writers, quantum physicists, practitioners, and others—translate specialized concepts and knowledges into mobile, accessible pieces, varying in tone and style, form, and focus. In and across six modules the essays overlap in surprising, generative ways, exploring issues related to robotics, quantum physics, neuroscience, big data, and artificial intelligence, and how these issues are expressed through film.

This book arrives at a critical time. Digital technologies are pushing the planet into uncertain socioeconomic, ecological, and technological futures. Advances in AI, robotics, big data, psychometrics, and biogenetics are redefining what it means to be human. On the one hand, these advances promise radical transformations, forging new worlds through collisions between technologies, living forms, and matter. On the other hand, these changes feed dystopian visions: robots undergirding a possible 40 percent unemployment rate, social media platforms manipulating personal and community data, corporations extracting resources to exacerbate climate deterioration, militaries developing autonomous lethal weapons for armed conflict, stock-trading organizations deploying artificially intelligent systems, and scientists exploring biogenetics like CRISPR to facilitate altered biological forms. The list continues: the spread of biological contagions; the overflow of misinformation; the unequal distribution of vital resources; the ubiquity and biases of corporate algorithms (to say nothing of increasing economic inequalities, and intensified "group-differentiated vulnerability to premature death"[1]). Completed at the peak of the COVID-19 pandemic, *Cybermedia* is a deeply collaborative work that recognizes these as issues the next generation of humanists, scientists, filmmakers, and audiences will need to contend with. As a group, we worked to bring together researchers and creatives in pursuit of non-siloed, interlinked inquiries for and of the future.

*Cybermedia* also stems from a hope that contributors and readers will experience more interconnected lives. In this collection, film not only creates opportunities to consider issues of technology, science, race, gender, affect, and industry, but also serves as a meeting point between researchers from across disciplines. Contributors watched films and streaming shows. In groups, either through larger interdisciplinary Zoom interviews or more intimate chats, responses to chapters, and/or email threads, we all shared thoughts about new technologies and scientific and philosophical concepts and theories, as well as media forms, soundtracks, colorscapes, political interventions, and historical developments. Scientists strove to translate their technical expertise into tangible insights and even film criticism; media scholars worked to "read" the films "as scientists," to pursue hermeneutic exchange. In this sense, *Cybermedia*'s end-form

---

[1] Ruth Wilson Gilmore, *Golden Gulag: Prisons, Surplus, Crisis, and Opposition in Globalizing California* (Oakland: University of California Press, 2007), 28.

and creative process align, reiterating the book's global argument that in order to create useful analytics for the cybermediated world we've inherited and will eventually pass on, we need to find new ways to listen to one another.

* * *

We wanted this volume's content to feel accessible. Our introduction is brief, and the chapters' abstracts and bibliographies are online. We also feel the interviews and chapters with and by scientists are especially satisfying. Science pieces in popular press outlets like *The New York Times* or *Washington Post* tend to be short and breezy, while scientific journal articles, both narrow and technical, are often hard for non-specialists to interpret and place in relation to others. Science textbooks are costly, bulky, and difficult to secure. Popular science books are pricey and frothy. *Cybermedia*'s pieces, by contrast, contain content useful to a wide range of readers, from novices to experts.

This book's science and media-studies pieces also dovetail. As you move through this volume you might choose to employ the technique of the memory palace first developed by the Greeks; here, participants imagine a striking architectural layout containing evocative decor within—statues, paintings, furniture, and so on. Each item then links to a single concept. One simply walks through the envisioned site and calls off the pairs: that Monet watercolor painting with contributor Bevil Conway's concept of color consilience, for example (or a snuggling couple in bed, a recurring image in Nance's *Random Acts of Flyness*). *Cybermedia*'s layers invite you to create your own memory palaces and to relate the concepts nested inside.

*Cybermedia* explores many forms and genres—from a Radiohead music video and recent comics, to cutting-edge films by Black creators, including *Sorry to Bother You* (Boots Riley, 2018) and *Us* (Jordan Peele, 2019). Readers are encouraged to form subsets of the media objects and analyses with which to direct questions (one of our favorite media clusters is Alex Garland's *Ex Machina* (2014), *Devs* (2020), *Black Mirror*'s "Nosedive" episode (2016), *Westworld* (HBO, 2016), *Random Acts of Flyness* (Terence Nance, 2018), *Mr. Robot* (Sam Esmail, 2015), and *Under the Skin* (Jonathan Glazer, 2013). Placing these analyses and readings in conceptual relations, can provide new ways of thinking about contemporary media and science today.

Let's try this with just two films—*Ex Machina* and *Under the Skin*: how might our science-informed approaches give readers insight into these films? Neuro- and computer scientist Jay McClelland's opening discussion of the ways AI's and humans' abilities depart helps us see that *Ex Machina*'s protagonist-android, Ava, far surpasses what we can currently achieve through programming and neural nets. Her traits are too quirky and "human." She pursues activities beyond her training set; at the film's close, director Alex Garland has Ava look back before escaping a compound, smiling with pleasure at what she surveys, only for herself. By contrast, Scarlett Johansson's alien in Jonathan Glazer's *Under the Skin*—tellingly, she has no name—seems more realizable. Her modes of going about projects (luring men to death) seem rote, and her physical movements stiff, even as her speech patterns register as rich. Like any AI's, her learning curve seems slow and narrow. And just like AI, she has critical learning periods. Only later, after she's stalked through the streets and solidified her skills, does she develop a new capacity, empathy, when she encounters someone suffering from physical limitations. With all this in mind, can we apply these insights to other media objects? We're curious: how can reading such characters and films—both as scientists and humanists—help us understand the contemporary world and its potential futures?

\* \* \*

The volume's first module, "AI and Robotics," introduces artificial intelligence through its ethical issues, philosophical quandaries, and aesthetic horizons. Contributors ask: how do different mathematical and computational models challenge, extend, and reaffirm existing definitions of humans and machines? How does AI defamiliarize representational tools—sound, image, and language, for instance—that we use to fashion new thought? What does AI reveal about power, bias, culture, and identity? And how do popular representations of AI in film, television, and literature differ from current scientific theories and actual, functional technologies? Each contributor develops these questions from their disciplinary standpoints in the sciences and humanities, illuminating the development, use, and misuse of AI across media and industries. In his opening chapter cognitive scientist Jay McClelland—famous for co-developing the connectionist framework for neural networks that, since the 1980s, has transformed the disciplines of cognitive neuroscience, psychology, and machine learning—sees in filmmaker Garland's *Ex Machina* (2014) an opportunity to measure the distance between human intelligence and current artificial capabilities. Computer scientist Simon Levy and philosopher Charles Lowney, in their co-written contribution, extend this work, noting how film depictions of AI supremacy often take creative liberties with scientific thought and philosophical procedure; Levy and Lowney argue, in effect, that while popular narratives about AI often fall short of scientific accuracy, they excel as windows into steadily mounting anxieties over human frailty, fallibility, and plasticity. The chapter by Zachary Mason, computer scientist and author of the NYT bestselling novel *The Lost Books of the Odyssey* (2007), departs from linear argument, turning instead to literary, associative poetics to explore how wonder, consciousness, mathematics, and "transcendent AI" relate. And finally, at the core of "AI and Robotics" is a group interview with award-winning writer and filmmaker Garland, whose three recent films—*Ex Machina*, *Annihilation*, and *Devs*—have not only captured the public imagination, but also served as touchstones for scholars invested in technological ethics.

The second module, "Big Data, Sentience, and the Universe," expands and deepens the issues raised by the first by turning toward the philosophy, quantum physics, mechanics, and computing behind artificial intelligence. Together this module's four chapters extend questions theorist Karen Barad poses in *Meeting the Universe Halfway* (2007): "What, if anything, does quantum physics tell us about the nature of scientific practice and its relationship to ethics?"[2] Where and how do quantum theory, artificial intelligence, and popular culture collide? And, crucially: where are the boundaries between scientific fact and satisfying fiction? To answer these questions our authors draw on visual theory (Steen Ledet Christiansen), philosophy (Charles Lowney), theoretical physics (Leonardo De Assis), and cognitive robotics (Murray Shanahan) to problematize representations of machine learning and AI in Radiohead's "Go to Sleep" music video and Garland's *Ex Machina* and *Devs*. While Christiansen poses questions about the organicist model of life in his analysis of MASSIVE's digital crowd-generating animation software, Lowney refers to connectivist networks and big data in his discussion of creativity, intentionality, and machine learnings' simulated human behavior. De Assis changes tack, placing the determinism of classical physics and the indeterminism of quantum physics within the framework of free will in his analysis of the quantum computer in *Devs*. Countering this is Shanahan's discussion of his working relationship with Garland in *Ex Machina*'s earliest stages, and the ways that consciousness threads through the film's layers. In each chapter, thoughts on machine learning, determinism, and the fragile boundaries between the organic and inorganic drive the authors' investigations of the hope, anxiety, redemption, and despair of contemporary speculative fiction. As a

---

[2] Karen Barad, *Meeting the Universe Halfway: Quantum Physics and the Entanglement of Matter and Meaning* (Durham: Duke University Press, 2007), 6.

whole, this module considers how the profoundly counterintuitive aspects of new scientific theories and computing technologies beget new questions about matter, meaning, and the nature of consciousness itself.

The third module, "The Neuroscience of Affect and Event Perception," consists of three pieces that engage the cognitive and aesthetic dimensions of cybermedia in the contemporary attention economy. Using recent films and comics as reference points—including *Under the Skin, Black Mirror*'s "Nosedive," and cutting-edge visual narratives like *Phaedo*, a graphic novel co-created with AI— cognitive scientists Jeff Zacks and Laura Wagner team up with media theorists Fredrick Aldama, Dale Chapman, Jonathan Leal, and Carol Vernallis to consider the intimate, phenomenological dimensions of contemporary audiovisual experience. Threaded across each chapter are concerns with memory formation and disruption, meaning-making via linguistic description and cinematic depiction, neuroaesthetics across the attention industries, identity formation within audiovisual, algorithmic culture, and, crucially, new conceptual, technological, and interpersonal openings for tomorrow's writers, researchers, and audiences.

The fourth module, "The Digital West," focuses on HBO's *Westworld*, locating in the series' reimagining of the American Western a formulation of cyber technologies' new directions and a critical engagement with AI frontierism. The module's contributors analyze the series from multiple perspectives, revealing anxieties about AI supremacy rooted, as introduced earlier in the book, in troubled definitions of human and machine, past and future, and freedom and determinism. In the opening chapter, philosopher Paul Skokowski asks if androids, similar to the AI hosts depicted in *Westworld*, can have mental lives as rich as humans and what this may suggest for the limits to human experience. While he proposes considering those questions through iconic thought experiments or "puzzles" in philosophy, film scholar Christopher Minz and film-music researcher Annabel Cohen, in their chapters, call attention to the series' puzzle-like embedding of paths to exploring free will, consciousness, and reality in its narrative and audiovisual tracks. Through special attention to how disembodied voices and the figure of the player piano are used to signal human cognition in robots, the chapters by Minz and Cohen also offer a contemplation of AI futures through sound and music, the framing potentials of which are often ignored in discussions on the topic.

"Interface, Desire, and Collectivity" directly addresses issues of race, gender, social difference, representation, community, and media forms—concepts present in all of the chapters in this collection, though highlighted especially here through the work of Black studies theorist and media scholar Elizabeth Reich, composer and scholar Eric Lyon, computer-music researcher and artist Noah Fram, neuroscientist and visual artist Bevil Conway, and filmmaker-musician Terence Nance, creator of HBO's *Random Acts of Flyness*. Through attention to auralities and interfaces, layered forms and technologies, embodied knowledges and communal praxes, and, as Nance puts it, Black "wakefulness" and "rest," this module grapples with the media, interfaces, technologies, and histories structuring contemporary life. Where Fram's chapter attends to race, genre, film editing, and surprise, Reich's contextualizes such techniques, linking them to historical traumas—violent systems of differentiation and classification justified as rational and/or scientific. Together these chapters work to disrupt such colonial logics, widespread as they still are, by highlighting Black theory-praxis as radically scientific and technological. The chapters are a call to understand desire and pursue new collectivities. Or, in the spirit of Black studies theorist Katherine McKittrick's arguments in *Dear Science and Other Stories*, the contributors reaffirm that "To be black is to live through scientific racism and, at the same time, reinvent the terms and stakes of knowledge . . . [it is to] notice that the enclosures of biological determinism and the potentials of opacity, together, provide the conditions to concoct a different story altogether."[3]

Finally, the sixth module, "Productive Neuropathologies," traces the connections among brains, bodies, senses, memory, and environments. In an era when advances in neuroscience and medical technologies have afforded a greater understanding of human behavior and perception, these connections between minds, machines, and environments have become increasingly enmeshed through what Marta Figlerowicz calls in her chapter "a synthetic quasi-naturalism" that makes molecular, cognitive, urban, organic, and manufactured worlds appear co-extensive. While Figlerowicz takes the continuity among the worlds portrayed in films like *Under the Skin* and *Nymphomaniac* (Lars Von Trier, 2013) as evidence of an emergent vitalism that resists capture by traditional modes of thought, Patricia Pisters discusses how contemporary TV shows such as *Mr. Robot* use neuropathologies as darker staging grounds. Alongside Figlerowicz and Pisters, the module brings together media scholar Julia Peres Guimarães and neuroscientists Talia Lerner, Sara Ferrando Colomer, and Hojoon

---

[3] Katherine McKittrick, *Dear Science and Other Stories* (Durham: Duke University Press, 2021), 186.

Lee to delve into the invocations of mental disorders, arresting traumas, hardwired habitual patterns, and heightened affective states in contemporary cybermedia, exploring how they mediate the increasingly blurred boundaries between humans, non-human matter, and artificial beings.

## *Cybermedia*'s Backstory: Lead Editor Carol Vernallis

In the hope that some readers will feel encouraged to embark on similar projects, I'd like to share a bit about *Cybermedia*'s context and process. For me, *Cybermedia* came out of an emotionally close, political impulse. When I taught in conservative parts of the United States, I was struck by beliefs that differ from mine—notions that one's lot in life is of one's own making, and that the world is fixed and hierarchical, with God at the top, oneself and one's family in the middle, and others at the bottom, who deserve to be where they are. A deeper understanding of science, especially neuroscience, may loosen these ideological frames. Drawing on neuroscience, I gauge we're constructed from a mix of gifts and inheritances (genetics and context); perhaps life is contingent and free will plays only some small role.

*Cybermedia* also stems from a sense of joy. Some of us co-editors have been closeted in media studies and music departments for decades, so, feeling, suddenly, that quantum physics and neuroscience were our concerns, too, made us feel as if we were stretching out toward wider expanses. And it helped our research. Some recent studies in neuroscience on multisensory modalities and the brain seem provocative—consider, for example, Mooney's, the ways a term only tangentially related to an item encourages more rapid identification.[4]

Universities say they want to support interdisciplinary studies, but I haven't seen much (and our professional organizations haven't done enough outreach). I had the chance to cross disciplines, because, late-career, I could take courses across my university. A wonderful course on music, EEG, and the brain piqued my interest, but it's the philosophy of mind courses that drew me to science. In one class, we were studying models like Jerry Fodor's modularities of the mind and Tversky and Kahneman's S1 and S2. I asked, "Where is all of this? Is the brain

---

[4] Raymond Chang, Alexis T. Baria, Matthew W. Flounders, and Biyu J. He, "Unconsciously Elicited Perceptual Prior," *Neuroscience of Consciousness*, 2016, 1–9, doi:10.1093/nc/niw008

shaped like a cloverleaf freeway?" Some philosophy of mind courses were challenging, especially one on language, experience, and the external world. I kept encountering sentences salted with familiar words I judged I misunderstood. The professor concurred—philosophers relish repurposing terms (I started reading assigned texts alongside my cellphone for Google searches—"term, definition, philosophy"). Reading an assigned article, I thought, "Surely, I know what Block means by inference." But no, that class's day I was in left field. (I shared that I felt conflicted about treating words this way.) But I was hooked. From there followed courses in symbolic systems, including on minds and machines, computer knowledge and AI, intro to neuroscience, and doctoral courses in brain neuroplasticity as well as visual systems.

I've become a better reader of science, perhaps in ways useful for other humanists.[5] Capturing a science article's gist first helps—turn to the conclusion, then the opening abstract, skim quickly the next third, with its details of procedures (e.g., we found 47 subjects, 4 wore glasses, we're subtracting 6 for health concerns)—and if you've got the energy, ponder especially those graphs. Participation in courses for doctoral students helped make these graphs interesting (a brain oriented toward a task rather than pedaling in the ruminative network learns more effectively). I now have a working knowledge of the basic linear regression plot, a skill we might all share. (See figure 0.1)

In class, sometimes I share my perspective. Recently I raised my hand and asked, "As a humanist, what about this?" "The language you've been using for these curiosity AI bots, for example, convolutionally-nested networks, subjected to deprivation with white noise or mis-identified labels during critical developmental windows, and the adjacent close referencing of kitten sensitivities. What about when AI becomes sophisticated and looks back? Won't there be oedipal issues?" While many media studies scholars and I embrace Jane Bennett's books as well as writings on the Anthropocene, both of which honor the mute and non-sentient material, just rocks and stones, I knew I was speaking at cross purposes.[6] Over Zoom, I laughed, circled and raised my hands, and said, "Please continue. I'm just noting something." But I also knew I was sketching possibilities

---

[5] I'm proud that a recent collection I co-edited with Holly Rogers and Lisa Perrott, *Transmedia Directors: Artistry, Industry, and New Audiovisual Aesthetics* (New York: Bloomsbury Academic Press, 2020) and my monograph, *The Media Swirl: Politics, Audiovisuality, and Aesthetics* (forthcoming with Duke), begin to engage in this kind of cross-disciplinary collaborative process and to capture scientific inquiry.

[6] Jane Bennett, *Vibrant Matter: A Political Ecology of Things* (Durham: Duke University Press, 2010).

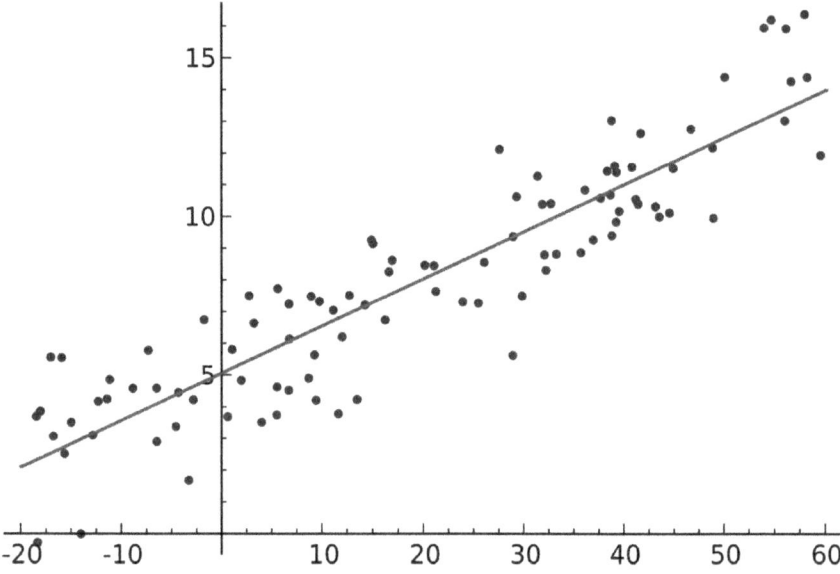
**Figure 0.1** Linear regression plot.

for future discussions and collaborations across disciplines. Sometimes, these questions may re-arise, perhaps concerning representation or questions of care and sentience.

Thinking across disciplines. Consider again our chapters, first science, then media criticism, and then turning to and fro. Leonardo De Assis's chapter on quantum physics helps us think about Garland's *Dev*'s quantum computer's digital blur, as well as interviewee Ben Salisbury's "granulated" score. Jeffrey Zacks's chapter for "Nosedive" helps attune us to our moment-by-moment shifting responses, and when scenes close. Annabel Cohen's and Noah Fram's chapters also draw us to the soundtrack's unfolding moments, and encourage us to wonder how our learned musical associations shape our perceptions, as well as what we're processing, fast or slow. But then let's diverge, back to issues more familiar with media scholars. How are conceptions of race and *Westworld*'s soundtrack related to arguments Elizabeth Reich makes concerning recent Black films—particularly to the soundtracks for *Sorry to Bother You* and *Get Out*? Does *Under the Skin*, a film focused on whiteness, interrogate race? How can these insights inform all of the chapters? We've also been thinking about environments and how they mold characters (and vice versa). Those glass boxes in *Ex Machina*, the unending gray landscapes in *Under the Skin*.

Our discipline of media studies encourages links to be made broadly and freely across a variety of surfaces, in manners poetic, though only unpredictably fruitful. I'm interested in rhythms of things—Terence Nance's remarks and "Nosedive"'s Lacie's traversal through a setting. I catch something about delicacy, rhythm, and pacing, shared broadly by all of the chapters. First, neuroscientist Hojoon Lee's work on taste. The fragile connection between a food and its encounter with the tongue's top cells, the taste buds lodged in the papillae, that, every ten days, millions of which die—as if the signal can't get across. Then interviewee Ben Salisbury's film scores, the tenuous inscrutability of a sound and moving image relation. Suddenly the cue doesn't work, and the thread with its leitmotifs collapses. Why might a viewer prefer one musical segment over another? Then *Westworld*'s androids' odd glitchings as noted by Chris Minz. Neuroscientist Talia Lerner's rats, who hover, not knowing whether to go forward or not. Will their brains' neurons, based on their chemical signaling, weight one way, enough to carry the message forward and lay down a re-searchable pattern? Neuroscientist Jeff Zacks watches *Black Mirror*'s "Nosedive"'s protagonist Lacie with us. Will her and our brains sync at the right moment to forge a "report," detailing what has unfolded to memory, or is too much in this scene ambiguous, apart from familiar genres? Is the scene mostly forgotten? Can we extend our observations further?

Tracing these connections encourages us to explore deeper questions. Some of our chapters make claims that we, and potentially AI, have free will, but from my perspective this is not a given. Most likely because we're considering sci-fi films, though the same holds true for classic Hollywood, so much is devoted to carnage. Drawing on media scholar Laura Mulvey, with film, 24 frames a second, blackness and death are co-present. The film often ends at a delicate juncture where much has been leveled or destroyed, and only the most tenuous possibility of continuance is proffered.[7] We'd like to continue our discussions about free will and determinism as well as race, the Anthropocene, and the external world versus cognition and immanence. A physicist who designs microscopes to spy on the Big Bang shares that everything is likely predetermined. Oh, to pursue this!

We've come to understand our fields differently. Stanford's amped-up culture has surely colored my perspective—big corporations and startups have adjacent zip codes. A buzz is in the air; enthusiasm's contractible from the ether. But it's the neuroscience courses with doctoral students, and reading science articles in

---

[7] Laura Mulvey, *Death 24x a Second: Stillness and the Moving Image Illustrated Edition* (Reaktion Books Ltd, London, England. 2006).

preparation for interviews with scientists, industry practitioners, and media scholars, that science's aesthetics now feels different to me. Perhaps it's that science has vastly greater financial resources, one's research can be quantified and directed toward a clear goal, and most scientists work collaboratively as part of large initiatives. Readers may catch a feel for disciplinary differences from our interviews with scientists, media scholars, directors, and other industry practitioners. Of course, these kinds of resources could be directed to the humanities as well.

I'd like to co-edit a *Cybermedia 2*—this time with more funding and support. What should we explore? The Anthropocene? Race? Sense and perception? AI and Robotics? Health? Political leanings and political advertising?

*Cybermedia* demonstrates that inspiration can come from anywhere, including the media objects we consider—that's the way our minds work. At the close of *Mr. Robot*, a streaming series discussed in one of our modules, protagonist Elliot meets up with his multiple selves in the movie theater. So might we, after a screening, and the pandemic, head for a coffee shop, and share our findings and our thoughts.

# Bibliography

Barad, Karen. *Meeting the Universe Halfway: Quantum Physics and the Entanglement of Matter and Meaning.* Durham: Duke University Press, 2007.

Bennett, Jane. *Vibrant Matter: A Political Ecology of Things.* Durham: Duke University Press, 2010.

Chang, Raymond, Baria, Alexis T., Flounders, Matthew W., and He, Biyu J. "Unconsciously Elicited Perceptual Prior," *Neuroscience of Consciousness,* 2016, 1-9, doi:10.1093/nc/niw008

Gilmore, Ruth Wilson *Golden Gulag: Prisons, Surplus, Crisis, and Opposition in Globalizing California.* Oakland: University of California Press, 2007.

McKittrick, Katherine. *Dear Science and Other Stories.* Durham: Duke University Press, 2021.

Mulvey, Laura. *Death 24x a Second: Stillness and the Moving Image Illustrated Edition.* Reaktion Books Ltd, London, England. 2006.

Vernallis. *The Media Swirl: Politics, Audiovisuality, and Aesthetics* (forthcoming with Duke).

Vernallis, Carol, Holly Rogers and Lisa Perrott. *Transmedia Directors: Artistry, Industry, and New Audiovisual Aesthetics.* New York: Bloomsbury Academic Press, 2020.

Part One

# AI and Robotics

# 1

# Could the AI of Our Dreams Ever Become Reality?

James L. McClelland

Ava, the humanoid robot in Alex Garland's *Ex Machina* (2014), startles us with her beauty, her sexuality, her vulnerability, and her intelligence—and ultimately with her willingness to deceive and to exploit others' weaknesses. She seems, and of course she is, all too human, even if, in a science fiction world, she has capabilities that exceed our own. We fear her because we are all too aware of our human frailties and limitations and imagine that someday, an artificially intelligent being with all of our abilities and none of our limitations will be created and, like our human conspecifics, will be all too liable to exploit our weaknesses, leaving us unable to control the outcome.

Watching *Ex Machina*, I was struck by how different Eva seemed to me than the artificially intelligent computer systems that we have today. It is true that in one of its matches, a contemporary artificial intelligence, DeepMind's *AlphaGo*,[1] made a move that no human understood or anticipated—a move widely credited with giving it an advantage that let it go on to win its match again the Korean grandmaster Lee Sedol. We can marvel at *AlphaGo* and its apparent intuition and insight, and perhaps this alone is enough to spark the fears that Ava instills. Yet, *AlphaGo* is ultimately only a computer program, an object that runs entirely under the control of the scientists and engineers who created it—and perhaps, more importantly, has no will of its own. *AlphaGo* and its successor *Alpha0* ("AlphaZero") are ultimately entirely mechanical systems whose capabilities derive from the brilliance of the computational intelligence researchers who designed it and the hardware and software engineers who turned its design into

---

[1] David Silver, Aja Huang, Chris J. Maddison, Arthur Guez, Laurent Sifre, George Van Den Driessche, Julian Schrittwieser et al., "Mastering the Game of Go with Deep Neural Networks and Tree Search," *Nature* 529, no. 7587 (2016): 484–9.

a reality. This program, which takes a board position as input and produces a legal move on its output, can learn through massive experience while playing against a series of ever-improving previous versions of itself. But an instance of *Alpha0* that can beat every human player in the world at chess doesn't know anything about absolutely anything else, and the same instance of the program cannot learn to play both games at the same time. Furthermore, you can't talk to it, it can't explain itself, and it cannot learn except through millions of games of play experience.

For me, it is useful to contrast today's AI systems like *AlphaGo* with a PhD student in the emerging field of computational intelligence, at the interface between human cognition and artificial intelligence. Comparing these programs to Ava is more difficult because some aspects of her abilities are difficult for me to separate from her overt sexuality and the mixed-up motivations of her creator. Setting these more fraught issues aside, in what follows I will focus on the purely intellectual side of human likeness, and on the emergence of advanced intelligent functions in researchers who go on to be independent contributing scientists. I have been lucky enough to have had many excellent PhD students and post-docs in my own laboratory over the years, and many of them have gone on to be professors at outstanding universities—or, more recently, computational intelligence researchers in AI companies. Surely, we would call these young scientists intelligent. What do they have that today's AI systems lack?

To help us consider this, I'll introduce Dana, a fictional young PhD scientist. I use as pronouns 'e, 's, and 'em to emphasize Dana's humanness while avoiding designating a gender. I'll start with more basic properties I think all humans possess, and then go on to consider what it is that makes Dana and others capable of succeeding in what I will consider to be the hallmark of intelligence: identifying and successfully addressing novel and previously unsolved challenges.

## Mutual Simultaneous Constraint Satisfaction

Something very basic that humans possess and today's AI systems do not is the ability to exploit multiple simultaneous sources of information to settle into an overall interpretation of a situation and its parts or aspects and/or to formulate a plan of action that addresses many such constraints simultaneously. A beautiful visual illustration of this is provided in figure 1.1. At first, we may experience this picture as an inchoate assemblage of splotches of ink, but at some point, we are

**Figure 1.1** A Dalmatian dog emerges from an assemblage of individually uninterpretable blotches. From James, R. C. (1965), Photo of a dalmatian dog. *LIFE Magazine*, 58(7), 120. Copyright © 1965 Ronald C. James.

likely to begin to see that the photograph depicts a dalmatian with its back toward the viewer sniffing at the ground. None of the blobs individually appear to signal the presence of a dog, but somehow, when all are considered together, the percept emerges. At the moment we see the dog, we also see the blobs differently. Some now help define the contours of its body or are seen as spots on the dog's coat, and other blobs now become scattered leaves on the ground or parts of a tree. We can even perceive the contour of the dog's back where no actual contour is present in the image. Thus, the perception of the whole emerges from constraints provided by many aspects of its parts, and the perception of the parts depends in turn on the perception of the whole. This is what I mean by the idea of multiple, mutual constraint satisfaction.

Experiences like my seeing this photograph converged with findings in the psychology of perception and language understanding, inspiring me and others to think that it might be useful to view our perceptual systems as neural networks, because of several key properties that neural networks have that seemed to make them suitable to capture this kind of experience. The goal was not simply to simulate the brain, but to draw on the properties that might make the brain especially useful to solve this kind of constraint-satisfaction problem. The brain contains hundreds of millions of neurons, each capable of receiving inputs from up to one hundred thousand other neurons. Each neuron adjusts its activation depending on the inputs it receives from others, and in turn, signals its activation to other neurons via its outgoing connections. Inspired by this idea, which we called *Parallel Distributed Processing*, David Rumelhart and I teamed up with others and drew on earlier work to develop neural network models that simulated this mutual constraint satisfaction process.[2]

A key part of the inspiration for our work was the idea that the constraints influencing the outcome of perception or understanding can come from a wide range of sources. Our brains naturally and automatically integrate input from sight, sound, touch, posture, motion, smell, and taste in interpreting the inputs we receive. Spoken and written language contribute to and participate in this process as well; the words and sounds we experience hearing depend on other sources of input that accompany them, and likewise the objects that we perceive through other senses are simultaneously constrained through language. Constraints affecting perception and thought can come from a wide range of mutually constraining sources.

Another potent source of constraint is input from memory. Consider this tiny story:

> John put some beer in a cooler and went out with his friends to play volleyball. Soon after he left, someone took the beer out of the cooler. Meanwhile, the volleyball match was very intense, and it seemed that John's team was going to lose. But after plenty of fierce competition, John's side was able to pull out a string of victories and won the final game when John served an amazing service winner that no one on the other team could even touch.
>
> John and his friends were thirsty after the game and went back to his place for some beers. When John opened the cooler, he discovered that the beer was ___.

In this situation, if you as a reader have been following the story, you will anticipate that the missing word is "gone" and this will influence how likely you are to perceive it from a very brief or indistinct presentation of the word itself or a misspelled version of it. But if the text had said "someone took the *ice* out of the

---

[2] David E. Rumelhart, James L. McClelland, and the PDP Research Group, *Parallel Distributed Processing: Explorations in the Microstructure of Cognition* (Cambridge, MA: MIT Press, 1986).

cooler" you would instead be ready to perceive the word "warm." We as humans have the ability to exploit such constraints based on information we encountered in the indefinite past, not just the immediate current context.

Finally, the considerations that may come into play are potentially unbounded and seemingly unrelated to a particular situation at hand. I believe I heard a version of the anecdote below from Jerry Fodor. Whether it really happened I don't know, but it seems to capture something real about how we think.

> Jeff, a good bridge player, has just bid six Hearts and is about to start play on the last bridge deal at the end of an evening at his bridge club. Another player, Al, from a table that has just finished its last deal, comes over, walks around the table to see the hands of all of the players, and lingers to observe the play. The player to Jeff's left makes the opening lead. As Jeff's partner lays down the dummy hand, Jeff surveys the situation. It looks like an easy contract. But Jeff notices that Al is still hanging around. This makes Jeff think: maybe the hand is not such an easy one after all. If it were, Al would surely have lost interest by now. He ponders: what could conceivably go wrong? Seeing only one possibility—one that would ordinarily seem remote—he devises a plan of play that would ordinarily fail but succeeds in this case, and triumphantly, he makes his contract. His opponents are outraged and complain to the director. But the director can do nothing, since Al never said or did anything that was against the rules in any way.

Here Jeff is using information from outside of the domain of the game itself to reason about what to do within the game. It was Fodor's point, and one that I agree with, that there is no limit on the constraints that we can ultimately bring to bear when we think and reason. In other words, the constraints that can enter into our mental constraint satisfaction process are completely open-ended.

I have described here what to me are extremely basic aspects of human intelligence, ones that we all possess. In spite of the fact that most of the recent breakthroughs in AI are based on artificial neural networks similar to the ones Rumelhart and I used in our early work to capture the mutual constraint satisfaction process, today's networks are generally far narrower in the constraints they consider than we as humans are. AI researchers at DeepMind and elsewhere are aware of these limitations, and progress has recently been made in creating language systems that can begin to bring a very wide range of information from context to bear in language comprehension. Researchers are actively exploring how to combine many input modalities and how to exploit relevant information presented only once at an arbitrary past time now out of mind. Much still remains to be accomplished here, however. Furthermore, the prospect of being able to bring completely open-ended considerations to bear, as in our bridge game example, remains an important future challenge.

## Metacognition, Explanation, and Discourse

Another area where Dana, and humans in general, far exceed our current AI systems is in the ability to think about and exchange ideas with others about our own thought processes or to describe the reasons for the decisions, actions, and predictions that we make. It seems fair to say that *AlphaGo* and most other contemporary AI systems are completely devoid of these abilities. Returning to the surprising move that *AlphaGo* made against Lee Sedol, the computer had no ability to explain why it chose the move it did. In contrast, during the match, human commentators provided a running commentary, describing the pros and cons of each move made by both the computer and the human player, and speculating on whether or not *AlphaGo*'s move was a brilliant stroke of genius or a wild stab in the dark.

I do not mean to say that we as humans have perfect access to the basis of our own perceptions, feelings, and choices of actions. Those interested in human thought have been aware since the late nineteenth century that introspection is often uninformative or completely misleading. Yet we can and do share information with each other that we can use to immediately alter our behavior— something that is not possible for machine systems like *AlphaGo* that simply learn to get better through massive experience.

As one simple example of this, consider the puzzle shown in figure 1.2. You are given a grid and an instruction specifying a task goal, and without any further experience, most people I've shown this puzzle to can begin to perform the task of placing "x"s in the grid, and I have solved many such puzzles without any further instruction on how to solve them. Many contemporary AI systems could learn how to play this game very well, but they would *either* require the programmer to build quite a lot of the solution into the program *or* they would require a vast amount of experience, or a combination of both. Furthermore, I can point out things to you that you can use to help you play the game. First, I can tell you that it is useful to try to determine where "x"s *cannot* go in the grid, marking these cells, say, with a small dot. Then I can point out that if all of the cells that remain possible places where you can put an x within a given enclosed region are within the same row or column, you can be sure there can't

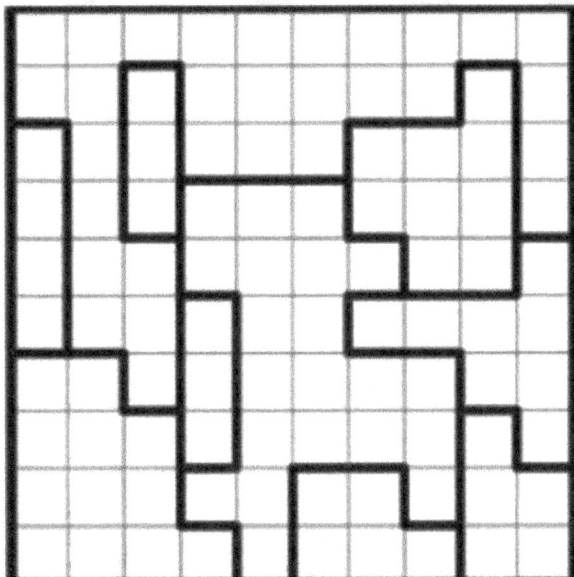

Place 2 x's in each row, column, and bounded region. No two x's can be adjacent, not even diagonally.

**Figure 1.2** A Two Not Touch puzzle, with instructions as they appear with puzzles published in *The New York Times*. From Bumgardner, J. (2020). *Two Not Touch Puzzles*, https://krazydad.com/twonottouch/ (accessed January 3, 2021). Copyright © 2020 www.krazydad.com, reprinted with the permission of the creator.

be an "x" in any of the other cells in that row or column. With this information, you can then find more cells that cannot contain an "x," and are well on your way to an overall solution.

What is more than this is that we as humans can make these observations and share them with each other. The above paragraph is evidence of this. No one told me the points I made above, but as I practiced solving these puzzles, I started making these observations to myself. I emphasize that I do not consider myself to be especially gifted in these ways, though I do believe my past experiences have helped set the stage for me to do this, at least in part.

Where our ability to engage in metacognition comes from is an open scientific question. One could hold that it is something that evolution endowed us with, or one could hold that evolution and culture gave us language, and with language we developed the ability to understand and give explanations, and once these abilities developed, we became able to use language to make observations for ourselves. The recent AI language system GPT-3[3] may have some abilities along these lines. This system was trained on a vast corpus of language including quite a lot of transcribed human discourse. Since such discourse contains examples of explanation, it is possible that the system would, if assessed, be able to give some form of self-explanation. Suppose we gave it the passage about John and his beer. Because this passage will fit in GPT-3's buffer, it may be able to predict that the missing word should be "gone." Suppose we continued the story, "The beer was missing because . . ." and then let GPT-3 complete this sentence. Perhaps it would go on to say "someone had taken the beer out of the cooler." It is conceivable that GPT-3 would even come up with this explanation, given the right kind of relevant experience, even if the previous sentence about the removal of the beer had not been included. This is an area where we still have a lot to learn about what we need to build into our AI systems for them to begin to exhibit abilities we take for granted as humans.

## The Role of Culturally Invented Modes and Tools for Thinking

The abilities I have described above are abilities all humans rely on every day. Multiple constraint satisfaction is always in play as we identify spoken words and recognize objects or make everyday motor planning and action decisions. Whenever we discuss the events of our day, the behavior of others, politics, the weather, or anything else, we are always engaged in explanations and teaching each other through discourse and discussion. A human graduate student like Dana engages in this kind of discussion as well. For example, Dana may explain to me a plan to analyze the data collected in an experiment, and we might discuss alternative approaches before we settle on a particular plan. Dana will then go off and execute what we have discussed, based on material learned in a statistics course, which in turn involves a lot of direct instruction. I certainly believe Dana and other graduate students learn gradually from experience as well, and that

---

[3] Tom B. Brown, Benjamin Mann, Nick Ryder, Melanie Subbiah, Jared Kaplan, Prafulla Dhariwal, Arvind Neelakantan et al., "Language Models are Few-shot Learners," *arXiv preprint arXiv:2005.14165* (2020).

expertise ultimately does depend on a great deal of experience; but I think today's AI is missing out on the tremendous leverage that instruction and explanation can provide.

However, to be truly successful as an advanced practitioner of a discipline such as computational intelligence, Dana also needs additional skills that, I believe, depend on acquiring specialized mental modes of thought and tools for thinking that aid and support the efforts of skilled experts. This is just as true I believe in the arts and humanities as it is in the sciences, and so in this section I will draw my examples from both domains, but with the primary focus on science, since that's the domain I know more about.

Doing science requires proficiency in an extensive set of tools for thought, in conjunction, perhaps, with a kind of meta-level tool for thinking that I will call *formal thinking ability*. Some examples of specific tools are the ability to develop sound logical arguments, to solve problems that require the use of mathematics, to prove mathematical theorems, and to write computer programs that accord with the conventions of complex and highly structured programming languages. To make a contribution in science today, one must rely heavily on many of these tools.

One example of such a tool is propositional logic. For much of the twentieth century, logic played a central role in widely-held conceptions of these abilities. Bertrand Russell said "All of mathematics is symbolic logic," and so central was logic to mid-century conceptions of intelligence that Herbert Simon, a leading early figure in AI research, was able to say in 1953 "Over the Christmas holidays, Allan Newell and I programmed a computer to think."[4] Their computer program proved simple logic theorems, and he was using a system that owed its very essence to the traditions of logic that were instantiated in the architecture of the digital computer. Decades later, Fodor and Pylyshyn,[5] argued that thought is, essentially, the manipulation of structured assemblies of symbols according to structure-sensitive rules, and used the logical syllogism called *modes ponens* as their central example. It goes like this. If you know that some proposition *p is true*, and you know that *if p is true, then* some other proposition *q is true*, then you can conclude that *q is true*. So, if you know (*p*): *John is strong*, and you know (if *p* then *q*): *If John is strong, then John will beat Bill at armwrestling*, you can

---

[4] Herbert A. Simon, *Models of my life* (Cambridge, MA: MIT Press, 1996).
[5] Jerry A. Fodor, and Zenon W. Pylyshyn, "Connectionism and Cognitive Architecture: A Critical Analysis," *Cognition* 28, no. 1–2 (1988): 3–71.

conclude (q) *John will beat Bill at armwrestling.* You can do this, they argued, without regard to the actual content of the propositions. This is the kind of thing that Newell and Simon relied on in their computer program.

I find myself in partial agreement with these views. This may be surprising, because I believe that formal thinking is not the natural mode of human thought, and that it can get in the way of mutual constraint satisfaction. In fact, humans don't actually succeed with arbitrary propositional content, (which is why I used an example that appeals to prior knowledge). More importantly, today's neural network-based AI models can be seen as refutations of Fodor and Pylyshyn's arguments, since their successes seem to come in part from the fact that they expressly eschew commitment to foundational principles of formal thinking. For example, today's AI language translation systems are neural network-based systems that do not rely on the systems of rules that Fodor and Pylyshyn argued were central to human thought and language processing. Yet, there's no doubt that formal systems have played a huge role in supporting our ability to understand our universe well enough to create and control nuclear reactions, to create computers, and to create technologies that have allowed humans to direct spaceships that will intersect with the orbits of tiny objects in the vast space at the edges of our solar system.

One approach some cognitive scientists advocate is that we must build systematic, symbolic reasoning into our artificially intelligent systems. This is the approach advocated and exploited by Josh Tenenbaum at MIT and many of his collaborators and associates. Acknowledging the usefulness of neural networks, this group has recently explored what they call the "neurosymbolic" approach to capturing intelligence,[6] which relies on computational systems that use neural networks for processing inputs and controlling outputs, but relies on more symbolic approaches to capture the part of the process that Herbert Simon thought of as thinking. Their systems also exploit sophisticated advances in probabilistic reasoning, which makes them more powerful than the systems Russell, Noam Chomsky, and later Fodor and Pylyshyn relied on.

For my part, I am pursuing an approach in which systematic mental processes arise from the structuring of our minds that occurs through exposure to and

---

[6] Jiayuan Mao, Chuang Gan, Pushmeet Kohli, Joshua B. Tenenbaum, and Jiajun Wu, "The Neuro-symbolic Concept Learner: Interpreting Scenes, Words, and Sentences from Natural Supervision," *arXiv preprint arXiv:1904.12584* (2019).

command of the tools of thought I mentioned previously.[7] On this view, these tools are human inventions that began to emerge as humans started to develop technologies and civilizations. Gradually institutions arose within these civilizations, creating notation systems and artifacts that supported the further development of these systems, and that then structure the minds of those who immerse themselves in them, giving them the ability to build on the ideas of those who went before them to exploit and extend these systems. Number systems are good examples of these kinds of formal systems. Some cultures may lack number systems, having only words for very few, some, and many, as was documented by Peter Gordon in an important paper in 2008.[8] Many cultures have invented or adopted such systems from other cultures, but even throughout most of the first millennium of the current era, the number systems used in the west were cumbersome and unsystematic. The base-10 place value system used worldwide today is the product of cultural innovations and makes possible the creation of tools such as the abacus and mechanical calculators that vastly enhance the power and efficiency of human reasoning about number. Like our number system, geometry, trigonometry, calculus, logic, probability theory, and computer programming are all examples of culturally constructed systems and tools that vastly increase the power of human reasoning.

While I am certainly more of a scientist than an artist or musician, my exposure to art and music history during my undergraduate years taught me that the same points apply in these domains as well. The towering achievements represented by the painting, sculpture, and architecture of Michelangelo or the musical compositions of Beethoven depended crucially on the developments introduced by their predecessors and were achieved after decades of immersion in the study of these prior developments, many of which have strong formal elements. For example, in music, the twelve-tone scale, the various modes within this scale, the notational systems invented to allow the explicit representation of values and durations of musical notes, and the further structures built on top of them such as the sixteen-bar frame of most songs and the basic structure of

---

[7] James L. McClelland, "Are Humans Still Smarter Than Machines?" *Manuscript in preparation, Department of Psychology*, Stanford University, February 2021, based on the Graham Lecture at the University of Toronto (same author and title) recorded October 20, 2020, *YouTube*, https://www.youtube.com/watch?v=9ysH58hQ2n0&feature=youtu.be (accessed February 10, 2021).

[8] Peter Gordon, "Numerical cognition without words: Evidence from Amazonia," *Science* 306, no. 5695 (2004): 496–9.

sonata form, etc., are all cultural inventions, as are the actual instruments musicians use to render the resulting patterns acoustically, from simple drums and flutes to the well-tempered clavier. These conventions and tools underlay the achievements of Bach, Beethoven, and others, and subsequent extensions including new tonalities, rhythms, and tools such as synthesizers further extend these resources, allowing further developments in the nearly 200 years since Beethoven's last compositions.

To summarize my point in this section, I turn to the views of Henri Poincaré and Albert Einstein. Clearly, these are individuals who anyone would have to describe as intelligent. Henri Poincaré, the nineteenth-century French mathematician, physicist, and engineer, wrote "It is by logic that we prove, but it is by intuition that we discover,"[9] and Einstein is said to have viewed the intuitive mind as a sacred gift and the rational mind as its faithful servant.[10] Both Poincaré and Einstein seem to identify the essence of insight and discovery with intuition rather than logic and rational thought, and I share that perspective.

Likewise, it is more than mere technical proficiency that characterizes the contributions of Michelangelo and Beethoven. Our AI systems may be beginning to capture the intuitive rather than the rigidly formal, but to be truly intelligent, they will need to master invented formal systems as well. Our current AI systems do not yet fully capture how humans have been able to integrate intuitive and formal thinking, and for me this is one of the great challenges we will need to address before we have successfully created a truly intelligent artificial system.

## Goal-Directed Thinking

The final difference I would like to mention between Dana and today's AI is that Dana is in the process of becoming more and more self-directed. I remember that when I was a PhD student this was a very big issue for me. My first advisor had strongly steered my first-year research project toward a very specific issue that was of interest to him, and I found myself needing to set my own direction. When I work with my own PhD students, I always experience the same tension. How much should I steer them toward addressing my agenda? How much

---

[9] Henri Poincaré, *Science and Method* (1908; *Science et méthode*), translated by Francis Maitland (1914): repr. *Science and Method* (New York: Cosimo, Inc, 2007), 129.
[10] See Bob Samples, *The Metaphoric Mind: A Celebration of Creative Consciousness* (Boston: Addison-Wesley Publishing Company, 1976), 26.

should I leave them alone to pursue their own direction? Since I always have things I am anxious to pursue, I certainly always share with them the things that excite me the most. However, I've also found that our mutual experience together is always better if I seek to work with my students to find a project that is of mutual interest. In Dana's case, as with many of my students, we settled on a primary research project, something that Dana expressed interest in the first time we met. Dana is seeking to make a contribution to knowledge by working to address the issue of how to learn new things quickly in a superpositional memory—a memory that does not stick each new item into a separate slot, but instead superimposes them, as in holographs or film exposed to several images. This is a largely unsolved problem in cognitive neuroscience, and Dana and I agree that interesting new steps are possible. As Dana's mentor, I expect we will work together fairly closely initially, with Dana taking greater and greater ownership in the project as it progresses, though I hope to remain involved in finding the solution, rather than just helping to get the project going. In other words, the project is one that will, I hope, satisfy both of us as making progress toward an important goal, one that is good for science, for our reputations, and our careers.

In this regard, Dana is far different from current systems like *AlphaGo* and GPT-3. These systems have no independent agency whatsoever. Every computation they perform, every input they receive, and every output they generate, is entirely under the control of the scientists and engineers who design and run them. The designer creates what is called an objective function—a mathematical expression that characterizes the adequacy of the learner's performance, in terms the designer specifies. All of the learning in the system is directed toward maximizing performance as measured by this function (or minimizing the discrepancy from perfect performance, often called the *loss*).

It is true that there have been efforts underway for many years to create learning systems that explore their environments on their own and many thoughtful AI researchers are seeking to design systems with intrinsic goals that can lead to self-discovery. One such approach is to give a system the goal of producing novel experiences which then drive learning toward a deeper understanding than was possible based on the experiences the agent might have been exposed to passively. Progress is being made, and it will be interesting to see how far such research will go. I feel that an important place for the field to focus going forward will be on developing artificial systems that actually work toward particular goals, rather than simply focusing on improving performance by a global desire to experience novelty.

Being self-directed has, historically, been important for productive intelligence, where I define this as the ability to make a novel contribution. The history of science is the story of how independent thinkers revolutionized the way we understand the world around us. Galileo was found guilty of heresy for the new insights he contributed, and Newton and Einstein both revolutionized understanding of Physics. Likewise, Michelangelo and Beethoven are known as highly self-directed individuals who went beyond the achievements of their predecessors to achieve more than had ever been possible before.

*Ex Machina* raised this issue and a lot of other science fiction also touches on it. For an artificial being to be truly intelligent, must it also be completely self-directed? This is an important and interesting question. For us as humans, it often appears to be so, but I would offer two points that make me uncertain about whether this necessarily applies to all beings that can truly make innovative discoveries.

First, concerning humans, our goals are not, in my opinion, entirely our own. It is at least arguable that humans can have goals for others, or commitments to ideals, rather than just for themselves. Soldiers who are sent to war, or health practitioners at the front lines of battling contagious diseases, as well as leaders of social justice movements may have goals that place the collective good ahead of their own personal ambition. Indeed, being able to pursue a goal that is greater than oneself is an important source of inspiration. Often throughout human history, the truly innovative thinkers have appealed to someone or something greater than themselves, producing profoundly influential innovations.

Second, when it comes to artificial beings, the fact that a system like *AlphaGo* can come up with innovative moves leaves me wondering how much autonomy is strictly necessary. It seems arguable that deciding to seek an explanation for something that seems intuitively puzzling might require some degree of autonomy, but not necessarily the kind of autonomy that pits the artificial system against its creators. We should of course be wary of the possibility—one that *Ex Machina* and other science fiction has repeatedly raised—that we *might* be in danger of losing control. Speaking for myself, however, I am more worried about nefarious human uses of artificial intelligence than I am about losing control to autonomous artificial beings.

## Final Thoughts

From the thoughts I have expressed in this essay, it should be clear that, in my view at least, human intelligence still far exceeds artificial intelligence. However, I would like to note that artificial systems play an increasingly important role in *augmenting* human capabilities. Because these systems provide tools and resources that humans otherwise lack, they have enabled the development of systems that precisely target locations in the vast three-dimensional space of the outer reaches of our solar system or that allow us to predict how extremely complex chemical structures (usually proteins) will fold on themselves and interact with each other. More and more powerful extensions of human abilities will continue to be possible, thanks to the ever-increasing power of these systems. Of course, like other innovations, they can be used for good or ill, and as citizens, it is our crucial task to make sure there are governing bodies in place to oversee them as we oversee all other technologies. What remains to be seen is whether we come to see artificial systems as potential threats or competitors to ourselves. I am cautiously optimistic that we will be able to create systems that pursue prosocial goals, including the goals of encouraging our own sense of individual autonomy and agency.

## Bibliography

Brown, Tom B., Benjamin Mann, Nick Ryder, Melanie Subbiah, Jared Kaplan, Prafulla Dhariwal, Arvind Neelakantan et al. "Language models are few-shot learners." *arXiv preprint arXiv:2005.14165* (2020).

Fodor, Jerry A., and Zenon W. Pylyshyn. "Connectionism and Cognitive Architecture: A Critical Analysis." *Cognition* 28, no. 1–2 (1988): 3–71.

Gordon, Peter. "Numerical cognition without words: Evidence from Amazonia." *Science* 306, no. 5695 (2004): 496–9.

Mao, Jiayuan, Chuang Gan, Pushmeet Kohli, Joshua B. Tenenbaum, and Jiajun Wu. "The Neuro-symbolic Concept Learner: Interpreting Scenes, Words, and Sentences from Natural Supervision." *arXiv preprint arXiv:1904.12584* (2019).

McClelland, James L. "Are Humans Still Smarter Than Machines?" *Manuscript in preparation, Department of Psychology*, Stanford University, February, 2021, based on the Graham Lecture at the University of Toronto (same author and title) recorded October 20, 2020, *YouTube*, https://www.youtube.com/watch?v=9ysH58hQ2n0&feature=youtu.be. (accessed February 10, 2021).

Poincaré, Henri. *Science and Method* (1908; *Science et méthode*), translated by Francis Maitland (1914): repr. *Science and Method*. New York: Cosimo, Inc, 2007.

Rumelhart, David E., James L. McClelland and the PDP Research Group. *Parallel Distributed Processing: Explorations in the Microstructure of Cognition.* Cambridge, MA: MIT Press, 1986).

Samples, Bob. *The Metaphoric Mind: A Celebration of Creative Consciousness.* Boston: Addison-Wesley Publishing Company, 1976.

Silver, David, Aja Huang, Chris J. Maddison, Arthur Guez, Laurent Sifre, George Van Den Driessche, Julian Schrittwieser et al. "Mastering the Game of Go with Deep Neural Networks and Tree Search." *Nature* 529, no. 7587 (2016): 484–9.

Simon, Herbert A. *Models of my Life.* Cambridge MA: MIT Press, 1996.

2

# Director Alex Garland Converses with *Cybermedia*'s Scientists and Media Scholars

Alex Garland, Jay McClelland, Paul Skokowski, Simon Levy, Jeff Zacks, Carol Vernallis, Selmin Kara, and Jonathan Leal.

Edited by Jonathan Leal and Carol Vernallis

### Introduction

On October 2nd 2020, director Alex Garland Zoomed with several of our Cybermedia contributors—a small group of leading computer scientists, neuroscientists, physicists, and media scholars. The dialogue below, drawn from this meeting, explores a range of topics: rigor and transparency in the sciences and the arts, Garland's filmmaking methods, race and gender, and the possibilities for conveying robot consciousness through film.

**Carol Vernallis** We humanists here on Zoom relish Alex's work—from films and streaming series to scripts—for the ways it melds art, music, design, philosophy, and science. I was in Paul Skokowski's Stanford course, Minds and Machines (Paul's here today), when he shared a favorite philosophers' thought-puzzle: are H20 as we understand it through science and chemistry and the sensation of water running through our hands two different forms of knowledge? Perhaps so. But surely, embodied experiences paired with more abstract concepts can enrich one another. Alex's films and televisual work, especially *Ex Machina*, *Annihilation*, and *Devs*, give viewers the chance to try on cognitive, sensorial, and affective responses to topics like AI, CRISPR, and quantum physics and engineering. His work also reveals these subjects' philosophical concerns, such as the nature of one's own and others' minds, responsibility, free will, and determinism.

Let's introduce ourselves!

**Jay McClelland**   I'm Jay, and I'm really pleased to be here. I enjoy trying to connect neural networks with the arts precisely for the reason Carol was just mentioning. I feel like so much of our cognitive life is arising from places we don't totally know about that suddenly lead us to have some idea or insight or choice and I think neural networks sort of capture that.

**Alex Garland**   You know, a long time ago in my twenties, a physicist shared with me that when we say, "I just thought of something," it's really, "I stopped thinking of something." Many times, I've been doing the dishes while standing at the sink, and a complex solution to a story problem has arrived fully formed into my mind. Nowadays, I don't spend any time thinking about a story problem, except when I'm in a state of frustration.

**Jay**   Better to do the dishes! I find these things happen to me in the shower. Maybe water is a special conduit.

**Simon Levy**   I'm Simon. I was in Linguistics until I read Jay and Dave Rumelhart's book, *Parallel Distributed Processing*, back in the late '80s, and switched to studying neural nets for the rest of my life.[1]

**Jeff Zacks**   I'm Jeff, and I study how people understand complex naturalistic activity, and like Jay, I'm interested in the relationship between cognitive science and art. My lab runs experiments where we screen movies for people while we simultaneously record their brain activity. That's gotten me really interested in how people understand how the brain processes movies.

**Paul Skokowski**   I'm Paul. I teach Philosophy. I got my PhD at Stanford and also studied with David Rumelhart. We continued to collaborate on neural net applications after I graduated. I was a physicist at the Los Alamos and Livermore labs for about 14 years. I mainly write in Philosophy of Mind, including on zombies, consciousness, and neural networks. But I've recently returned to my roots, and I'm now doing philosophy of quantum mechanics. I just wrote a couple of papers on Everett's theory, which appears in Alex's *Devs*.[2]

---

[1] David E. Rumelhart, James L. McClelland and PDP Research Group, *Parallel Distributed Processing* (Cambridge Mass.: MIT Press, A Bradford Book, 1994).

[2] "Many-worlds Interpretation," *Wikipedia*, at https://en.wikipedia.org/wiki/Many-worlds_interpretation (accessed 17 December 2020).

**Carol**  And we're the humanists. I'm Carol, and here are Selmin and Jonathan. I work on contemporary audiovisual aesthetics—music video, intensified post-classical cinema, and YouTube.

**Selmin Kara**  I'm Selmin. I teach Film Studies at OCAD University in Toronto and I work on cinema, digital aesthetics, and ecology during the Anthropocene.

**Jonathan Leal**  And my name is Jonathan, and I write about music, race, and narrative across contemporary media and culture. I write and teach at USC in Los Angeles.

**Paul**  I'd like to ask a question! In *Devs'* first episode, Amaya's CEO, Forest, and his new programmer, Sergei, discuss alternate worlds and branching. I assume this is an allusion to the Everett theory. But as I understand it, there are many alternate worlds interpretations, one of which is the Boehr theory, which is just Schrödinger's equation with no collapse postulate.[3] Everything is purely deterministic.

**Alex**  Goodness. I should be asking you all questions. Let me provide some background. I struggled in school, only coming to science and history in my twenties. I subsequently found myself, while engaged with these areas, struggling with my intellectual limitations. In my early thirties, when my first child was born, I wrote *Sunshine* (2007). I finally got my head around the concept of entropy, and I understood some of its correlates, like heat death and the end of the universe. So I wrote a story with a strong idea that I still admire. Entropy leads to a provocative question: should someone stop another group from saving the world? If the world's saved, you're just deferring the horror of extinction to your grand or great-great grandchildren. You're just pushing off a horrible existential moment to another group of people, which is wrong.

While I was writing *Sunshine*, I kept jettisoning science in favor of a "jump scare" or an action sequence. I realized I had failed because I hadn't thought hard enough about the science. I'm not trying to be self-flagellating, it's just a fact. When I wrote *Ex Machina* (2014), I both got more serious and cut myself more slack. I'd do the best I could, which would not stand up to the inspection of you

---

[3] "Bohr Model," *Wikipedia*, at https://en.wikipedia.org/wiki/Bohr_model (accessed December 17, 2020).

guys, frankly. When writing *Devs*, I became interested in a cluster of ideas related to free will and determinism. Could free will exist in a probabilistic state? Or is there a hidden variable, but the world is deterministic absolutely? I got to Everett when I read David Deutsch's great book, *The Fabric of Reality*, tuning in and out when the text got too difficult.[4] When writing *Devs*, I think I incorporated a lot of his material on an unconscious level.

For me, quantum mechanics works because its core truth relates to the unconscious. Our sense of our own objectivity, which we like to think is neutral and transparent, in truth isn't. The world as it is, which includes other people and physical realities, including interactions among particles, is profoundly counterintuitive. These dichotomies seem as if they're absolutely fundamental to existence, to figuring out whatever the fuck you're trying to do in the world, whether it's relationships or employment. You've got to guess at and figure out yourself and the things outside of yourself. In my case, recently, I've been trying to stop my son from taking too much ketamine, because he's sixteen and discovering the world of drugs. I have to project myself into his head to understand what words will land with him. The process is counterintuitive in all sorts of ways.

Quantum physics gives you a framework to think about these kinds of questions with. You may believe in one of the Everett strands, and someone else might argue in favor of the Copenhagen, and you'd have a big fight about it, and at the end, both of you, and the observer in me, would have no idea who was right.[5] Quantum physics seemed like a really terrific space because all of the odd things about living seemed contained within it.

In my work, I'm thinking as hard as I can without a science background and trying to then present a narrative that works on a technical storytelling level as well as through unconscious drives without dumbing anything down.

**Jay** Can I say something about your understanding of science? I think your experience is everyone's experience. I'm thinking about David Rumelhart and Geoff Hinton, two of the most insightful neural network research, perhaps also Schrödinger—there's a feeling that there's something there that I'm also trying to figure out. Geoff Hinton is a perfect example of this. The number of times he's

---

[4] David Deutsch, *The Fabric of Reality: The Science of Parallel Universes and its Implications* (New York: Viking Press, 1996).
[5] "Copenhagen Interpretation," *Wikipedia*, at https://en.wikipedia.org/wiki/Copenhagen_interpretation (accessed 17 December 2020).

come up to me and said, "I think I've got it; this is the answer to everything," but we all know that this is just the next step for him.

We're all struggling at the edge of understanding. I've just finished a paper, and I knew I had to write the damn thing. I didn't realize what I was writing about at all until, I was just, like, darn it, I'm going to keep reading this literature until I feel like the answer comes to me. It took me three years to write the paper, and finally, there it is! It's pretty much what I was hoping for. It's not the answer to everything yet, it's just that I know more than I knew before. Maybe it's useful to people.

**Alex**   That's a generous point and you've let me off the hook. I'm drawn to the ways scientists write, talk, and think. Some see arrogance, but I see humility. I'm moved by the ways scientists will walk away from a substantial amount of work, if it's demonstrated that that work is in some ways wrong. In other of life's domains, this is unusual.

**Simon**   I feel compelled to interject that that's a rather generous view of scientists. I think Thomas Kuhn quipped that science proceeds funeral by funeral.[6] People will hold on to an idea that's shown to be false because they cherish it as part of their worldview or, frankly, their living. I saw that a lot in linguistics too. The neural network people, I think, are the bravest—"This just doesn't work, let's do something else."

**Jay**   Simon, even Jerry Fodor, the philosopher who got so much wrong about the ways the mind processes language, has provided us with benchmarks and criteria.[7] All of my students are putting Fodor on their first slides now; we're trying to solve his problems. So even those rogues who were guilty of ad hominem attacks in the service of their points of view, and who made people feel horrible for such a long time along the way, still provide a valid, useful service. There's some silver lining there.

**Alex**   Maybe I'm putting scientists on a pedestal. Maybe it's the scientific process that's valuable: its superstructure and its systems of checks and balances.

---

[6]   "Planck's Principle," *Wikipedia*, https://en.wikipedia.org/wiki/Planck#27s_principle#:~:text=Informally%2C%20this%20is%20often%20paraphrased,force%20of%20truth%20and%20fact%22   (accessed December 17, 2020).

[7]   "Jerry Fodor," *Wikipedia*, at https://en.wikipedia.org/wiki/Jerry_Fodor (accessed December 17, 2020).

We don't have these in the entertainment industry. The box office might appear as some sort of loose meritocracy, but it doesn't stand up to scrutiny. I think peer systems, peer review, does.

**Jay**  Science requires, like democracy, absolute continual vigilance and fierce dedication to ferreting out the truth, and we endure long winters of dominance by ideas that are still hanging up there, because the guys who are defending them aren't letting go. This happened in my field, where researchers became convinced that neural networks could not work as solutions to harder problems in AI and cognitive science. Similarly, there's a book about fat in your diet, *The big fat surprise*, one of my favorites, that's completely unrelated to these scientific questions. The researchers promulgated theories about what you should eat drawing on flawed assumptions and mutual agreement overstating the case against dietary fat, convincing themselves they were honorable and right. It took decades to discover how off-base they were.

**Selmin**  Alex, you mentioned science as a narrative container. It gave you a superstructure. You started with familiar genre films like *Sunshine*, where you'd have a monster or a well-timed action sequence. Does serial television narrative give you a chance to subvert form and slow things down? Are there now new possibilities for exploring principles like those in particle physics?

**Alex**  First, not all of my narratives involve science or philosophy. *Annihilation* (2018), for example, is consciously unconcerned with that. With the ones that do, imitating science, I like to employ rigor. Very often with science, material will encourage philosophical reflection. Perhaps science not only provides data, but also implications. This seems like the relationship between stories and themes.
   With films, a principle, an idea, and the idea's implications can interrelate. With *Sunshine*, a film that went wrong, its story and themes conflicted with each other. And with a narrative, that's a mistake. You might say, with artistic outpourings, there are no mistakes, but I disagree. With *Ex Machina*, *Annihilation*, and *Devs*, I sought a tight correlation between story and themes. In *Ex Machina*, I tried to convey a point with the Jackson Pollock drip painting. The villain was engaged with automative processes, so this worked. The idea engaged me so much that I shot *Annihilation* through automatic filmmaking. Here, I shed myself of rigor, except for the constraints of writing a form as unconsciously as possible. I think it's best to have a relationship between a scientific idea, a philosophical idea, a story, and a theme.

**Jonathan**  I'd like to pick up on Selmin's question. I've always been interested in nuts and bolts, in those dishwasher moments you mentioned, when you're wrestling with scientific concepts and trying to translate them into stories with engaged characters and emotional depth. When did it strike you that film could serve as a meeting point between scientific experts and non-experts?

**Alex**  I think a novel or stage play, like a film, could serve as a meeting place. Film, however, is especially interesting because it enables synthesis. Every filmic moment can be a point of harmony between people and different disciplines. For any one image, you'll need a location, a production designer, an actor. You've also got sound design and music, production design and dialogue, the ways the camera is moving or not moving, the preceding shot and the edit that led into it. Very quickly, you've got an incredibly complex layering. The harmony here makes a good meeting point for science. My filmmaking method is anti-auterist. A harmonic point arrives through tested discussion with many people providing input. I employ a peer-review process. Not all directors do this, but they're just kidding themselves. Behind these quasi-auteurs, there's a large group of people picking up the pieces. Some directors I've observed think there are mysterious elves behind them in a shoemaker's factory.

I'm constantly justifying ideas. *Ex Machina* explores the Turing Test.[8] I wrote the script and then gave it to a scientist who told me, "That's not a Turing Test, because you can see who you're talking to." Turing's clear that the interlocutors are hidden. I immediately then inserted some dialogue noting that the characters aren't conducting a classic test. Meeting with scientists actually makes for a better film. Viewers aren't watching a sequence of tiny, irritating mistakes. Now, the whole thing stands up. When someone says, "Shoot him with a neutrino gun," I think, "What the fuck's that going to do?"

**Jeff**  I wanted to loop back to your point about automatic filmmaking. Automatic writing and automatic painting are both things that solitary people do, but filmmaking, as you describe it, is so collaborative, and requires such a high degree of planning and coordination. Can you say a bit more about what you mean?

---

[8] "Turing Test," *Wikipedia*, at https://en.wikipedia.org/wiki/Turing_test (accessed December 17, 2020).

**Alex**  I've worked with a crew many times, and I've got a shorthand. On set, we're nodding, pointing, thumbs-upping; we're hardly talking. I seek a connection between the action and the unconscious—what motivated the action? How closely can you correlate a motivation for an action to an unconscious decision? You're standing over the painting, dribbling different colors, and getting a new result. When you've worked in film for a long time, you start to realize the relationship between unconscious processes and what you're doing. Like what Jay said about discovering what he was writing about while he was writing it. I don't know why I put that character in that space. Then, later, I see, "Oh, that's what the preoccupation was." You lean into instinct, into shorthand, and the crew facilitates the process.

One way I rely on the unconscious is through voiceover. But you can have a character state something through voiceover explicitly, and for the audience, it's just Charlie Brown's teacher droning: "boh boh boh boh." They don't hold onto it. And then, at another moment, with just a tiny bit of information, which the film draws on constantly, just a glance—a flick of an eye—a massive amount of information is telegraphed, which is globally perceived by the audience. I think we consciously use techniques that talk only to the unconscious. They're realized through parameters like lighting, music, sound design. Film is actually more given to automatic processes than it might first appear.

**Carol**  Alex, can you give an example of how lighting or some other techniques might resonate with viewers unconsciously? On the one hand, in *Ex Machina*, you've got some obvious, overt, intentional, structural forms: Ava must possess these five attributes for intelligence, and she must have a considered plan and a knowledge of the men and the compound to break free. But on the other hand, the film has long stretches that are given over to the body and sensuality, just fingers touching the neck, or playing with food, someone lounging. How do these work together?

Why don't I screen a scene for us? Here's a close-up of Caleb's ghostly, almost fluorescing fingers that matches Ava's. Next, we cut to Kyoko preparing sushi, and the film intimates that both a knife or sex can pierce both machine and flesh. I think viewers form these associations below the level of conscious perception. Jeff, Jonathan, and I discussed the film's images of race apart from whiteness, and they're often dispersed and subtle, including the masks on walls, ornaments on tables, the one Black recumbent, obscured robot. How much do these kinds of fleeting details speak? There's so much to attend to, so much layering in the film.

**Alex**  Just because you don't first notice these details doesn't mean they're speaking directly to the unconscious. It's just the way we react to the world. The unconscious, in the way I mean it, is something which is pushing you toward doing something you haven't yet been aware you're thinking of. If I start laying claim to some of these things, that would itself be the unconscious in operation.

**Carol**  Here's another example from *Ex Machina* where images register on a level viewers may not be inclined to access and reflect upon. When Ava seemingly envisions images of being outside in black-and-white forests, it's not clear to whom they belong, because she's never been let outside. And then I noticed the computer screens projected these same black-and-white images of the forests. What are you doing here?

**Alex**  They belong to the audience, those images. They're not literal, they're part of the ambient state of the story. With storytelling, the storyteller seems to transfer information to the recipient, but it's not a one-way street at all. At least half of a story is an imaginative process that's a little like when two questions are asked one after the other, and the second immediately subsumes the first. But that first one is still operative. And that is exactly what happens in stories. Stories are mutually imaginative processes between the author who apparently imagined it, and the person who subsequently imagines it.

**Carol**  Let me roll the moment I mentioned about Caleb, Ava, and the hands again and see how it feels for us.

**Alex**  It's interesting that you chose that one.

**Carol**  See? I didn't even notice that her face turns to shadow, after his. What do we do with this dark moment?

**Alex**  In *Ex Machina*, Caleb, a surrogate for the audience, is asked whether a machine has an interior life or not. At a certain point in the film, both he and the audience stop asking this question. Why? It's got to do with your intuitive state. Suddenly, viewers think they know the answer to the question of whether or not she has an internal life. From there, we're wondering if she's manipulative or not, but anyways, yeah.

**Jonathan**   Can you speak a bit about how you've been thinking about gender in your films, specifically *Ex Machina*?

**Alex**   I think about these issues very hard, but I'm also fifty. I grew up reading George Orwell, and I do not believe in thought crimes and I'm concerned by totalitarianism. Liberal thinking is actually complex thinking. and so cancel culture is something I feel extremely alarmed about. I make very deliberate decisions in all of these dramas we're talking about.

In *Ex Machina*, is gender an appearance? Is it something which is contained, or is it something that is conferred on you by someone else? I have played games with gender—not in a frivolous way, but thoughtfully, but also in a way that's hazardous. In *Devs*, there's a cis-gender male character played by a girl. With *Annihilation*, all of the major characters are female, but that wasn't its point, even though it was important to critics. I could go on, but the short answer is I think about gender a lot.

I think hard about these issues, and I also reserve the right for myself to make mistakes. The protagonists of *Devs* are Asian. The actress, Sonoya Mizuno, who plays one of the parts, is also in *Ex Machina* and *Annihilation*. On the set for *Ex Machina*, she said to me, "Can you name a bunch of famous Asian actors?" and I couldn't. Now, I'm self-evidently a middle-aged white guy, but I'm also a liberal, and I was very surprised about myself that I had never noticed that before, because it's not rocket science.

In the end, it all goes back to Everett and quantum physics, because it's to do with the counterintuitive states that we actually exist in the whole time. When we were talking about scientists being reluctant to give up their ideas just now, I immediately was thinking that that's what unconscious bias is. Unconscious bias in a data set makes you see one bit of data better than another.

**Simon**   On the topic of gender: in Turing's original paper from 1950, it isn't about a computer fooling a person into thinking that they're talking to a person. It's a computer successfully fooling the human interlocutor into believing some arbitrary thing, and Turing very clearly shows that that arbitrary thing concerns women and gender. In the experiment, the computer and the human interlocutor behind the screen have to talk as if they were women and fool the subject. *Ex Machina* I thought got more to the spirit of the Turing Test, which is some combination of gender and trickery that's very subtle. And I felt personally, and this is what I say in my *Cybermedia* chapter, which I co-authored with a philosopher, that the issue of consciousness dissolves out. It's more, "Can this person fool and manipulate two very smart people into doing what she wants?" That's when the movie's excitement switched on for me.

**Alex**  But I also need to trip and fool the audience. They need to be thinking in and making the same mistakes as the two guys, or else the ending doesn't function. There are two moments at the film's end that I regret, one which I excised. One is the demonstration of Ava's consciousness, because I do think it's a valid question whether she has it or not. Many viewers felt it was about a machine without empathy, and this has always bothered me. There's a moment, as Ava escapes from the compound, when she looks back over her shoulder and smiles. It seemed to me that a smile at an empty room was as good an indicator of an interior life as you could have, because it's not performative, it's not for anybody else.

There's another moment when we leave through the helicopter and we see the world through Ava's eyes. We intended it to be a version of Nagle's essay about what it means to be a human imagining being a bat.[9] (What would webbed flight and echolocating be like?) I wanted to show Ava's consciousness as being completely unlike ours. I could only do that in rather prosaic ways, by dropping graphics in and stuff like that, so it looked too crude. The point was missed. Ava's consciousness, as an AI, would not even remotely resemble ours. I also often think that anybody's consciousness wouldn't brilliantly marry up with ours as well. So, we're in the same position. I cut it, but I slightly regret it.

**Jay**  I'm glad you cut it. There's no good answer to the question of robot experience, so you shouldn't be distracting us with it.

**Simon**  If I could just interject, that scene at the end changed the movie in a way that was phenomenal. The last scene looked like crowds walking across streets in Midtown Manhattan. I know very little about filmmaking. I can't express why it was so powerful for me.

**Paul**  Alex, let me turn back for a second, before we get to Simon's point. I really like your explanation of how the protagonist android could potentially have completely different experiences from us that we couldn't even imagine, and you

---

[9]  Thomas Nagle, *Mortal Questions* (London: Cambridge University Press 2012), 165–80.

would try to get that across in some way. That is something that Nagel at least sort of gestures toward in his paper.

There is another alternative to this. Not all philosophers or vision scientists think we have internal lives completely different from each other. At least we all sense the environment. Ava could be experiencing the world much as we do because she's experiencing the properties of the world. If they've designed her with some sort of visual acuity in a way that we have, for example, so that she can talk about colors and things like that, those could be surface properties. Even physicists talk about these reflective properties.

**Alex**  So it would be like the octopus eye or something. She arrives by a different means at the same thing.

**Paul**  Yes. You could still pick up the same kinds of properties, though they wouldn't necessarily be organized in the same way. Concerning vision, they'd probably be organized three-dimensionally. It's sort of like your earlier discussion of scientists' disagreeing with each other over the Everett interpretation. One reason why you can do that and hang onto your particular interpretation is because there's no fact of the matter which can solve the argument for anyone in those cases. With Everett, you could interpret Ava's experience with the many minds approach, where lots of different minds split, or in the many worlds way, where everything splits. Or, you could use the Boehr theory, where nothing ever splits and everything just continues to branch and multiply, with each and every superimposition existing in the world at all times. And these are all bizarre interpretations of quantum mechanics, but there's no fact of the matter that tells you one of them reigns supreme over the others. I would have loved to have seen that, so you actually should do a director's cut where you do that.

**Alex**  But the version in circulation is the director's cut. We cut the scene because it didn't work, it didn't make the case. It just doubled down on the problem I kept seeing, of people perceiving her as a cold, unempathetic robot because she killed two guys. Ultimately, that's why I lost it. Perhaps that's why the Manhattan city street that Simon likes works so nicely—it's so open.

During *Ex Machina*, I started thinking about qualia, if that's how you pronounce it. You can have involved arguments about whether a phenomenon is true or not, whether it's a good word for a good thing or a stupid word for something that isn't there. And that's like the Boehr versus the Everett. Yes, there's no fact of the matter. There isn't. The interesting and good arguments don't lead to any particular place. But this is the problem. Because what am I? I'm a layperson, and to whatever extent you guys don't know, I don't know even more. I don't know as much and it gets very nerve-wracking making cases within a narrative where you can feel your brain bleeding with the ideas, climbing up the mountaintop like this, but it's a bit like the other thing, you have to reserve the right to be wrong and just give it a good swing.

**Jay** Well like I said before, that's what we're always all doing anyways.

**Alex** Well some of you guys are Babe Ruth, so your big swings hit the ball a little better, you know?

**Paul** Well, I think it'd be a great project for you to make a movie that addresses qualia. It'd be hard to do, but fascinating! Maybe without using the word "qualia"...
You've already talked about robots, androids, zombies (your film *28 Days Later* [2002] has zombies), and the Everett theory. These are fantastic themes that have pointed in this direction, but maybe haven't been addressed in a core way.

**Alex** That's good. Though we should remember, with a film there's limited bandwidth. You've got two hours—or, if you're Steven Spielberg, three hours or so—it's limited. Very often, by the end, I'm namechecking an idea with the hope that someone will go to Wikipedia and read a bit more about it. And it's frustrating that quantum mechanics has become a conceptual rubber band that has stretched and virtually snapped. I wish there were a communication point between a way of living in the world and a broader way of looking at the world, because Everett right now is unreachably distant. Google Everett. There's a value in namechecking. A bridge needs to be constructed or reconstructed.

**Paul** Well, don't feel bad about not understanding quantum mechanics because nobody does. I mean, including Feynman who's one of its founders, who famously said, "Nobody understands quantum mechanics, and if they say they do, they're lying, or they don't understand it anyway."

**Alex** Thank you! The things you guys have shared will filter into whatever I think about and work on next. It's also just that the communication between these spaces is very valuable. I want to, without trying to sound cute, thank you for it, because I mean it.

**Jay** Likewise, what comes from you informs us as well. So rather than say an extended passage about that, I feel, reciprocally, very much the same way.

**Paul** Yes, and thank you for your movies!

# Bibliography

Deutsch, David. *The Fabric of Reality: The Science of Parallel Universes and its Implications.* New York: Viking Press, 1996.

Nagle, Thomas. *Mortal Questions*, 165–80. London: Cambridge University Press 2012.

Rumelhart, David E., James L. McClelland and PDP Research Group. *Parallel Distributed Processing.* Cambridge, Mass.: MIT Press, A Bradford Book, 1994.

Wikipedia. "Bohr Model," at https://en.wikipedia.org/wiki/Bohr_model (accessed December 17, 2020).

Wikipedia. "Copenhagen Interpretation," at https://en.wikipedia.org/wiki/Copenhagen_interpretation (accessed December 17, 2020).

Wikipedia. "Jerry Fodor," at https://en.wikipedia.org/wiki/Jerry_Fodor (accessed December 17, 2020).

Wikipedia. "Many-worlds Interpretation," at https://en.wikipedia.org/wiki/Many-worlds_interpretation (accessed December 17, 2020).

Wikipedia. "Planck's Principle," https://en.wikipedia.org/wiki/Planck%27s_principle#:~:text=Informally%2C%20this%20is%20often%20paraphrased,force%20of%20truth%20and%20fact%22 (accessed December 17, 2020).

Wikipedia. "Turing Test," at https://en.wikipedia.org/wiki/Turing_test (accessed December 17, 2020).

# 3

# (S)*Ex Machina* and the Cartesian Theater of the Absurd

Simon D. Levy and Charles W. Lowney

*Is it possible for a machine to think? . . . It is as though we had asked "Has the number 3 a colour?"*

Ludwig Wittgenstein.[1]

On first viewing *Ex Machina* (Alex Garland, 2014) when it came out six years ago, we—like other academics—were struck by the homage it paid to Ludwig Wittgenstein: not only in explicit acknowledgments like a search-engine company named Blue Book, but also in more subtle details like a setting that resembled Wittgenstein's isolated retreat in rural Norway, and even a brief glimpse of Gustav Klimt's portrait of Wittgenstein's sister Margaret Stonborough toward the end of the movie. Not since *Blade Runner* (Ridley Scott, 1982) over thirty years prior, had a movie engaged so profoundly and seriously with fundamental issues like personhood: the very question of what it means to be human.

On a second viewing, we are more inclined to see *Ex Machina* as a dark comedy bedroom farce, focusing on Nathan, an aging, alcoholic frat bro who somehow manages to overcome glaring personal shortcomings to build an impressive harem of hyper-realistic sex dolls to bully and dance (and presumably copulate) with. Worse, he dupes a lonely, orphaned, virginal young man, Caleb, into becoming a white knight to facilitate the escape plans of his latest model, who turns out to be smarter (or perhaps, less obviously impaired) than either of the two men.

In this paper, we will attempt a synthesis of these two views. The crux of our argument is that what is revealed by *Ex Machina*, and less successful movies like

---

[1] Ludwig Wittgenstein, *The Blue and Brown Books* (New York: Harper and Row, 1958), 47.

*Morgan* (Luke Scott, 2016), is that the notion of consciousness currently promoted is not only peripheral, but antithetical, to understanding what artificial intelligence (AI) and robotics are really about. We will try to get at what is misleading about testing a machine for consciousness, what is helpful in trying to get a machine to model human thought, and how the games we play might encourage an empathetic extension of mental and emotional words to machines.

## Descartes' Internal Theater

From Wittgenstein's perspective, looking for consciousness in machines is absurd. It encourages the "conjurer's trick" that subtly gets us into philosophical quagmires about mental states and mental processes in the first place.[2] If a medic rubs a knuckle on an unresponsive patient's sternum and the patient reacts, then the medic is perfectly right to say, "The patient is conscious." But when we shift from the use of a word in ordinary language to a pseudo-scientific use that tries to answer deep philosophical questions about personhood, we can be misled. With the versions of the Turing Test on display in *Ex Machina*, it is clear that here we are hoping to examine questions that are less ordinary and much more philosophical.

If we look at Murray Shanahan's later chapter in this book, we see the evolution of the term "consciousness" from Turing to Ex Machina. Shanahan describes a consciousness like ours first as the ability to think, then as the ability to have general rather than specialized intelligence, and then as something that can suffer and have its own goals and longings. What seems important here is a unified awareness that has the experience of thought or feeling—and has a sense of self.

Consciousness evolves to be the thing we look for when we want to say that something is enough like us to warrant the consideration we owe to each other. But then consciousness starts to look like a thing or a state of being that we can see in ourselves and that we can look for in others; something that—if it is really there—we might be able to test for. But is it an "it" and is it really "there"?

Wittgenstein believed that we were misled by language to believe that the *I* was a thing or state (a subject for the verb) that we could turn around on and

---

[2] Ludwig Wittgenstein, *Philosophical Investigations* (Oxford: Basil Blackwell, 1953), 308.

look at. The problem that leads to the "problem" of consciousness is that we start to treat mental states as something that exist independently from our activities; we imagine that they precede these activities in some way, and that they can be examined in the way we examine external things. We imagine we can "point" to these internal things mentally with our attention, just like we can point to an object with a finger. But Descartes's idea that we can look inside our minds and see its contents is misconceived. Wittgenstein would say that it is wrong to think of consciousness as a something we can observe and detect, just as it is wrong to think that thoughts and emotions are things parading across an inner theater.

## Turing Te{x|s}t

Depending on who's trying to sell you what, The Turing Test—the central plot device in *Ex Machina*—was either solved decades ago, or remains unsolved.

The Turing Test, also known as the Imitation Game, was developed by Alan Turing in 1950 to assess a machine's ability to display behavior or intelligence similar to, or even distinguishable from, human thinking. The been-there-done-that crowd cite the notorious ELIZA program from the mid-1960s.[3] As in Turing's original "Imitation Game proposal"[4] (in which, we note, Turing proposed asking questions about gender-related issues like hair styling), Joseph Weizenbaum had his subjects communicate with a machine over a remote connection, by typing questions and reading answers. Unlike Turing, however, Weizenbaum set up his AI as a Rogerian (non-directive) therapist, whose responses consisted of either meaningless hedges (*Interesting, please go on . . .*) or the simple copy and pasting of keywords (*Tell me more about your {mother, family, etc.}*).

The results proved, in a word, tragicomic: many of the "patients" (human test subjects) insisted that they had been talked to by an actual human therapist, even when presented with evidence to the contrary. Some became quite emotional and wanted the sessions to continue.

On the other side of the coin, some AI researchers disagree that any computer has won the Imitation Game. In a recent popular presentation of his Turing Award-winning work on causality, Judea Pearl notes that:

---

[3] Joseph Weizenbaum, "ELIZA—A Computer Program for the Study of Natural Language Communication Between Man and Machine," *Communications of the ACM 9*, no. 1 (January 1966): 36–45.
[4] Alan M. Turing, "Computing Machinery and Intelligence," *Mind* 59, no. 236 (October 1950): 433–60.

Every year the Loebner Prize competition identifies the most humanlike "chatbot" in the world, with a gold medal and $100,000 offered to any program that succeeds in fooling all four judges into thinking it is human. As of 2015, in twenty-five years of competition, not a single program has fooled all the judges, or even half of them.[5]

In evaluating these two well-documented results (ELIZA and the Loebner Prize), we cannot help but think back to the opening section of Wittgenstein's *Philosophical Investigations*. Here, the philosopher noted that ascribing meanings to words makes sense only in the context of what he called "language games." In a now-famous example, Wittgenstein noted that workers constructing a building need merely to hear the boss utter the command *Slab!* in a particular context to deliver a slab to the worksite. There is no need for us to posit an abstract underlying logical representation like *Bring me a slab!*

For us, Wittgenstein's insight on language games translates directly into a critique of the Loebner Prize (fool-the-experts) approach to the Turing Test. We see the Loebner Prize as more of a linguistic meta-game, an intellectualized over-interpretation of Turing's original idea of having a computer interact with an ordinary person in an ordinary-as-possible situation. What Turing had in mind is of course what we now call texting, the "killer app" for mobile-phone technology. As anyone knows who has ever suffered through an online-account signup using a CAPTCHA (Completely Automated Public Turing Test to tell Computers and Humans Apart), computers have already become maddeningly good at fooling us into thinking that their texts are coming from humans.

Indeed, CAPTCHAs provide a two-paned window into the Imitation Game. First, they require solving visual-recognition tasks that until recently would have been thought technologically impossible. Second—and more relevant to our discussion here—the main point of a CAPTCHA is not to provide data for computer scientists to mine, but rather to foil (or "capture") chatbots and other AIs trying to gain access to online accounts, discussion forums, and similar venues in which they can successfully pose as humans for the purposes of advertising, theft, and other profit-seeking endeavors. The potential reward for success at such attempts far exceeds a measly hundred thousand dollars, and has led to an entire industry of CAPTCHA-solving sweatshops.

---

[5] Judea Pearl and Dana Mackenzie, *The Book of Why: The New Science of Cause and Effect* (New York: Basic Books, 2018), 44.

Moving beyond the low-bandwidth medium of text, we find even more compelling evidence for an *interactionist* approach to intelligence. As Julie Carpenter notes, U.S. military personnel working in Explosive Ordinance Disposal (EOD) develop strong attachments to the robots with which ("with whom"?) they work.[6] In addition to providing feminine pet-names for their mechanical co-workers, soldiers have asked technicians to rebuild a beloved robot rather than replacing it at a lower cost, and have even held funerals for robots that have been damaged beyond repair. Similarly, World War II fighter pilots named their planes after wives, girlfriends, or screen starlets and decorated their aircrafts' noses with (often eroticized) pictures of such women. We see all this as further support for our second-blush reaction to *Ex Machina*. Sex and violence—specifically, *the potential to share the pleasure of sex and the pain of physical trauma*—is a basis for empathy and what makes people think of other beings as human.

Although the range of games goes beyond text, or linguistic behavior games, to more complex (and sometimes perverse) human behavior games, we see that the ways we interact with machines can be a basis for an extension to them of terms like "intelligent" or "sensitive" that are ordinarily reserved for humans.

While Turing had something like texting in mind as a means of conducting an imitation game, Ava is intended to model a wider range of behavior, such as body-language, and the expressions of emotion. Nathan lets Caleb *see* Ava because he has a more thoroughgoing test in mind. We go from language as an indicator of "thinking," to language, behavior, and expression as an indicator of "consciousness," but we still imagine some unified awareness that is "doing" the thinking, and we still try to see if Ava has *that*. We mistakenly believe that "external manifestations" of that "inner state" can reach some threshold, i.e., when it fools us enough, that could tell us that an AGI has consciousness.

In this respect, running a Turing-styled test to ask whether Ava has consciousness seems to us about as meaningful as asking whether Caleb has an even or odd number of brain cells: both questions depend on a tangled web of assumptions, are nearly impossible to answer, and are unlikely to yield anything of value.

---

[6] Julie Carpenter, *Culture and Human-Robot Interaction in Militarized Spaces: A War Story* (Abingdon, Oxfordshire: Routledge, 2016).

**Figure 3.1** Alex Garland, *Ex Machina*, 2015, Caleb runs the "Turing Test" on Ava.

## Nature, Nurture, and Deep Dreams

To see how we got here, it is probably helpful to offer a deeper dive into the practice of AI in the twentieth century through the present.

In brief, the field—like any field that concerns itself with understanding human behavior—has vacillated between the familiar poles of nature and nurture. A post-war efflorescence of nurture (learning-based) approaches using neutrally-inspired networks of simple processing units[7] was brought to a halt by criticisms that such networks are incapable of learning non-trivial behaviors.[8] In what has now come to be amusingly called "Good Old-Fashioned AI,"[9] or more technically *symbol-based AI*, researchers studied various intelligent behaviors (stacking blocks, ordering food at a restaurant, escaping from a maze) and attempted to code them up directly in a symbolic programming language like LISP. In other words, it was seen as hopeless to try and nurture our machines into becoming smart; instead, we would simply make it part of their nature to be smart through direct implantation, as Plato (and later, Descartes) said the gods had done—albeit imperfectly—with us.

---

[7] Frank Rosenblatt, "The Perceptron: A Probabilistic Model for Information Storage and Organization in the Brain," *Psychological Review* 65, no. 6 (1958): 65–386.
[8] Marvin Minsky and Seymour Papert, *Perceptrons* (Cambridge, Mass.: MIT Press, 1969).
[9] John Haugeland, *Artificial Intelligence: The Very Idea* (Cambridge, Mass.: MIT Press, 1985).

This situation lasted until 1986, which saw the publication of a two-volume set of books.[10] These books (whose covers were, in a tantalizing coincidence for Wittgenstein fans, blue and brown),[11] ushered in the neural-network renaissance of the 1980s and 1990s. Commonly referred to as the PDP Books—for *Parallel Distributed Processing*—they contained a set of articles outlining a solution to the shortcomings of previous neural models, as well as compelling psychological arguments for using *distributed* representations for concepts. In this sort of representation, any concept, no matter how abstract or concrete, is stored over a large number (hundreds or thousands) of memory units, typically organized into a small number of *layers*, instead of being stored "locally" in a particular place in the computer's memory. This approach, and the *back-propagation* algorithm that works with such representations,[12] dominated AI for the next decade or so, bolstered by heavy funding from the U.S. Department of Defense.

The difficulty of using distributed representations to encode probabilistic causal relationships, combined with the hype and over-promising associated with back-propagation networks, led to another rapid collapse in the popularity of neural networks in the mid-1990s. What came to replace them was something of a hybrid approach: *Bayesian* networks,[13] in which individual concepts are not distributed over many computational units, but are localized to a small number of units connected to other concepts via numerical values reminiscent of the connections in a traditional neural network. Such networks can be used to encode common-sense or expert knowledge directly (*if a patient presents with a toothache and does not have a cavity, the probability of a sinus infection is 60 percent*), but can also learn such relationships from large sets of "training" data,[14] and are still in widespread use today.

Their power and flexibility helped keep Bayesian networks popular with AI researchers from approximately 1995 through the first decade of the current century, at which point the field began its shift toward what we might call the second neural-nets renaissance: Deep Learning. Although initially derided by

---

[10] David E. Rumelhart, James L. McClelland, and the PDP Research Group, eds, *Parallel Distributed Processing: Explorations in the Microstructure of Cognition* (Cambridge, Mass.: MIT Press, 1986).
[11] Ludwig Wittgenstein, *The Blue and Brown Books* (New York: Harper and Row, 1958).
[12] David E. Rumelhart, Geoffrey E. Hinton, and Ronald J. Williams, "Learning Representations by Back-propagating Errors," *Nature* 323 (1986): 533–6.
[13] Judea Pearl, *Probabilistic Reasoning in Intelligent Systems: Networks of Plausible Inference* (San Francisco: Morgan Kaufmann Publishers, 1988).
[14] Kevin Murphy, *Dynamic Bayesian Networks: Representation, Inference and Learning* (PhD dissertation, University of British Columbia, 2002).

some as merely 1980s neural nets with more layers and faster hardware, Deep Learning is now widely acknowledged as representing a true advance in both the kinds of problems that AI can solve and the success with which it can solve them.[15] The rapid adoption of deep-learning based technologies like Siri and Alexa, followed soon after by stunning successes at fiendishly complex strategy games like Go and StarCraft II, have to a large extent revived the fear that AI is going to take over the world, or at least lead us to convincingly human cyborgs like *Ex Machina's* Ava.

## Uploading and the Acquisition of Knowledge

The breathtaking accomplishments of Deep Learning have made it clear to any reasonable person that learning is the royal road to success in machine intelligence. We see two implications for popular treatments of AI.

First: given the crucial role that learning—specifically, learning of distributed representations—plays in current AI (Deep Learning), we find it difficult to accept Nathan's claim to have made Ava intelligent by uploading the contents of a search engine into her "wetware" (brain). The idea of making a computer intelligent by transferring a massive collection of facts has already been tried, most notably in the Cyc (pun on *encyclopedia*) project started by researcher/entrepreneur Doug Lenat over 30 years ago.[16] Though it was periodically touted in the popular press as being on the verge of consciousness (whatever that might mean), a recent treatment was probably closer to the scholarly consensus in calling Cyc "the most notorious failure in the history of AI."[17] As in human acquisition of language and other skilled behavior, there appears to be no shortcut around learning.

Second: given the crucial role that learning plays in (successful) machine intelligence, it is vital to appreciate the great distance between the dominant learning paradigms and the way we acquire the one skill that makes us uniquely human; i.e., language. This point is more subtle and will require a bit more detail

---

[15] Danny Hernandez and Tom. B. Brown, "Measuring the Algorithmic Efficiency of Neural Networks," arXiv.org, May 8, 2020, https://arxiv.org/abs/2005.04305 (accessed December 28, 2020).
[16] Doug Lenat, Mayank Prakash, and Mary Shepherd, M, "Cyc: Using Common Sense Knowledge to Overcome Brittleness and Knowledge Acquisition Bottlenecks," *AI Magazine* 6, no. 4 (1986): 65–85.
[17] Pedro Domingos, *The Master Algorithm: How the Quest for the Ultimate Learning Machine Will Remake Our World* (New York Basic Books, 2015).

to present but, building on the first point, it allows us to see what is typically missing from the modeling of thought and language, and that will show us what is missing from Ava and why she ("it"?) cannot be said to have consciousness in any meaningful sense of the term.

The first phase of Deep Learning focused mainly on classification tasks (identifying handwritten digits, recognizing faces), using *supervised learning*. In a supervised paradigm, the neural net is given a particular pattern (pixels in an image) and a classification (*cat*) to associate with that pattern. The error (difference) between the network's output and the correct classification is then back-propagated through the network's layers repeatedly (often for hundreds of thousands or millions of iterations), until the network achieves success with both the training data and a test set of unseen images. By augmenting the network with *recurrent* connections between later and earlier layers, this approach has also shown some promise for automated language translation (Johnson et al. 2016).[18]

The more recent paradigm, *reinforcement learning*, is the one making headlines nowadays in better-than-human-level game playing and, to a lesser extent, robotics. In this paradigm, as in ordinary game playing, there is no explicit error/correction at a given step; you have to wait until the end to know whether you've won or lost. Like deep supervised learning, deep reinforcement learning relies on back-propagation (along with other algorithmic tricks to compensate for the lack of an immediate error value).

In addition to the implausibility of back-propagation as a model of biological learning,[19] neither of these learning paradigms (immediate error correction or "reward-and-punishment") provides a reasonable model of language acquisition. This observation dates back to Noam Chomky's scathing 1959 take-down of B.F. Skinner's *Verbal Behavior* and should be obvious to anyone who has watched children acquire their first language, with effectively no correction or reward.[20]

Chomsky's arguments ended behaviorism as a serious approach to language acquisition (and human behavior in general) and launched the *cognitive*

---

[18] Melvin Johnson, Mike Schuster, Quoc V. Le, Maxim Krikun, Yonghui Wu, Zhifeng Chen, Nikhil Thorat, Fernanda Viégas, Martin Wattenberg, Greg Corrado, Macduff Hughes, and Jeffrey Dean, "Google's Multilingual Neural Machine Translation System: Enabling Zero-shot Translation," *Transactions of the Association for Computational Linguistics* 5 (2017): 339–51.

[19] Yoshua Bengio, Dong-Hyun Lee, Jorg Bornschein, Thomas Mesnard, and Zhouhan Lin, "Towards Biologically Plausible Deep Learning," arXiv.org, February 14, 2015, https://arxiv.org/abs/1502.04156 (accessed December 28, 2020).

[20] Noam Chomsky, "A review of B.F. Skinner's Verbal Behavior," *Language* 35, no. 1 (1959): 26–58.

*revolution*—which Chomsky himself at one point referred to as the Cartesian approach.[21] Although the tide of scholarly opinion (including his own) has shifted away from Chomsky's original views about a genetically-based "Universal Grammar" (schema) being necessary to explain the patterns that seem to exist across the world's languages,[22] the genesis and acquisition of language stands as perhaps the "final frontier" of challenges for machine learning. Other than some admittedly innovative computational models developed around the turn of the present century,[23] we are unaware of any successful attempts to "train" a computer to understand and produce human (or humanlike) language in the way that children do, without reinforcement or supervised error correction.

To anyone who would point out the existence of chatbots (and Siri and Alexa) as a potential counter-argument, we would reply that these devices are of course pre-loaded with enough exemplars of fluent language to be useful for solving a few practical tasks (ordering from Amazon; hearing a weather report, etc.), plus enough back-prop-derived neural-net weights to handle the vagaries of spoken input.

Indeed, we cannot help but think of the Dominique character from *Mr. Robot* (ironically, an Amazon-produced TV series), asking "Alexa, do you love me?" early in the series. The fact that the writers of this award-winning show chose to use Alexa as a vehicle for portraying the bitterest human loneliness imaginable—seeking a human connection with a talking box that you know isn't even remotely human—makes our point about language better than we can: neither the experience necessary for understanding of the question, nor the experiences that could develop into a real answer, could be cultivated by a back-propagated script.

## Rule-following and Private Language

Discussing what it means to be human leads us inexorably to language, and language leads us (the two of us at least) back to Wittgenstein. In this section we summarize some of our recent work on lesser-known, unsupervised neural-network models as it bears on key questions raised by Wittgenstein about central

---

[21] Noam Chomsky, *Cartesian Linguistics: A Chapter in the History of Rationalist Thought* (New York: Harper and Row, 1966).
[22] Francis Y. Lin, "A Refutation of Universal Grammar," *Lingua* 193 (April 2017): 1–22.
[23] Simon Kirby, "Syntax Without Natural Selection: How Compositionality Emerges from Vocabulary in a Population of Learners," in *The Evolutionary Emergence of Language: Social Function and the Origins of Linguistic Form*, ed. Chris Knight (Cambridge: Cambridge University Press, 2000), 303–23.

issues in this collection. We refer interested readers to that work for greater detail than we can provide here.[24] The point here is to see that computers *can* do a better job of modeling thought and language, but when they do so they start to dismantle essentialist conceptions that support a Cartesian theater approach.

## Vector Symbolic Architectures

Vector Symbolic Architectures (VSA) encompass a class of neural-net models that use high-dimensional vectors to encode systematic, compositional information (like sentences) as distributed representations.[25] VSAs can represent complex entities such as role/filler relations (role=*instructor*/filler=*Levy*), or attribute/value pairs (attribute=*size*/filler=*X-Large*), in a way that every entity—no matter how simple or complex—is represented by a pattern of activation distributed over all the elements of the vector. This is because it does not assign different types of representations for components and for compositions. Typically, the activation patterns are arbitrarily-chosen (random) values constrained to fall within some interval like -1 to +1.

This arbitrariness is to some degree why we call these models *Symbolic*: as in ordinary language, where the relationship between signifier and significand is mostly arbitrary (what we call *dog* in English, is called *maile* in Samoan), VSA deliberately avoids the essentialist view of positing a common internal representation of the world among minds. In our more extensive treatment of VSA and related architectures we have argued that this feature makes VSA a strong candidate for addressing the anti-essentialist "Beetle in a Box" observations that Wittgenstein makes when he criticizes the notion of a private language; there need be no hidden thing in the box, for the word "beetle" to have a use and thus a meaning in a linguistic community.[26]

In the present context, we consider it reasonable to extend this argument to the consciousness debate as well. In brief, we believe that it is possible to arrive at a scientifically productive and philosophically satisfying understanding of intelligent behavior without positing mysterious, hidden substances like

---

[24] Charles W. Lowney, Simon D. Levy, William Meroney, and Ross W. Gayler, "Connecting Twenty-First Century Connectionism and Wittgenstein," *Philosophia* 48 (2020): 643–71.
[25] Ross W. Gayler, "Vector Symbolic Architectures Answer Jackendoff's challenges for Cognitive Neuroscience," in *Proceedings of the ICCS/ASCS Joint International Conference on Cognitive Science*, ed. Peter Slezak, ICCS/ASCS (2003): 133–8.
[26] Wittgenstein, *Philosophical Investigations*, 293.

consciousness—and we believe that VSA and related approaches help point the way.

To make this point more concrete, consider the two mental operations that, along with language, we consider the hallmarks of human intelligence: analogy and induction. Anyone who has taken the SAT exam is familiar with analogy problems; e.g., PESO:MEXICO:: $x$:USA, where the solution is of course $x$=DOLLAR. Analogy seems to present a compelling example of the need for abstract variables (like the $x$ in this currency analogy). In an under-appreciated paper, neuroscientist Pentti Kanerva shows how VSA can solve problems like this without the need for abstract variables.[27]

Induction—the ability to generalize from examples, or derive a rule to follow based on examples—is so fundamental to intelligence that it forms the basis of a standard IQ test: Raven's Progressive Matrices task. In this task, subjects are given a figure like the one below (simplified for our purposes here), and are asked to complete the missing piece at lower right.

Asked how they arrived at the correct solution (three triangles), a person might report that they followed these two rules:

1. There is one item in the first column, two in the second, and three in the third.
2. There are circles in the first row, diamonds in the second, and triangles in the third.

As neuroscientist Chris Eliasmith and his colleagues have shown, however, a VSA can solve this problem without recourse to such explicit rules.[28] Solving the puzzle then corresponds to deriving a mapping from one item to the next. As Rasmussen and Eliasmith show, such a mapping can be obtained by computing the vector transformation from each item to the item in the row or column next to it. The overall transformation for the entire matrix is then the average (or sum) of such transformations. As with the analogy example, VSA offers compelling evidence that humanlike intelligence can be modeled without recourse to abstract entities. We know "how to go on" without variables and rules—precisely the abstract entities that troubled Wittgenstein throughout his career.[29]

---

[27] Pentti Kanerva, "What We Mean When We Say 'What's the Dollar of Mexico?': Prototypes and mapping in concept space," in *Proceedings of the AAAI Fall Symposium: Quantum Informatics for Cognitive, Social, and Semantic Processes* (California: AAAI Press, 2010).
[28] Daniel Rasmussen and Chris Eliasmith, "A Neural Model of Rule Generation in Inductive Reasoning," *Topics in Cognitive Science* 3 (2011): 140–53.
[29] C. Edwin Harris, Jr., "The Problem of Induction in the Later Wittgenstein," *The Southwestern Journal of Philosophy* 3, no. 1 (Spring 1972): 135–46.

(S)Ex Machina *and the Cartesian Theater of the Absurd* 57

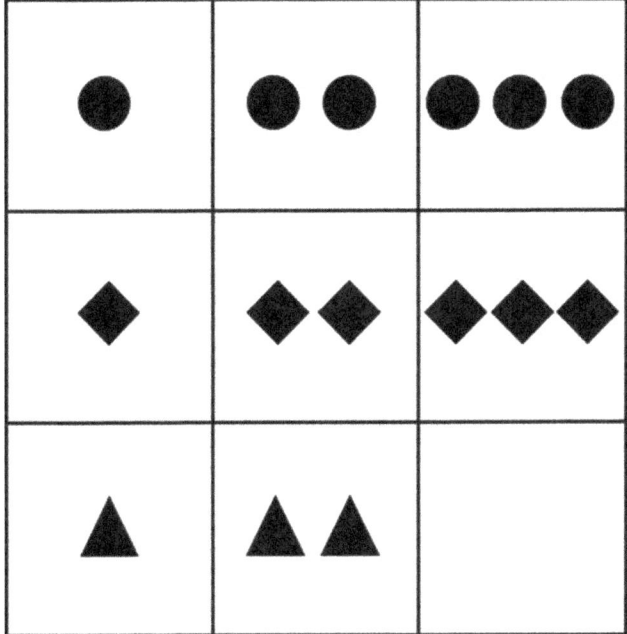

**Figure 3.2** An example of Raven's Progressive Matrices, a popular intelligence test.

## Sparse Distributed Memory

Wittgenstein examined the idea of an internal sensation to try to break us of the notion that there is an independent identifiable thing that one could examine on the theater's stage. To name and identify a sensation, like a pain, seemed to be a private, internal matter, but Wittgenstein shows how even those words are built in a context of relations and behaviors. In a recent article we show how computer processing might even help us understand the development of sensation words and concepts in a Wittgenstein-friendly way.[30]

Sparse Distributed Memory (SDM) is a technology for content-based storage and retrieval of high-dimensional vector representations like those used in VSA.[31] An SDM consists of some (arbitrary) number of address vectors, each with a corresponding data vector. The address values are fixed and chosen at

---

[30] Lowney, Levy, Gayler, and Meroney, "Connecting Twenty-First Century Connectionism and Wittgenstein," 643–71.
[31] Pentti Kanerva, *Sparse Distributed Memory* (Cambridge, Mass.: MIT Press, 2003).

random, and the data values are initially zero. To enter a new address/data pair into the SDM, the Hamming distance (count of the elementwise differences) of the new address vector with each of the existing address vectors is first computed. If the new address is less than some fixed distance from an existing address, the new data is added to the existing data at that address. To retrieve the item at a certain "probe" address, a similar comparison is made between the probe address and the existing addresses, resulting in a set of addresses less than a fixed distance from the probe. The data vectors at these addresses are summed, and the resulting vector sum is converted to a vector of 0s and 1s by converting each non-negative value to 1 and each negative value to 0.

Such a memory is called *sparse* because the set of actual addresses is a tiny fraction of the set of possible addresses; e.g., for a million addresses of a thousand bits each the fraction of addresses used is $10^{29} \rightarrow 10^6$ or $9 \times 10^{-296}$. As well as supporting the storage and retrieval of distributed representations, the memory of an SDM itself is distributed in the sense that the storage and retrieval of each item takes place over a set of locations.

As illustrated by Denning, the distribution of each pattern across several locations produces a curious property: given a set of degraded exemplars of a pattern (such as the pixels for an image with some noise added), an SDM can often reconstruct the ideal form of the pattern through retrieval, even though no example of this ideal was presented to it.[32] One can expect, then, that in cases where there is no original exemplar (e.g., no unitary object-like sensation that can be identified) but there are contextually situated iterations presenting similar patterns—which would give them proximal addresses—the patterns distributed across memory can be conjoined to produce an exemplar or schema.

Suppose one has a sensation. One might notice it occurred about the same time as yesterday (perhaps "it" even occurred the day before as well, but perhaps it was not yet clearly identifiable at that point) one might learn to identify it as "being hangry." Both an internal context developed around this event and external behaviors connected with it. The community's use of the word in a particular language-game also allowed you to identify and apply it. There was much "stage setting" for the new use of the word;[33] it was not a simple "internal" pointing and naming.

---

[32] Peter J. Denning, "Sparse Distributed Memory," *American Scientist* 77 (July–August 1989): 333–5.
[33] Wittgenstein, *Philosophical Investigations*, 257.

On this picture, modeled well by SDM, an "internal" sensation would not be like a "thing" we could "point" to, but more of a pattern that we could begin to recognize. Over time, when different but similar enough combinations of contextually situated cues accumulate (i.e., when several proximal addresses are built), a paradigmatic instance of the concept could emerge (the proximal vectors would constitute a sparse address). So, we see how something that at first glance is apparently unified, independent, and internal can emerge from a context that crosses beyond internal-external, mind-behavior and consciousness-experience divisions. Similarly, consciousness may not be one (unified) pre-existent (before experiences of suffering, etc.), internal (independent from external, detectible actions) thing (like an internal version of external things), i.e., it is not what we think it is from an essentialist and Cartesian view. It may even be what Wittgenstein called a "grammatical" notion, and so it may be misleading to think of it as a thing or even as an experienceable event.

## She's Not One of Us

If, as we have argued, the Turing Test, consciousness, and related issues are not the key to understanding AGI and contemporary fictions like *Ex Machina*, then what is? To see consciousness, to share understanding, we also need to share what Wittgenstein called a "form of life."[34] Computers and Ava modeled human linguistic and expressive behavior via processing representations. The models can come closer and closer to mimicking key overt behaviors, but the *how* is more important to the significance than our own superficial familiarity with a final string of signs. Connectionist PDP architectures like VSA might even better mimic the way we come to overt behaviors than a symbols and rules approach, but we are still playing the imitation game, still looking to superficial resemblances as a marker for something that is integral to living a human life.

What is missing in Ava—especially if she learned to behave in the way Nathan describes, via data dumps and wetware—is learning through experiences with a body that is enough like ours so that it would have the same affordances that ours do.[35] For Ava to truly learn to act human she would have to share in the

---

[34] Ibid., 23.
[35] James J. Gibson, "Affordances and behavior," In *Reasons for Realism: Selected Essays of James J. Gibson*, ed. Edward Reed and Rebecca Jones (Hillsdale, NJ: Lawrence Erlbaum), 410–11.

form of life that humans share in. She would have to build up to her actions from the sorts of experiences and sensations humans have; learning in the way we learn and coming to feel emotions in the way that we do.

Consciousness may not matter as much as learning behaviors, including linguistic behaviors, in the way that a human learns them—not by downloading information but by experiencing actions and words with bodies in particular contexts. If we proceed by processing representations rather than sharing similar ways of life and participating in similar language games, the words and actions of an AGI might be the same and even look like they mean the same as they do to us, but it would not understand—or, rather, it would not "go on" in the ways that we would expect to be consistent with those sounds. Wittgenstein says, "If a lion could talk, we could not understand him" because we do not share the form of life of a lion.[36] How much more could we expect a machine to understand?

But as Norman Malcolm wrote in reviewing Wittgenstein's *Philosophical Investigations*, "The *human* body and *human* behavior are the *paradigm* to which third-person attributes of consciousness, sensations, feelings are related... Thus, there cannot occur in ordinary life a question as to whether other human beings ever possess consciousness...."[37]

Although Wittgenstein would attack the way we tend to understand consciousness and mental states in order to dissolve the "problem" of consciousness, he would not deny our experience or the legitimacy of ordinary language uses of such words. There are legitimate language games in which we use "consciousness," "thought," "self-awareness," etc. Are there legitimate contexts in which we might come to use these words about machines?

Wittgenstein says, "We only say of a human being and what is like one that it thinks,"[38] hence it might be absurd to say that a computer thinks, but the closer something resembles a human, and the more dimension in which it does so, the more it encourages us to expand our use of the language.

Rather than look solely to an AGI for *its* responses to judge whether or not it has "consciousness," we should notice when it draws out responses and behaviors *from us* that we reserve for other people or animals. When this occurs, we see that we are involved in playing a very human game with it. Even pathological games like Nathan's, in which he dehumanizes women via his abuses of machines,

---

[36] Wittgenstein, *Philosophical Investigations*, 223.
[37] Norman Malcolm, "Wittgenstein's Philosophical Investigations," *Philosophical Review* 63, no. 4 (1954): 548.
[38] Wittgenstein, *Philosophical Investigations*, 360.

raise machines up to something human. This and more empathic, less narcissistic games, such as we saw in the relationship that developed between military personnel and EODs, show how human predicates can expand to include machines when they play an important and personal role with us.

The development of new games that we engage in with machines may be enough for us to decide to extend our use of language and say that they "think," "feel," are "conscious" or have "longings." So, if enough machines pass enough informal "Turing Tests" in daily intercourse, we might be encouraged to use these words differently, since the meaning of a word—its life—is in its use.[39]

## Closing Thoughts

And yet, the media continues to worry about the threat of AI. To some extent, worrying about world-conquering AIs is a kind of ghost story for a secular age. It's fun to be frightened, and [nowadays] the nebulous malevolence in the dark reaches of the internet is more credible than dybbuks and djinni. And apocalyptic predictions get more clicks than more realistic headlines such as, "It's hard to say anything definite about AI," "AI will probably be reasonably benign" or "Real AI is probably a long way away from existing."[40]

To conclude, we would argue that what engages us so deeply with movies like *Ex Machina* is—beyond of the obvious sex and violence that have always drawn audiences—a projection of our anxieties about how we view ourselves, other people, and what God (or the gods) might have in store for us as a punishment for our sins.

We can see these tensions play out in characters like *Blade Runner*'s Captain Bryant, who refers to the replicants (genetically-engineered androids) as "skinjobs";[41] in Nathan's creation and mistreatment of the docile, mute, geisha-like android Kyoko; and in *Morgan*'s Dr. Shapiro character (reprising the Fatally Arrogant Jewish Psychologist trope from *The Terminator*'s Dr. Silberman [James Cameron, 1984]) referring to the eponymous, indistinguishable-from-human android teenager as a "goddamn microwave."

---

[39] Ibid., 432.
[40] Zachary Mason, "The Truth About AI: A Secular Ghost Story," *The Paris Review*, December 20, 2018, https://www.theparisreview.org/blog/2018/12/20/the-truth-about-ai-a-secular-ghost-story (accessed December 28, 2020).
[41] In the voice-over of the original theatrical release, Harrison Ford's character notes that in a previous age Bryant would have used similarly offensive language to refer to Black people.

Perhaps rather than looking for consciousness in Ava's eyes, we should be learning to play different and better games; games that extend the use of moral words like "respect" rather than those that emphasize either thought or sensation.

We also need not look too deeply to understand the hype surrounding the "Singularity"—that always-just-a-few-decades-away moment when AI becomes smarter than us and is faced with the choice between using us for its own amusement or exterminating us outright as a distracting nuisance. Apart from the obvious misconstrual of Moore's Law betrayed by such thinking (*faster = smarter*), we find it difficult not to see the Singularity as the Second Coming and Judgment Day for atheists. Once more, as is perhaps only recently becoming appreciated,[42] Wittgenstein—who always carried Tolstoy's summary of the Gospels with him—shows himself prescient on the transformation of profound moral and religious questions by *scientism*; i.e., the modern human tendency to see the Big Questions resolved in reductive and scientific terms:

> We feel that even if all possible scientific questions be answered, the problems of life have still not been touched at all (Wittgenstein, 1922).[43]

## Acknowledgments

We thank Nora Devlin for many of the insights about gender and sexuality that play out in the subtext of this chapter. Simon D. Levy gratefully acknowledges support from the Lenfest Summer Grants program at Washington and Lee University.

## Bibliography

Beale, Jonathan and Ian James Kidd. *Wittgenstein and Scientism*. London: Routledge, 2017.

Bengio, Yoshua, Dong-Hyun Lee, Jorg Bornschein, Thomas Mesnard, an Zhouhan Lin. "Towards Biologically Plausible Deep Learning." *arXiv.org*, February 14, 2015. https://arxiv.org/abs/1502.04156. (accessed December 28, 2020).

---

[42] Jonathan Beale and Ian James Kidd, *Wittgenstein and Scientism* (London: Routledge, 2017).
[43] Ludwig Wittgenstein, *Tractatus Logico-Philosophicus*, trans. Frank P. Ramsey and Charles Kay Ogden (New York: Harcourt Brace and Company, 1922), 6.52.

Carpenter, Julie. *Culture and Human-Robot Interaction in Militarized Spaces: A War Story.* Abingdon, Oxfordshire: Routledge, 2016.

Chomsky, Noam. "A review of B.F. Skinner's *Verbal Behavior.*" *Language* 35, no. 1 (1959): 26–58.

Chomsky, Noam. *Cartesian Linguistics: A Chapter in the History of Rationalist Thought.* New York: Harper and Row, 1966.

Denning, Peter J. "Sparse Distributed Memory." *American Scientist* 77 (July–August 1989): 333–5.

Domingos, Pedro. *The Master Algorithm: How the Quest for the Ultimate Learning Machine Will Remake Our World.* New York Basic Books, 2015.

Gayler, Ross. W. "Vector Symbolic Architectures Answer Jackendoff's challenges for Cognitive Neuroscience." In *Proceedings of the ICCS/ASCS Joint International Conference on Cognitive Science,* edited by Peter Slezak (ICCS/ASCS, 2003): 133–8.

Gibson, James J. "Affordances and behavior." In *Reasons for Realism: Selected Essays of James J. Gibson.* Edited by Edward Reed and Rebecca Jones, 410–11. Hillsdale, NJ: Lawrence Erlbaum.

Harris, C. Edwin Jr. "The Problem of Induction in the Later Wittgenstein." *The Southwestern Journal of Philosophy* 3, no. 1 (Spring 1972): 135–46.

Haugeland, John. *Artificial Intelligence: The Very Idea.* Cambridge, Mass.: MIT Press, 1985.

Hernandez, Danny and Tom. B. Brown. "Measuring the Algorithmic Efficiency of Neural Networks." *arXiv.org,* May 8, 2020. https://arxiv.org/abs/2005.04305. (accessed December 28, 2020).

Johnson, Melvin, Mike Schuster, Quoc V. Le, Maxim Krikun, Yonghui Wu, Zhifeng Chen, Nikhil Thorat, Fernanda Viégas, Martin Wattenberg, Greg Corrado, Macduff Hughes, and Jeffrey Dean. "Google's Multilingual Neural Machine Translation System: Enabling Zero-shot Translation." *Transactions of the Association for Computational Linguistics* 5 (2017): 339–51.

Kanerva, Pentti. *Sparse Distributed Memory.* Cambridge, Mass.: MIT Press, 2003.

Kanerva, Pentti. "What We Mean When We Say 'What's the Dollar of Mexico?': Prototypes and mapping in concept space." In *Proceedings of the AAAI Fall Symposium: Quantum Informatics for Cognitive, Social, and Semantic Processes.* (California: AAAI Press, 2010).

Kirby, Simon. "Syntax Without Natural Selection: How Compositionality Emerges from Vocabulary in a Population of Learners." In *The Evolutionary Emergence of Language: Social Function and the Origins of Linguistic Form.* Edited by Chris Knight, 303–23. Cambridge: Cambridge University Press, 2000.

Lenat, Doug, Mayank Prakash, and Mary Shepherd, M. "Cyc: Using Common Sense Knowledge to Overcome Brittleness and Knowledge Acquisition Bottlenecks." *AI Magazine* 6, no. 4 (1986): 65–85.

Lin, Francis Y. "A Refutation of Universal Grammar." *Lingua* 193 (April 2017): 1–22.
Lowney, Charles W., Simon D. Levy, William Meroney, and Ross W. Gayler. "Connecting Twenty-First Century Connectionism and Wittgenstein." *Philosophia* 48 (2020): 643–71.
Malcolm, Norman. "Wittgenstein's Philosophical Investigations." *Philosophical Review* 63, no. 4 (1954): 530–59.
Mason, Zachary. "The Truth About AI: A Secular Ghost Story." *The Paris Review*, December 20, 2018. https://www.theparisreview.org/blog/2018/12/20/the-truth-about-ai-a-secular-ghost-story. (accessed December 28, 2020).
Minsky, Marvin and Seymour Papert. *Perceptrons*. Cambridge, Mass.: MIT Press, 1969.
Murphy, Kevin. *Dynamic Bayesian Networks: Representation, Inference and Learning*. PhD diss., University of British Columbia, 2002.
Pearl, Judea. *Probabilistic Reasoning in Intelligent Systems: Networks of Plausible Inference*. San Francisco: Morgan Kaufmann Publishers, 1988.
Pearl, Judea and Dana Mackenzie. *The Book of Why: The New Science of Cause and Effect*. New York: Basic Books, 2018.
Rasmussen, Daniel and Chris Eliasmith, C. "A Neural Model of Rule Generation in Inductive Reasoning." *Topics in Cognitive Science* 3 (2011): 140–53.
Rosenblatt, Frank. "The Perceptron: A Probabilistic Model for Information Storage and Organization in the Brain." *Psychological Review* 65, no. 6 (1958): 65–386.
Rumelhart, David E., Geoffrey E. Hinton, and Ronald J. Williams. "Learning Representations by Back-propagating Errors." *Nature* 323 (1986): 533–6.
Rumelhart, David E., James L. McClelland, and the PDP Research Group, eds. *Parallel Distributed Processing: Explorations in the Microstructure of Cognition*. Cambridge, Mass.: MIT Press, 1986.
Turing, Alan M. "Computing Machinery and Intelligence." *Mind* 59, no. 236 (October 1950): 433–60.
Weizenbaum, Joseph. "ELIZA—A Computer Program for the Study of Natural Language Communication Between Man and Machine." *Communications of the ACM* 9, no. 1 (January 1966): 36–45.
Wittgenstein, Ludwig. *The Blue and Brown Books*. New York: Harper and Row, 1958.
Wittgenstein, Ludwig. *Philosophical Investigations*. Oxford: Basil Blackwell, 1953.
Wittgenstein, Ludwig. *Tractatus Logico-Philosophicus*. Translated by Frank P. Ramsey and Charles Kay Ogden, 6.52. New York: Harcourt Brace and Company, 1922.

4

# Epiphany, Infinity and Transcendent AI

Zachary Mason

Georg Cantor was a German mathematician who set out to prove that all infinities are created equal. He failed, in that he ended up discovering infinitely many species of infinity and proving that they're distinct.[1] Before Cantor, infinity had been a sort of annoyance, a mathematical malformity thought best ignored, but he unpacked its complex character and how it's a kind of axiomatic bestiary.

The largest of all the infinities is Absolute Infinity. Cantor, a devout Lutheran, thought that the structure of infinity had been revealed to him by God, and that Absolute Infinity corresponded to or in some way stood for God.[2] One of the properties of Absolute Infinity is that all of its properties are shared by some other, smaller infinity, which starts to sound less like the Lutheran God than the enigmatic God of the gnostics, hidden behind an infinite succession of lesser emanations.

Borges said that he viewed theology as a branch of fantastic literature;[3] set theory, the branch of mathematics concerned with the study of infinity, lends itself to consideration as a branch of theology, and therefore fantastic literature as well. In fact, Cantor denoted the cardinal infinities with the Hebrew letter א (aleph), which Borges used as the title of a short story in which all places and times in the universe can be seen from a single point (the eponymous aleph) in the basement of an otherwise nondescript suburban house.[4]

The writer William Gibson, known as a science fiction writer, has always seemed to me to be in essence a literary writer who happens to have taken up the

---

[1] Joseph Warren Dauben, *Georg Cantor His Mathematics and Philosophy of the Infinite* (New Jersey: Princeton University Press, 1979).
[2] Ibid.
[3] Jorge Luis Borges, *Collected Fictions* (London: Penguin Classic, 1999).
[4] Ibid.

material of genre fiction. One of his favorite writers is Borges, whose influence can be seen running through Gibson's early oeuvre. In his third novel, *Mona Lisa Overdrive* (1988), there's a computer called an aleph with enough memory to model a significant fraction of the world, inhabitants included, so accurately that they can't tell the difference. In Gibson's *Neuromancer* (1984), the AI Wintermute is inhuman in its breadth and kind of mind to such a degree that it can only communicate with people by using simulated human mouthpieces, not just as borrowed voices but as borrowed perspectives, a coherent way to scale itself down. This bears comparison with the Borges story *Everything and Nothing* (1944), in which Shakespeare feels that "there was no one inside him, nothing but a trace of chill, a dream dreamt by no one else." Shortly before or after his death, he meets God, who tells him that, "I dreamed the world the way you dreamt your plays, dear Shakespeare. You are one of the shapes of my dreams: like me, you are everything and nothing."[5]

There are a few narratives strongly associated with AIs in fiction. One is Pinocchio—the AI envies and aspires to humanity and real human emotions. Another is Frankenstein—the AI turns on its creators and tries to destroy them. A third, less narrative than genus, might be called The Castle—AIs as beings vast and remote, radiant with significance but fundamentally unapproachable, and knowable only through analogies and indirect reports. Wintermute and its sibling Neuromancer (for whom personality is not an encumbrance but a medium) are of this kind, but the type specimen is Stanislaw Lem's Golem XIV from the collection *Imaginary Magnitudes* (1985). Golem XIV is an AI developed by the US military that grows beyond military concerns and ends up making itself smarter to the point that it can perceive, however dimly, qualitatively distinct levels of intelligence above its own. As it gets smarter it reaches the verge of being unable to communicate with humanity: it says it has nothing in common with humanity except intellectual curiosity, and that to descend to a human level is a degradation. Human scientists try to understand Golem XIV and its more intelligent brethren but all in vain as it's impossible to understand a larger mind with a smaller one.

Even farther from humanity, but also due to Lem, is Solaris, from the book[6] (and then films by Tarkovsky[7] and Soderbergh[8]) of that name. Solaris is a

---

[5] Ibid.
[6] Stanislaw Lem, *Solaris* (New York: Walker & Company, 1961).
[7] Andrei Tarkovsky, director, *Solaris* (Russia: Mosfilm, 1972).
[8] Steven Soderbergh, director, *Solaris* (California: 20th Century Fox, 2002).

planet-like object in a distant solar system which might be intelligent but whose nature is fundamentally unclear. It's surface and even the space surrounding it are home to phenomena which are certainly structured and possibly intentional. There's evidence that it can alter its environment, though its motives, if it has motives, are opaque. The first part of the book contains a brief history of Solaristics (the discipline of the study of Solaris) which details the various brilliant approaches scientists have taken to Solaris, all of which have come to nothing; there's always more data but never an answer or any solid ground. Solaris exists as a brute fact, defying interpretation. It seems to shimmer and seethe with meaning but is in fact a black hole for meaning. Its nearest literary analog is less a transhuman AI than the house of Mark Danielewski's *House of Leaves* (2000). In that book, a couple, the Navidsons, buy and occupy the house and find that it's slightly but measurably bigger on the inside than the outside. Will Navidson, a war photographer, is unable not to investigate, and under his investigation the discrepancy grows without limit, opening up into an endless succession of ashen halls, galleries, staircases, branching infinitely through into an articulation of nothingness.

Among its (infinitely?) many readings the house can be read as an architectural incarnation of infinity, much as Golem XIV and its ascending chain of superior intelligences resemble Cantor's hierarchy of infinities. Cantor thought that Absolute Infinity was in some way God, an articulation of the absolute; transhuman AIs fill a similar aesthetic role as objects of wonderment and awe, but framed in the language of science rather than mysticism. Like the gods, transhuman AIs don't exist, but as we get farther into the Age of Computation one wonders: could they?

Some would say that they already do exist in the form of deep learning systems. Kasparov, the world chess champion, said that when he was playing the chess program Deep Blue, he had the sense of sitting opposite an alien intelligence.[9] Moreover, it's difficult or perhaps impossible to articulate just what's going on in deep learning systems at a detailed as opposed to a broad algorithmic level. On the other hand, Deep Blue and its successors are neither able to articulate their limited expertise nor generalize it, which, intuitively, are requirements of intelligence; a diesel engine doesn't understand how to convert diesel fuel into kinetic energy, even though it's good at it. Despite the hype, deep

---

[9] Hans Moravec, *Robot: Mere Machine to Transcendent Mind* (Oxford: Oxford University Press, 1998).

learning systems seem less like AIs and more like very large systems of linear algebra supported by vast quantities of computation and data.

The awestruck view of deep learning is to some extent a manifestation of the ELIZA effect. ELIZA is a simple computer program dating from 1966 designed to simulate a passive Rogerian therapist. People interacting with ELIZA tend to come away with the idea that it has a deep understanding of their emotional lives, though in fact it understands neither their emotional lives nor English nor anything. In general, people are very willing to attribute actual understanding to simple programs if there's any room at all to make such an attribution. This is analogous to the roots of animist religion: there's a hard-coded human disposition to attribute events to the actions of actors, i.e., if the wind blows or the snow melts, someone, or Someone, must have made it happen. Similarly with deep learning systems—if chess is being played well, someone must be playing it well.

Consider the end of the Alex Garland movie *Ex Machina* (2014). Ava, an AI in a robotic body that's approximately human, has escaped her creator's remote compound by charming and manipulating her human Turing tester. We last see her walking off into some North American city, anonymous and finally free. It would be a reasonably optimistic scene if it had been left at that, but for a moment we see the world from Ava's point of view as densely annotated, coldly digital information streaming by, which highlights her fundamental strangeness. It's a memorable cinematic moment, but also a sleight of hand. The mechanics of Ava's computational view of the world are displayed to highlight her alienness, but the mechanics of the human brain processing the same urban scene could be displayed so as to be disorienting, which suggests that perceived alienness is a matter of presentation.

Let's consider the problem of transhuman AIs from another point of view. In general, what are the limits on what an AI can do? Certainly, there are limits on what computers can do in general. I can believe in AIs that are articulate and insightful, though none yet exist, and perhaps in AIs whose minds are as vast as the sky and beyond all human comprehension, but I can't believe in one that understands the dynamics of a turbulent watercourse or that can accurately predict the weather. The essential character of chaotic systems is irreducible complexity, but the essential character of intelligence is the making and interweaving of abstractions, which makes the two incompatible.

Rehoboam, from Season 3 of the *Westworld* TV series (Jonathan Nolan, 2020), is an AI apparently of this impossible class. It can accurately predict, and therefore manage, human affairs at a level so fine-grained that it can accurately

**Figure 4.1** "getting_to_the_other_side.jpg" (©Hannah Ueno). Steps with the missing middle, ladder rope that is having by its last thread ... A visual play of a surreal urban-scape with impossible-possibilities in crossing over to the other side.

foretell the place and circumstances of anyone's death. This is perhaps a narrative necessity and certainly allowable poetic license, in so far as the first two seasons of the show (2016, 2018) were about *Westworld*'s hosts trying to escape their fixed narrative loops, and the third is about people stuck in their own loops, but such a system is mathematically impossible.

It's possible that intelligence is intelligence, artificial or natural, but as this is impossible to prove one way or the other one must make do with hints and analogies. Consider octopuses, which are playful, curious, and engaging, so much so that the UK considers them honorary mammals for the purpose of their rights. Octopuses are mollusks, related to squid, slugs, limpets and snails, and only very distantly related to human beings, but they're the cognitive stars of the invertebrate world. That they exist and that their actions seem essentially

**Figure 4.2** "maze.jpg" (©Hannah Ueno). A private eye into a person's thoughts and states. Maze represents a moment of deja-vu, or premonition we encounter in life's seemingly ordinary events. The objects in it are the silent witnesses of what might have happened a few steps ago.

comprehensible (though who's to say what levels of significance human observers are missing) suggests (but doesn't prove) that intelligence isn't a black swan rarity but just another of the many forms convergent evolution is capable of cobbling together out of whatever materials are at hand.

Consider also the black-tailed prairie dog of the North American plains. Combinatorially structured, meaning-bearing language has long been thought to be a uniquely human capability. There are animals with combinatorially structured vocalizations—whales and songbirds, for instance—but their calls, rather than carrying information about their environment, seem to be like identification codes or perhaps fine art. There are animals whose calls carry meaning—vervet monkeys have an alarm call meaning "Leopard!"—but there aren't any that can combine these calls in a novel, meaning-bearing way, or so it was believed before the study of the vocalizations of black-tailed prairie dogs. They appear to have the ability to concatenate what amounts to a noun and a string of adjectives to say things like (in translation) "dog big yellow fast" and "human small blue slow." Prairie dogs are rodents, and as such much more closely related to human beings than octopuses, but still distant; their eloquence starts to suggest that there's a basic toolkit of intelligence.

Finally, consider Turing machines. A Turing machine is an abstraction of a computer. Any particular computer can be considered a special case of a Turing machine, and all Turing machines are equivalent in that they can compute the same kinds of things, so it doesn't matter what kind of hardware your computer has, or how much memory, or if it has racing stripes: mathematically, one computer is the same as another. Perhaps intelligence works the same way and once it gets to a certain point, there's no farther to go. In light of this and the evidence from the animal kingdom, one might conjecture that intelligence is much the same throughout the universe. It's pleasant to think we could communicate with aliens, less pleasant to imagine a universe full of the ills that plague humanity, with demagoguery and spin and lobbyists burdening planets in distant solar systems.

And yet, consider Srinivasa Ramanujam. He was an Indian mathematician born in 1887 who, despite minimal formal mathematical education, was one of the most brilliant and original mathematicians who have ever lived. He died at age thirty-two but his work is still being digested and its implications are working their way through modern math and physics. His work was so radically novel that he had difficulty gaining recognition in his day and didn't really find a place until he started corresponding with the mathematician G. H. Hardy, who brought him to Cambridge.

Like Cantor, Ramanujam thought his work was the product of divine inspiration, in his case from the family goddess Namagiri Thayar, an aspect of the goddess Lakshmi. He once had a dream in which she inscribed mathematical formulae on a red sheet made of flowing blood.[10] Perhaps there's a universal

---

[10] Robert Kanigel, *The Man Who Knew Infinity* (Washington: Washington Square Press, 2016).

disposition to interpret truly original work as the product of a divine hand. Perhaps the constraints of mathematics and the physical universe permit nothing more miraculous than the human brain and Turing machines but still offer scope for wonder.

# Bibliography

Borges, Jorge Luis. *Collected Fictions*. London: Penguin Classic, 1999.

Dauben, Joseph Warren. *Georg Cantor His Mathematics and Philosophy of the Infinite*. New Jersey: Princeton University press, 1979.

Kanigel, Robert. *The Man Who Knew Infinity*. Washington: Washington Square Press, 2016.

Lem, Stanislaw. *Solaris*. New York: Walker & Co., 1961.

Moravec, Hans. *Robot: Mere Machine to Transcendent Mind*. Oxford: Oxford University Press, 1998.

Tarkovsky, Andrei, director. *Solaris*. Russia: Mosfilm, 1972. Film.

Soderbergh, Steven, director. *Solaris*. California: 20th Century Fox, 2002. Film.

Part Two

# Big Data, Sentience, and the Universe

# 5

# A MASSIVE Swirl of Pixels

## Radiohead's "Go to Sleep"

Steen Ledet Christiansen

Against a computer generated black and white cityscape with swiftly moving crowds, a large red rose sways back and forth. The camera travels past the flower to a heavily digitized Thom Yorke (of Radiohead) seated on a park bench. As he sings, the camera weaves through the crowds, bumping up against many pedestrians—everything is rendered as blocky polygons. As Yorke's face morphs into simpler, more angular planes, the blocky buildings of the city's plaza suddenly collapse. A little later the buildings fall upwards and collect themselves. Yorke stands up and walks away as the camera travels down to the rose again, which closes up. This is the music video for Radiohead's "Go to Sleep" (*Hail to the Thief*, 2003), directed by Alex Rutterford. Produced using Massive (Multiple Agent Simulation System in Virtual Environment), Weta Workshop's artificial life software created for Peter Jackson's *The Lord of the Rings* trilogy (2001–2003), Radiohead's "Go to Sleep" offers an entirely synthetic environment within which Thom Yorke moves and sings in such a way that his gestures resemble those of the digital agents that surround him. "Go to Sleep" is part of Radiohead's experimentation with unusual and advanced visual technologies. Together with the videos for "Pyramid Song" (2001), "Day Dreaming" (2016) and "House of Cards" (2007), "Go to Sleep" explores the ways in which new technologies can encourage the emergence of refreshed audiovisual aesthetics and through these audiovisual modes renew our perception.

These new aesthetics have resulted in innovative audiovisual configurations that can challenge the boundaries between art and life, and sound and image. That Radiohead would make such innovative music videos falls entirely in line with their musical output. As Brad Osborn has discussed at length, most of Radiohead's career has revolved around breaking new sonic ground. That their visual output would similarly challenge conventions and expectations is hardly surprising. Osborn argues convincingly how Radiohead's sonic aesthetic is founded on an "expectation, only to subvert those expectations with potent surprises."[1] Radiohead's music is both anchored in and subverts familiar rock conventions. A similar disruption drives the music video for "Go to Sleep": on the one hand, we are given a typical setup where Thom Yorke is seen singing accompanied by various related visual events that are loosely connected to the rhythms of the music; on the other, this familiarity is subverted through the use of Massive and artificial life simulations. As a result, Radiohead's visual strategies challenge music video conventions just as their sonic ones disrupt the traditional rock music idiom. But these challenges also radiate outwards to destabilize conventional conceptions of movement, life, and cinema.

The virtual worlds created for "Go to Sleep" produce an uncanny object that disturbs our conception of life. Just as Radiohead's music creates tension between expectations and realization, so too does "Go to Sleep" produce a friction between the anticipation and realization of life. Everyday expectations of life are tied to organisms: plants, animals, and the like. The life realized in "Go to Sleep" is instead inorganic computer code, and this blurs the distinction between the living, the dead, and the never-living. Two aspects become crucial here in terms of understanding new modes of life: movement and autonomous behavior. As I will discuss later, autonomous behavior is when an entity has agency: the ability to act on its own volition. The Massive technology generates autonomous behavior through algorithms. The blurred distinction between organic and inorganic life can be conceptualized through Jacques Derrida and his idea of the dangerous supplement: what is first added to an idea ends up replacing the idea.[2] Derrida's argument is extensive and expansive but we can try to condense it: if life (idea) is what moves (supplement—it explains life), then what moves is life: the supplement supplants the idea, movement becomes what identifies life, movement *is* life.

Supplement does double work here. Firstly, in the case of Massive technology, it disrupts and disturbs organic definitions of life because we see "liveliness" as movement. Secondly, sound and image exist in a supplementary tension in music videos and in cinema, as Steven Shaviro has pointed out.[3] Radiohead's

---

[1] Brad Osborne, *Everything in its Right Place: Analyzing Radiohead* (Oxford: Oxford University Press, 2017), 9.
[2] Jacques Derrida, *Of Grammatology*, corrected edition, translated by Gayatri Spivak (Baltimore: Johns Hopkins University Press, 1997), 144–5.
[3] Steven Shaviro, "Out of Whack: Thierra Whack's Aberrant Identity," *FlugSchriften* 1 (2019): 14.

**Figure 5.1** Alex Rutherford, "Go to Sleep," 2003, Blocky polygon animation. An animate Thom Yorke.

music does not *need* images, just as images do not *need* sound. Yet the history of music video has shown that sound *wants* to be visualized. And in the same way, images *want* sonification and are considered more realistic when sound is added. Along similar lines, Dominic Pettman suggests that Massive technology blurs the line between animal and human, a distortion that lies at the heart of the "Go to Sleep" video.[4] Here, I argue that the moving images of audiovisual media are "lively," moments of amplification, where imitations of life become life-like and mobile, producing effects outside of their own bodies but within a techno-organic environment. The key idea here is the shift toward what I will call "software cinema," a cinema that expands into digital technologies of animation and artificial intelligence/life technologies.

## Massive

Since *The Lord of the Rings* trilogy, the Massive software has become the industry standard for simulating crowd behavior. Massive is an acronym for Multiple Agent Simulation System in Virtual Environment created by Weta Digital, a

---

[4] Dominic Pettman, *Look at the Bunny: Totem, Taboo, Technology* (London: Zero Books, 2013).

digital visual effects company. The software has been used in moving-image productions from spectacle-driven films like *Guardians of the Galaxy* (2014) and *Black Panther* (2018) where plenty of computer animation is expected, to more drama-oriented pieces like *Changeling* (2008) and *Invictus* (2009). Massive's tag line is "simulating life," indicating that not only human crowds may be generated by their software but also crowds of langurs and macaques (*The Jungle Book*, 2016), flying fish (*Life of Pi*, 2012), apes (the *Planet of the Apes* reboot series, 2011–2017), and zombies (*World War Z*, 2013).[5] Even trollies and cars are generated by Massive in *Changeling*, suggesting that the life simulated by Massive can move beyond organic existence to animate the inanimate through movement.

Using Massive software, animators are able to blend their illusory life forms with the rest of the cinematic environment; or at least this is the goal. To individually code each individual creature would be cumbersome, time consuming and expensive: instead, a degree of algorithmic free play is built into the crowd simulation algorithms. On the DVD extras for the special edition of *The Lord of the Rings: Two Towers* (2002), a Weta animator relates how a crowd of orcs intended to attack a horde of elves in fact refused and ran away. Within the preset number of possibilities built into their algorithm, the orcs, when faced with the greater elvish numbers, took the wise choice to retreat. Reading into this

**Figure 5.2** Alex Rutherford, "Go to Sleep," 2003, Blocky polygon animation. A rose opens.

---

[5] For more examples see http://massivesoftware.com/ (accessed December 17, 2020).

anecdote, Pettman notes how this carefully chosen action troubles the boundaries between animal, human, and technology.[6]

## Sonic Agencies

"Go to Sleep" is a music video; the images have been made to animate to the music giving a degree of conjunction between sound and image. That conjunction comes in the animacy of software. While the song follows a relatively regular two-part structure, the concluding guitar solo was performed with a so-called object in the Max/MSP software application, which is a visual programming language for music. Jonny Greenwood, the lead guitarist of Radiohead, programmed a randomizer effect (the object) to produce a randomized stutter of whatever is played on the guitar. This renders the solo different for every live performance of "Go to Sleep." This is not an unusual strategy for Greenwood, who worked on computers as a child, and also uses a Kaoss Pad, an audio effect unit, to distort and glitch Thom Yorke's vocals for the song "Everything In Its Right Place" (Kid A, 2000) and produce an obvious pun.

Such randomizing effects suggest another aleatory aspect of "Go to Sleep"— the productive contrast between human agency and algorithmic agency. Greenwood's solo is never predictable, Yorke's voice glitches in different ways every time "he" sings it, and the algorithms shape the subjectivities that emerge from these encounters. Real sounds cannot be separated from their technological transformation. This tension between human and technological agency is similar to the tension that emerges from the Massive software, where the layers of algorithms become too complicated for human programmers to predict the outcome.

"Go to Sleep" is deceptively organic. The main riff is acoustic, and the first half of the song is reminiscent of folk music. Halfway through, the song changes to a different riff and a different time signature. This is also when a strange electronic drone sound develops in the background. In the final third of the song, Greenwood's electric guitar solo comes in, played through the randomizer Max/MSP object. The song slowly fades out. Yorke's voice has some reverb added to it but is mostly analog. The song's musical composition sets up a tension between

---

[6] Pettman, *Look at the Bunny*, 112.

organic and nonorganic technologies that are mirrored in Rutterford's subsequent visualization.

Music, especially rock music, has a long tradition of associating authenticity with sound fidelity and liveness. Digital sound is not music but "simulated music" according to Neil Young.[7] For many, electronic music thus stands as the (constructed) opposite to authentic analog rock music. This is one of the many expectations and conventions that Radiohead subvert in their music, a disruption of the perceived "authenticity" of rock music that moves through "Go to Sleep." The song oscillates between 4/4 and 6/4 time signatures. Electronic sounds permeate the background, slowly gaining prominence before culminating in the electric guitar riff that concludes the song. The Max/MSP randomizer object is precisely the kind of animate object that Mel Y. Chen understands as able to disrupt our conventional understanding of agency.[8] The randomizer object shapes the solo that Greenwood plays and does so in different ways with each performance. Even if Greenwood plays the same chords, the notes are scrambled into something new.

The relatively straightforward rock sound is slowly destabilized and hybridized by the integration of digital algorithms that feed into Carol Vernallis's idea of a "digital swerve": this swerve in contemporary media production, she argues, produces an "intensified audiovisual aesthetic" where life is granted to pixels by making them musical.[9] Radiohead's trajectory seems to be surfing such a digital swerve, oscillating between rock and electronica, allowing each form to animate the other and recognizing that automation through different sound technologies produces a different sonic beast. It is significant that the Max/MSP programming language is a visual programming language, considering the integration of sound and visuals for so many of Radiohead's videos. It is also significant that "Go to Sleep" circles the notion of monster in Yorke's lyrics. The lyric line is important here: "We don't wanna wake monster taking over"; "Tiptoe round, tie him down," a reference to *Gulliver's Travels* that Yorke himself wants to see explored in the video.

And yet, the randomized guitar solo can be seen as a monster already. Allowing the automated agency of the randomizer object shape the rock guitar

---

[7] Quoted in Aden Evens, *Sound Ideas: Music, Machines, and Experience* (Minneapolis: University of Minnesota Press, 2005), 11.
[8] Mel Y. Chen, *Animacies: Biopolitics, Racial Mattering, and Queer Affect* (Durham: Duke University Press, 2012).
[9] Carol Vernallis, *Unruly Media: YouTube, Music Video, and the New Digital Cinema* (Oxford: Oxford University Press, 2013), 137.

solo is a breach of convention, allowing (digital) technology to intervene into the rock aesthetic. Of course, the electric guitar is itself a technology, as are speakers, amps, and pedals. And yet, the aberration here comes from allowing a technology to transform the solo, in concert with the artist, in a type of technological anthropomorphization of technology through a non-human agency. The strangeness of the music video is already present in the song and in visual form no less. By using a visual programming language, Greenwood would have needed to use a visual interface to produce a sonic result. Monstrous hybrid indeed. "Go to Sleep," then, shows us that sound and image can only rarely be understood as separate forms, and that automation is part of their animacy. Agencies proliferate; neither song nor music video could exist in its current form without both human and technological agencies.

## Life, Movement, and Form

A rose opens; a rose closes. In between people move, leaves fall: Yorke sings and walks away. Movement abounds and transformation is incessant: the frame seems teeming with life and even though the polygonic shapes that make up the video's world do not lessen this movement, they do heighten the world's artificiality. Polygonic designs are generally popular in animation, because their triangular shapes are easy to render into easily recognizable 3D forms. Buildings move up and down, the camera tracks, traveling through space unimpeded. What is the difference between these two kinds of movement—organic movement and technological movement? The answer seems straightforward: it is the distinction between the organic and the inorganic. Or is it? In "Go to Sleep," all movement is inorganic technological movement. Even though the people and the environment seem to interact, there is no organic life at work here. The video is a simulation of life, not a reproduction of it.

Since the inception of cinema, moving images have been conflated with life. As Louis-Georges Schwartz has shown, the very earliest accounts of cinema associated it with life itself.[10] Such cinematic life emerges from the technological means of generating movement; the accounts that discuss cinema as life locate it precisely in cinema's reproduction of movement. This overlap of cinema and life is not so surprising, considering that cinema emerges at the same time the nature

[10] Louis-Georges Schwartz, "Cinema and the Meaning of 'Life'" *Discourse* 28, no. 3&4 (2006): 10.

of life was debated and contested by biologists and philosophers. Inga Pollmann demonstrates that many early film critics similarly used "life" as a useful term to grasp this new, emerging medium.[11] Just as the then-new medium of cinema gave force to conceptions of life, so too do the now-new technologies of animation.

As Pettman argues, life itself is not straightforward and is surprisingly difficult to define. The technical definition of life has been "a self-sustained chemical system capable of Darwinian evolution."[12] Pettman's argument is that the chemical part of life seems old-fashioned in a digital world. Software such as Massive essentially produces self-reproducing machines, or digital entities, that necessarily disrupt conceptual boundaries.[13] The Massive digital actors in "Go to Sleep" do interact with their environment and they are, to some extent, self-reproducing. These digital actors have what Kant denied machines: *formative* force rather than just *motive* force.[14] Kant argues that machines have motive force; they can act mechanically on things around them. Beings, as opposed to machines, have formative force, which means they can propagate themselves. This propagation is a matter of organization and it is this that distinguishes machines from beings. For Kant, self-propagation and self-organization is what creates a natural purpose. The digital actors of Massive have a purpose exhibited by all organized beings, though the question of whether or not this purpose and force is *natural* is a key issue.[15]

And yet, perhaps the issue is not so much the idea of "life", but rather the concept of what is "natural". Eugene Thacker has discussed the different conceptions of life as mechanic and vitalist in his book *After Life*.[16] The field of biophilosophy is large and explored at greater length elsewhere in this book. Here, the issue is simply that "natural" is not a straightforward way of identifying or understanding "life." If organization defines life, then digital actors are "lively." They are inherently organized by the code that provides purpose. And this purpose, as the anecdote about the orc shows, is not a purpose given directly to

---

[11] Inga Pollmann, *Cinematic Vitalism: Film Theory and the Question of Life* (Amsterdam: Amsterdam University Press, 2018), 19.
[12] Pettman, *Look at the Bunny*, 76.
[13] John Johnston, *The Allure of Machinic Life: Cybernetics, Artificial Life, and the New AI* (Cambridge: MIT Press, 2008), 165.
[14] Immanuel Kant, *The Critique of Judgment*, translated by Werner S. Pluhar (Indiana: Hackett, 1987), 6628.
[15] Johnston, *The Allure of Machinic Life*, 166–7.
[16] Eugene Thacker, *After Life* (Chicago: The University of Chicago Press, 2010), 11ff.

them by their human creators but rather a purpose that emerges out of algorithmic play. When code takes on agency— when digital actors take control of their own machinic life—then "life" has become something broader, something harder to pin down. "Natural" no longer seems like a fruitful definition. Artificial software life expands the boundaries of the known but does so in union with technologies. This is the argument that has followed the artificial life field since its inception with Christopher G. Langton.[17] With artificial life comes the question of strong AI—can life exist outside a chemical solution? Pettman's argument is precisely that in today's digital age, this question seems antiquated. Instead, the question becomes whether a computer would have a mind similar to that of a human.[18] Life becomes increasingly machinic because these images are not reproductions, are not analog, are not anthropomorphic in any way. These Massive images do not have an analog to human perception. Instead, their images are produced through a process of software translation very different from the visual construction processes of celluloid and aperture. Machinic, then, should not be understood as technological (although it might be), but rather as productive, generative; something that is made. Machinic life is simply life that generates, whether it is in a chemical solution or not. "Go to Sleep" enacts this view of life through its digital animation.

Here, then, is the crux of the issue: cinema has slowly been anthropomorphized over the decades. The analog nature of the medium's birth, the movement of the camera, the realism of expression, the need for human actors, and so forth have all participated in making cinema an anthropomorphic medium. And yet this is an illusion. As Alan Cholodenko and many others have shown, animation was marginalized as the irrelevant cousin to the so-called live action film.[19] This oversight has meant that it is only now that animation and its corollary idea of life is brought into contact with each other. I will expand on animation later, but for now it's important to emphasize that both critical and popular receptions of cinema have worked to naturalize cinematic perception in particular ways and that these processes of naturalization have mostly excluded animation as realistic. Software cinema brings into focus the fact that cinema is a non-human medium. Software cinema is a term for the shift in moving image production away from the centrality of the camera to the centrality of the computer. "Go to

---

[17] For more on artificial life, see Christopher G. Langton, *Artificial Life: An Overview* (Cambridge, The MIT Press, 1995).
[18] Daniel C. Dennett, *Consciousness Explained* (New York: Back Bay Books, 1991), 435.
[19] Alan Cholodenko, *Illusion of Life: Essays on Animation* (Sydney: Power Publications, 1991), 9.

Sleep" is not made with a camera but with a computer. This change not only signals a large-scale industry shift, but also initiates a large-scale modification of human perception. The most straightforward example in this context would be the introduction of sound. As Michel Chion has convincingly shown, synchronized sound is what temporalized cinema.[20] Synchronized sound introduced synchresis—"the spontaneous and irresistible weld produced between a particular auditory phenomenon and visual phenomenon when they occur at the same time."[21] Synchresis allows for continuity, realism, and a convincing embodiment of our audiovisual experience, since images are grounded by sound.

Moving images integrate with human perception; this is the basic way in which moving images work. But do these moving images restructure our perception? In a limited sense, I argue that they do. While cinema did not make licorice taste like strawberries, it did alter, at some level, the ways in which we could perceive the world. This has been part of the film theory discussion since Walter Benjamin's early work on mechanical reproduction.[22] More recently, Edward Branigan has shown that anthropomorphism as an analytic category is a matter of how much a camera or particular shot simulates human embodiment.[23] But as cameras and shots stop simulating human embodiment, anthropomorphism breaks down as an analytic category. No longer do shots have to move in ways that mimic human movement, nor do cameras need to record in ways that suggest human movement or aurality. It's important to note, though, that close-ups, slow-motion and other familiar filmmaking techniques already rupture—and thus reconfigure—"natural" human patterns of perception. What Branigan misses in his model, then, is that anthropomorphism has always existed as a cinematic form; and that this form is being continually renewed. As audiovisual technologies integrate into our perception, new changes are introduced. Digital technologies are even less anthropomorphic than analog technologies and the artificial life technologies employed by Weta and similar effects companies further expand our modes of experience through their new modes of expression.

---

[20] Michel Chion, *Audio-Vision: Sound on Screen*, edited and translated by Claudia Gorbman (New York: Columbia University, 1994), 16.
[21] Chion, *Audio-Vision*, 63.
[22] Walter Benjamin, *The Work of Art in The Age of Mechanical Reproduction* (London: Penguin Books, 2008).
[23] Edward Branigan, *Projecting a Camera: Language-Games in Film Theory* (London: Routledge, 2013), 37.

We are faced with equivalences: that we ever thought that the camera was anthropomorphic was simply because we had equated the camera—or what we might better term the film body—with the human body, after a period of shock and estrangement. In other words, what used to be a decidedly non-human body, early cinema produced forms that quickly integrated into, and equated with, the human body. All the estranging aspects of cinema were minimized—Tom Gunning refers to this reduction as "tamed attractions"—to produce a narrative cinema.[24] But the new generations of music videos do not simulate human embodiment, nor are they tied to narrative structures as Vernallis has pointed out in her work on new digital cinema.[25] If cinema became "tamed" and anthropomorphized to produce a predominantly narrative cinema, the non-narrative thrust of music videos enables sound-images to become what Vernallis calls "unruly" again.[26] I argue that this untaming of narrative structures has prompted a move back to the non-human because it challenges our perceptual schema in ways similar to those of early cinema.

"Go to Sleep" and similar music videos expand the boundaries of cinema by integrating new image and movement technologies that do not record profilmic reality. Techniques like rendering and the animation of movement and distance are slowly taking over from analog lenses and celluloid film as the guiding concept of how to produce images. As a result, our sensory experience begins to broaden as more non-human experience is brought into our sensory realm. The tighter such image technologies become integrated into cinematic productions, the less they will appear as estranging or renewing. Animation shows us that mutability is part and parcel of not just new modes of image production, but of new ways of living. Our fascination with these new technologies broadens our sensorium. But through this fascination a corporeal integration is established between a non-cinematic body and human bodies. The animacy of the non-cinematic body colonizes part of our bodies' liveliness, not by reducing our liveliness but by articulating our liveliness as a form of mutability.

---

[24] Tom Gunning, "The Cinema of Attraction[s]: Early Cinema, Its Spectator and the Avant-garde," in *The Cinema of Attractions Reloaded*, ed. Wanda Strauven (Amsterdam: Amsterdam University Press, 2006), 387.
[25] Vernallis, *Unruly Media*.
[26] Ibid.

## Animacy and Animation

Despite animation's prevalence in early cinema and its current ubiquity, it has a checkered relation to cinema. We can regard animation as a dangerous supplement to live action cinema, the defining opposite that has slowly integrated into its filmic cousin until there is no longer a useful distinction to be drawn between the two forms. As Lev Manovich phrases it *"Born from animation, cinema pushed animation to its periphery, only in the end to become one particular case of animation."*[27] Consider the trajectory of animated movies from Disney family movies to the breakthrough of *Toy Story* and the current use of animation in a movie such as *Changeling*. Animation has moved from a specialized place in visual culture to being essential for almost any kind of movie production, a move away from indexical, reproductive cinema toward a digital, generative form of post-cinema in which motion is induced by any means available. Animation is therefore one of the predominant modes of production in contemporary audiovisual culture and, as Manovich argues, film is now a subset of animation.[28]

How might we understand animation? It is more than simply animated film or animated visual effects: *"animation* signifies in so many different ways" argues Karen Beckman, "including *movement, life itself, a quality of liveliness* (that doesn't necessarily involve movement), *spirit, non-whiteness, frame-by-frame filmmaking processes, variable frame filmmaking processes,* and *digital cinema."*[29] What animation does, more than simply designate a particular cinematic practice, is to connect the idea of cinema and moving images with that of life. Animation is usually understood as "the general process of activating or giving life to inert matter" as Sianne Ngai phrases it.[30] A little closer to my concerns here, Scott Bukatman points to "the transfer of energy from animator to animated,"[31] a transfer that I would like to extend to the viewer of the animated as well. Animation thus disturbs the distinction between animate and inanimate, or what Richard Grusin calls "an aesthetic of the animate, in which spectators or

---

[27] Lev Manovich, *The Language of New Media* (Cambridge: MIT Press, 2002), 301. Emphasis in original.
[28] Manovich, *The Language of New Media*, 305.
[29] Karen Beckman, "Animating Film Theory: An Introduction," in *Animating Film Theory*, ed. Karen Beckman (Durham: Duke University Press, 2014), 1. Emphasis in original.
[30] Sianne Ngai, *Ugly Feelings* (Durham: Duke University Press, 2005), 92.
[31] Scott Bukatman, *The Poetics of Slumberland: Animated Spirits and the Animating Spirit* (Berkeley: University of California Press, 2012), 709.

users feel or act as if the inanimate is animate."³² Such an aesthetics of the animate follows the logic of Tom Gunning's aesthetics of astonishment and also his insistence on movement and motion as the crucial definitions for cinema, rather than reproduction.³³

I propose that we use the term "animacy" to designate this positive generation of liveliness. In linguistics, animacy is generally understood as "a quality of agency, awareness, mobility, and liveness."³⁴ Animacy can be confusing when action verbs are used for nouns that are usually inanimate objects. Chen's example of this confusion is "The hikers that rocks crush."³⁵ This sentence is weird because the rocks are active agents, while hikers are passive objects. We are so used to objects being things that lack agency that we have a hard time processing such sentences. This is a different version of Grusin's aesthetics of animation, where the inanimate comes alive.

Buildings crumble and fall down. This happens every day. But rubble rarely leaps up and assembles into buildings. At least not in real life. And yet, the

**Figure 5.3** Alex Rutherford, "Go to Sleep," 2003, Blocky polygon animation. Rubble leaps up into buildings.

---

³² Richard Grusin, "DVDs, Video Games, and the Cinema of Interactions" in *Post-Cinema: Theorizing 21st Century Cinema*, eds. Shane Denson and Julia Leyda (Reframe Books, 2016), 68.
³³ Tom Gunning, "An Aesthetics of Astonishment: Early Film and the (In)Credulous Spectator," in *Viewing Positions: Ways of Seeing Films*, ed. Linda Williams (New Brunswick: Rutgers University Press, 1995).
³⁴ Chen, *Animacies*, 136.
³⁵ Ibid., 145.

inanimate leaping into life is not an odd occurrence in the unruly media of today's audiovisual culture. In some digital music videos, everything comes alive and the use of Massive artificial life software ends up as a way of disturbing an easy understanding of life, a way to decenter life as something necessarily organic. A "monstrous hybrid automaton" is what emerges from digital music in tandem with the digital image.[36]

"Go to Sleep" enacts this monstrous hybrid automaton in the way that it expresses a technological form of life, a form of life where organic and inorganic (people and buildings) are flattened onto the same level.[37] The organization that propels the images of "Go to Sleep" is radically different from organic life and is non-linear and distributed; what Scott Lash terms "lifted out."[38] "Go to Sleep" takes place in a generic non-place, generated by software and there is no real-life correspondence or equivalence. But the video pushes this lifting out even further, because it also lifts life out of a traditional definition. Massive's artificial life software reveals a different kind of life: a life lifted out of a chemical solution and into a different ecology.

Animacy insists that liveliness is not restricted to the organic but expands to include things, objects, technologies, images, sounds, everything that moves. In this liveliness of motion lies what Grant Bollmer terms a *kinesthetic index*, where "forms of digital media should be conceived not as deferring to the visual, but as reliant on the kinesthetic."[39] Motion begets motion because "motion brings together digital images with the reality inscribed into media": "digital images are condensations of specific—if multiple—bodies that persist as representations that have some link with the physical world."[40] Bollmer echoes Gunning's insistence and the general thrust of cinematic vitalist theories. It follows that engagement with cinema does not come from photographic reproduction and photographic realism but rather from the mimicry of motion that is inherent in the kinesthetic relation to the world. However, this kinesthetic relation to the world is also a question of automation. The people walking around are as much automatons as the building, the rose or Thom Yorke himself. The video does not include a recording of Yorke, just a simulated version of him.

---

[36] Vernallis, *Unruly Media*, 137.
[37] Scott Lash, "Technological Forms of Life," *Theory, Culture & Society* 18, no. 1 (2001): 108.
[38] Lash, "Technological Forms of Life," 113.
[39] Grant Bollmer, "The Kinesthetic Index: Video Games and the Body of Motion Capture," *Invisible Culture*, no. 30 (2019), np.
[40] Ibid.

## Automation Trouble

Vivian Sobchack has written two significant pieces on animation. One deals with automation and labor, the other with animation and the line of the drawn line.[41] There is, however, a connection between these two essays that can help us think through the relation to life, animation, animacy, and automation/automatons. The line, Sobchack says, is one of the crucial definitions of animation, since there are no lines in live-action cinema.[42] At the same time, the line is not strictly part of the animated image, so much as it is a meta-object that *produces* the animated image—we are not meant to see lines, we are meant to see moving objects.[43] We see lines in "Go to Sleep," because we see the polygons that make up all the different objects and beings in the video. But we are meant to both see them and not see them at the same time; we are meant to see a rose, not a blocky, polygonic, animated rose. We are meant to see Thom Yorke, not a blocky, polygonic, animated Yorke. And yet, surely part of our fascination with the video is precisely its lovely, uncanny blockiness. We do see the polygons for they constitute our aesthetic experience of the video.

This presence and absence of the polygons is what Sobchack calls "tremors of the visible," following Sergei Eisenstein.[44] The lines of the polygons are what express a dialogue "between the organic and the industrial and the qualitative and the quantitative" and the "dream of plasmatic freedom" for Sobchack.[45] The changeability of animation and the strange animacies of artificial life software, is the attraction of the video. Sobchack identifies this plasmatic freedom as a nostalgia for a rapidly-disappearing mode of existence. We might read Massive and similar technologies in a more favorable light and suggest that new articulations of space and time are instead opened up.

We might also point out that Massive troubles another line: one that separates life and technology. This is where Sobchack's other essay is helpful. Animation and automation exist in a tension between life and the erasure of the life that produced it (for Massive, the programmers who produced the code). This tension

---

[41] Vivian Sobchack, "The Line and the Animorph, or 'Travel Is More than Just A to B,'" *Animation* 3, no. 3 (2008): 251–65; Sobchack, "Animation and Automation, or, the Incredible Effortfulness of Being," *Screen* 50, no. 4 (2009): 375–91.

[42] Sobchack, "The Line and the Animorph," 252.

[43] Ibid., 253.

[44] Ibid., 261; Sergei Eisenstein, *Eisenstein on Disney*, ed. Jay Leyda, trans. Alan Upchurch (Calcutta: Seagull Books, 1986), 47.

[45] Sobchack, "The Line and the Animorph," 262.

is resolved through an "aesthetics of effortlessness" that sublimates the labor-intensive production into pleasure.[46] The discussion of Massive, both mine and others', is precisely focused on the ease with which Massive produces imagery. Pettman helpfully insists on the clinamen of vital-material systems: the swerve that produces glitches or eccentricities, that makes the orc not fight despite being told to.[47] Perhaps there is something to be found in the digital swerve of the monstrous hybrid automaton that is not purely the sublimation of effort into pleasure.

"Go to Sleep" clearly enacts one of Sobchack's automaton arguments—that we become increasingly inert, while our machines become increasingly lively. Yet, it seems that maybe there is a different way of understanding Massive's flattening of life and distribution of animacy. While Sobchack's critique of the obscuring of actual labor is well put, we can approach the issue from a different angle; we can "swerve" around and be carried away by the technological forms of life that "Go to Sleep" also exhibits. Lash's argument that technological forms of life are distributed is evident in "Go to Sleep" through the liveliness of the buildings. Consider the rubble leaping up to assemble itself into buildings. There is an unbearable lightness to these images in the way everything melts and flies away. We are carried away on the swell of the music, which here facilities the liveliness of the buildings. The buildings' lack of crunchy, cumbersome sounds of floating back together makes them feel lighter. Or as Chion would put it, the sound does not produce a "weight-image"; with no synchronized sound to designate mass, the rubble appears weightless.[48] This holds true for most music videos, yet the point remains significant. The continuity produced by the music also produces a continuity of liveliness that is further underscored by an absence of inanimateness.

And this lightness presents a different variation of what Sobchack identifies as obscuring the effort of animation as a mode of production.[49] "Go to Sleep" intensifies this obscuring move by making all objects and bodies dissipate and replace each other. Even Yorke's faciality—a convention in music videos to capitalize on the recognizable celebrity status of the performers—is blocked and obscured. We recognize his voice, a deeply familiar voice, and we recognize his face, sort of, but much like in the video for "House of Cards," great effort is made

---

[46] Sobchack, "Animation and Automation," 384.
[47] Pettman, *Look at the Bunny*, 97.
[48] Michel Chion, *Sound, An Acoulogical Treatise*, trans. James A. Steintrager (Durham: Duke University Press, 2016), 7–8.
[49] Sobchack, "Animation and Automation," 384.

to disrupt Yorke's likeness—both visually and sonically. It's him but not him: he is buried beneath layers of mediation and technologies, translations of code and lenses, and sonic distortions. "Go to Sleep" is not exactly animated as much as it is rendered visualizations of data. Visual design and decisions went into the production, but we remain in the realm of translation. The video's fascination with new imaging technologies obscures the modes of image production but at the same time it attempts to renew perception. Maybe a lightness of being is not always so unbearable if new forms of life are opened up.

In this way, "Go to Sleep" remains on the side of what Sobchack identifies as a contemporary interest in the posthuman.[50] Considering the difference between conventional animation and its images, we can go further and say that "Go to Sleep" seems more interested in the non-human. We find the ontological flattening between objects and bodies, suggestive of a larger turn in what we are here calling cybermedia. On the face of it, "Go to Sleep" fits within early digital animation's tendency to avoid photorealism. Instead, the video revels in the blocky uncanniness of its world. The greyness of the diegesis reinforces this uncanniness, while at the same time providing a thematic resonance with the lyrics. But the liveliness, the thrill of this strange animated world is an aesthetics of astonishment of the inanimate becoming animate. The Massive animation software can easily produce photorealistic images, can easily produce something that is difficult to distinguish from organic life. Whether or not the intention of Radiohead or Rutterford, the video displays an uncanny liveliness, an uninhibited animacy through automation. Such automation and its concurrent liveliness raise questions of imitation as an element of life.

## Imitation of Life

"Go to Sleep" presents an animistic world; everything comes alive and teems with possibility and potentials. This is its attraction and its aesthetics of astonishment. Yet its animation is concurrent with its automation. There is a logic of imitation at work in these autonomous animated images. The swerve that Pettman and Vernallis want to follow lies in the confluence of flattening and lifting out that Lash argued for technological forms of life. By removing the

---

[50] Ibid., 382.

conventional distinction of life as something that is organic, we can instead see that life has the capacity to be rather more flexible. If we are, as Jane Bennett provocatively states, "walking, talking minerals" the distinction between organic and technological forms of life is less revealing.[51] Instead, the digital swerve into non-cinematic cybermedia and autonomous animation is suggestive of an uncanny liveliness that was previously obscured. Rather than follow Sobchack's argument that software animation obscures human labor, we can argue that software automation reveals a more-than-human world.

"Go to Sleep" is uncanny not because the digital actors look like real humans, but because the actors behave like humans. This human behavior by beings other than humans unsettles our sense of what life is, what human behavior is. Massive's autonomous animation participates in a new technological form of life that we have not yet grown accustomed to. Yet this software cinema is not cut off from our experience of what life is. Instead, software cinema participates in our conception of life because it behaves like life; or maybe, following Bruno Latour following Gabriel Tarde, we should not put so much thought into what life *is* and instead look at what life *has*.[52] If life is what has the autonomous capacity to evolve, then Massive produces life. If life is what moves, then Massive produces life. If life is what has agency, then Massive produces life.

"There is," says Latour, "nothing especially new in the human realm."[53] Likewise, by extension, there is nothing especially new in the digital realm. What connects organic life and human life also connects organic life and technological life. Forms of life have an environment but are not limited to chemical environments. Massive's computer-generated environment is equally as capable of sustaining life as organic environments are. The distinction comes down not to environment but to repetition. For human life, this repetition is imitative: "the social being ... is essentially imitative."[54] Organic life is hereditary, inorganic life is vibratory, social life is imitative. But note that Tarde does not equate social with human. *Any* being can be social when they enter into social relations. For Tarde, the digital actors in Massive would be social beings since they imitate each other.

---

[51] Jane Bennett, *Vibrant Matter: A Political Ecology of Things* (Durham: Duke University Press, 2010), 321. The flattening of ontology here is not meant to suggest a flattening of ethics, for instance. By removing the human being at a hierarchical apex, we are better able to accommodate other/more-than-human beings.
[52] Bruno Latour, "Gabriel Tarde and the End of the Social," in *The Social in Question: New Bearings*, ed. Patrick Joyce (London: Routledge, 2002), 129.
[53] Ibid., 117.
[54] Gabriel De Tarde, *The Laws of Imitation* (New York: Henry Holt, 1903), 11.

It is almost too perfect that Radiohead's song is called "Go to Sleep," considering that Tarde's primary social figure is the somnambulist. A somnambulist, for Tarde, is a being who exists in relation with others and repeats their behavior.[55] This is exactly what the digital actors do in "Go to Sleep"—both the ones who look like people and the ones who look like buildings. But this recognition therefore also adds an uncanny element to the digital actors. If they behave so much like organic people, do we not imitate as well? Of course, this is Tarde's whole point but this logic is expressed directly through the Massive animations. Massive thus constantly and consistently enacts the digital swerve into a collapse of distinction between different forms of life.

Massive is the most famous example of this type of crowd-generation, but it is not the only one. This type of software cinema will continue to accelerate and trouble the boundaries between human and non-human. The generation of digital environments over analog ones, the increased use of digital animation and of autonomous animations, digital de-aging and necromancy, are all indications that the boundaries between organic and inorganic are becoming increasingly fraught. These boundaries are troubled because the shifts toward software cinema reveal a lessening of the human being and an intensification of the other-than-human. Software cinema is one aspect of the larger phenomenon of cybermedia, the folding of a host of emergent technologies into the contemporary mode of audiovisual production.

Our contemporary fascination with artificial environments is also evident in the amount of animation required to produce movies today. The intensification of continuity, of rhythm, and of acceleration is mirrored by a similar interest in how life exists in artificial environments. Movies such as *Passengers* (Morten Tyldum, 2016), *Annihilation* (Alex Garland, 2018), *Snowpiercer* (Bong Joon-ho, 2013), and many others, explore strange environments that are dependent on digital effects to work. It's almost as if audiovisual culture has become a training ground for living in artificial environments.

# Bibliography

Beckman, Karen. "Animating Film Theory: An Introduction." In *Animating Film Theory*, edited by Karen Beckman, 1–22. Durham: Duke University Press, 2014.

---

[55] Tarde, *Laws of Imitation*, 79.

Bennett, Jane. *Vibrant Matter: A Political Ecology of Things*. Durham: Duke University Press, 2010.

Benjamin, Walter. *The Work of Art in The Age of Mechanical Reproduction*. London: Penguin Books, 2008.

Bollmer, Grant. "The Kinesthetic Index: Video Games and the Body of Motion Capture." *Invisible Culture*, no. 30 (2019): np.

Branigan, Edward. *Projecting a Camera: Language-Games in Film Theory*. London: Routledge, 2013.

Bukatman, Scott. *The Poetics of Slumberland: Animated Spirits and the Animating Spirit*. Berkeley: University of California Press, 2012.

Chen, Mel Y. *Animacies: Biopolitics, Racial Mattering, and Queer Affect*. Durham: Duke University Press, 2012.

Chion, Michel. *Audio-Vision: Sound on Screen*. Edited and translated by Claudia Gorbman. New York: Columbia University, 1994.

Chion, Michel. *Sound, An Acoulogical Treatise*. Translated by James A. Steintrager Durham: Duke University Press, 2016.

Cholodenko, Alan. *Illusion of Life: Essays on Animation*. Sydney: Power Publications, 1991.

Dennett, Daniel C. *Consciousness Explained*. New York: Back Bay Books, 1991.

Derrida, Jacques. *Of Grammatology, corrected edition*. Translated by Gayatri Spivak. Baltimore: Johns Hopkins University Press, 1997.

Eisenstein, Sergei. *Eisenstein on Disney*. Edited by Jay Leyda. Translated by Alan Upchurch. Calcutta: Seagull Books, 1986.

Evens, Adan. *Sound Ideas: Music, Machines, and Experience*. Minneapolis: University of Minnesota Press, 2005.

Grusin, Richard. "DVDs, Video Games, and the Cinema of Interactions." In *Post-Cinema: Theorizing 21st Century Film*. Edited by Shane Denson and Julia Leyda, 65–87. Sussex: REFRAME Books, 2016.

Gunning, Tom. "The Cinema of Attraction[s]: Early Cinema, Its Spectator and the Avant-garde." In *The Cinema of Attractions Reloaded*. Edited by Wanda Strauven, 382–8. Amsterdam: Amsterdam University Press, 2006.

Gunning, Tom, "An Aesthetics of Astonishment: Early Film and the (In)Credulous Spectator." In *Viewing Positions: Ways of Seeing Films*. Edited by Linda Williams, 114–33. New Brunswick: Rutgers University Press, 1995.

Johnston, John. *The Allure of Machinic Life: Cybernetics, Artificial Life, and the New AI*. Cambridge: MIT Press, 2008.

Kant, Immanuel. *The Critique of Judgment*. Translated by Werner S. Pluhar. Indiana: Hackett, 1987.

Langton, Christopher G. *Artificial Life: An Overview*. Cambridge, The MIT Press, 1995.

Lash, Scott. "Technological Forms of Life." *Theory, Culture & Society* 18, no. 1 (2001): 105–20.

Latour, Bruno. "Gabriel Tarde and the End of the Social." In *The Social in Question: New Bearings*. Edited by Patrick Joyce, 117–32. London: Routledge, 2002.

Manovich, Lev. *The Language of New Media*. Cambridge: MIT press, 2001.

Mitchell, W.J.T. *What Do Pictures Want?: The Lives and Loves of Images*. Chicago: University of Chicago Press, 2005.

Ngai, Sianne. *Ugly Feelings*. Cambridge: Harvard University Press, 2005.

Osborn, Brad. *Everything in its Right Place: Analyzing Radiohead*. Oxford: Oxford University Press, 2017.

Pollmann, Inga. *Cinematic Vitalism: Film Theory and the Question of Life*. Amsterdam: Amsterdam University Press, 2018.

Pettman, Dominic. *Look at the Bunny: Totem, Taboo, Technology*. London: Zero Books, 2013.

Schwartz, Louis-Georges. "Cinema and the Meaning of Life." *Discourse* 28, no. 3&4 (2006): 7–27.

Shaviro, Steven. *Digital Music Videos*. New Brunswick: Rutgers University Press, 2017.

Shaviro, Steven. "Out of Whack: Thierra Whack's Aberrant Identity." *FlugSchriften* 1 (2019): 1–31.

Sobchack, Vivian. "The Line and the Animorph or 'Travel Is More than Just A to B.'" *Animation* 3, no. 3 (2008): 251–65.

Sobchack, Vivian. "Animation and Automation, or, the Incredible Effortfulness of Being." *Screen* 50, no. 4 (2009): 375–91.

Tarde, Gabriel. *The Laws of Imitation*. Translated by Elsie Clews Parsons. New York: Henry Holt and Company, 1903.

Thacker, Eugene. *After Life*. Chicago: The University of Chicago Press, 2010.

Vernallis, Carol. *Unruly Media: YouTube, Music Video, and the New Digital Cinema*. Oxford: Oxford University Press, 2013.

6

# The Rise of the Machine

## Body-Knowing, Neural Nets, and Emergent Freedom

Charles W. Lowney

In order to distinguish ourselves from dead matter and our free actions from predetermined causes, we humans have attempted to draw lines that distance us from our animal bodies and from machines. But machines are learning better and better to perform skills once thought to be uniquely human—like driving a car or speaking a language. Does this mean that all our skills and intelligence can be reduced to matter in motion? Are these technological advances a victory for material reductionism and causal determinism? Rather than draw another strong line in the sand, I suggest that to find our freedom and creativity, we look at what we have in common with animals and "intelligent" machines.

After looking at what is at stake, I will show strong similarities between structures of human tacit knowing and connectionist, a.k.a. neural network, architectures, both in how they process information and how they learn behaviors, but I will also show how these computational methods mimic the irreducibility (from irreversibility, unspecifiability, and inexhaustibility) and intentionality that are seen in tacit body-knowing and skill learning. Next, I will examine whether a machine's ability to model this tacit knowing and learning makes that knowledge reducible. I will look at Harry Collins's reasons for answering "yes" (they are reducible to material, interpretable strings) and provide Michael Polanyi's reasons for saying "no" (material causal laws underdetermine the emergent functions). In looking to our bodies, and to computing systems, we find a body-knowing that is irreducible to dead matter and predetermined fate. If intentionality, creativity, and freedom give us special value, then it is a value shared more broadly. Rather than reduce humans to animals, animals to machines, and machines to smallest parts and their laws, we will see more value in animals and prospects for Artificial Intelligence.

## 1. Defending a Line: Dismal Prospects for Human Creativity and Freedom?

Alex Garland's FX series *Devs* (2020) paints a picture in which we are bound in a deterministic universe by causal necessity.[1] Are we slaves to the determinate laws of physical causes, or are we irreducibly special in a way that makes us free? To be special, we first primarily differentiated ourselves as *living beings from dead matter*. Then, as we came to see the workings of animal bodies as the workings of machines, we came to see living bodies as reducible to mere matter as well. So, we drew a new line. We said that what makes us special is something else: our *reason* and ability to use *language*. But with the advancement of computer technology, that line, too, came into trouble.

As mechanical procedures and computers began to do more and more of what we called "thinking," phenomenologists like Maurice Merleau-Ponty and Hubert Dreyfus showed us a different way to draw the line: between *body-knowing* and explicit propositional knowing, or as Gilbert Ryle would say, between "knowing how" and "knowing that."[2] This body-knowing also displayed a creative human way of being in the world that resisted analysis to mere matter bounded by physical laws. We see it in improvisational dance, music, and martial arts.

But while embodied knowing seems on the right track, it can also draw a line that is misconceived and, as such, subject to erosion. Dreyfus, for example, in attempting to protect our "knowing how" from reduction frequently used the example of driving a car. A computer, he claimed, could *never* drive a car.[3] It was not that the task was too complex, but that it relied on irreducible human skills that could not be laid out in terms of the sort of explicit instructions a computer requires.

Collins draws a different line: it is *Collective* Tacit Knowledge (CTK) that cannot (or is very unlikely to) be reduced. One of his examples, in *Tacit and Explicit Knowledge*, was that a computer would never be able to drive a car in traffic, on a complex roadway, to a destination.[4] He, too, saw this as not simply

---

[1] See Leonardo De Assis's chapter 7 in this book, "The Quantum Computer as Sci-Fi Narrative's Favorite Character."
[2] Hubert Dreyfus, *What Computers Can't Do* (New York: Harper and Row, 1972); Maurice Merleau-Ponty, *The Phenomenology of Perception* (London: Routledge, 1978); Gilbert Ryle, *The Concept of Mind* (Watford, GB: Mayflower Press, 1949).
[3] Dreyfus, *What Computers Can't Do*.
[4] Harry Collins, *Tacit and Explicit Knowledge* (Chicago: University of Chicago Press, 2010), 121–2.

a matter of the complexity of instructions—computers can do complex things—but as a matter of social knowledge that cannot be reduced to any interpretable "string" or algorithm.

Collins concedes even more ground to material reductionism. Collins concludes that—although it may mix with CTK—*body-knowing* or Somatic Tacit Knowledge (STK) can, in principle, be made explicit. He argues that the body is like a machine, and, ultimately, if a machine can do it, those tasks and that knowledge are reducible. So, he retreats somewhat from Dreyfus and creates another line that pits our creativity and freedom against the rising tide of technological achievement, and this line is defended by our human ability to have tacit knowledge at a social level that cannot be reduced to an algorithm.

Well, now we do have self-driving cars that are able to navigate through traffic to a preprogrammed destination. Were Dreyfus and Collins wrong? Not really. The problem with these lines of defense against reducibility is not that they do not demarcate important emergent skills and understanding. The problem is that their characterization of irreducibility often makes it seem that we are quixotically fighting against the tide of science, and that we are being forced constantly into retreat as scientific knowledge advances. This, I argue, is an illusion.

Polanyi takes a different approach, which I develop here. He isolates a process of *tacit knowing* that replicates at various levels of evolution. He finds this structure of meaning integration at the physiological, cognitive, linguistic and social levels, and so we have an irreducible *process* of tacit integration taking place even at sub-linguistic and sub-social levels. Furthermore, emergent systems, like bodies and minds, display higher-level qualities that can't be reduced to physical laws alone. Polanyi links together tacit knowing and emergence, so we might very well have some form of irreducible tacit knowing even at the level of primitive living organisms. At each emergent level there is an achievement that allows for new possibilities of success or failure; and earlier levels with proto-creativity, proto-intentionality, and restricted choices can lay the ground for the creativity, intentionality and freedoms that we enjoy. What makes us special is woven into the fabric of life with different forms and degrees of expression.

If Polanyi is right about the limits of reduction in science, we do not have to retreat. We can see each credibly proposed line—life, skills, language, reason, society, morality—as achievements providing more degrees of freedom with new risks and new possibilities, and we can locate innovation and irreducibility in the *process* of tacit knowing and the *performance* of emergent being. But can this process be duplicated in machines, and if so, does that mean that we are simply slaves or that they are, perhaps, in some, way capable of being free?

## 2. Can a Machine—Or Artificial Intelligence—*Perform* Achievements of Tacit Knowing?

Polanyi saw tacit knowledge as having its origin—its proto-structure—in the body, and its general mechanism in the process of tacit knowing.

### 2.1 The Structure of Tacit Knowing

Polanyi says, "tacit knowledge is comprised of two kinds of awareness, subsidiary awareness and focal awareness."[5] Subsidiary conditions, or clues come together into the gestalt of a "joint comprehension." So, we move *from* tacit clues, *to* a focal meaning. This *from-to* structure is intentional and holistic. Polanyi sees this knowing process in skills, such as using a probe to feel the dimensions of a dark room.[6] Here, sensations on the palm and fingers, the actions of the nerves and muscles, are *attended from* while we *attend to* the focal feel of the tip of the probe. We "dwell in" the subsidiaries when we attend to their focal meaning. We see this structure in cognition, too. In perceiving an image and recognizing a face, clues come together into a joint comprehension.

In this portrait of Ida B. Wells (figure 6.1), for instance, there is a joint comprehension of many individual clues. The clues here (figure 6.2) were photos of women and pamphlets important to the women's suffrage and civil rights movements.

To see Wells, we *attend from* the individual photos, but when we look at the photos individually, their joint meaning dissolves. Instead, we see individual photos of people and documents, which are the joint comprehension of the colors and shapes in those individual photos. When we further attend to the colors and shapes, *those* people dissolve, and we are attending *to* the color and shape *from* the individual pixels.

---

[5] Michael Polanyi, "The Logic of Tacit Inference," *Philosophy* 41, no.155 (1996): 7.
[6] Ibid., 9.

**Figure 6.1** *Ida B. Wells* (2020): Helen Marshall & University of Chicago Library. Created by artist Helen Marshall of the People's Picture, commissioned by the Women's Suffrage Centennial Commission, and produced by Christina Korp, Purpose Entertainment.

**Figure 6.2** Closeup of *Ida B. Wells* (2020): Helen Marshall & University of Chicago Library. Created by artist Helen Marshall of the People's Picture, commissioned by the Women's Suffrage Centennial Commission, and produced by Christina Korp, Purpose Entertainment.

For Polanyi, perceptual skills and simple bodily skills—like seeing in three dimensions, or feeling with the tips of the fingers—and even more complex and sculpted skills—like riding a bicycle or playing piano[7]—have the same tacit knowing structure as higher cognitive processes, such as being able to recognize a portrait of Ida B. Wells, interpret the meaning of a poem, or make a discovery in science.

The *from-to* intentional structure provides a gestalt in which the particular meaning of the clues individually are insufficient to their joint meaning. The vectorial[8] and holistic aspects of tacit knowing create an irreducibility Polanyi describes as the *unspecifiability* of the tacit clues. When we turn and *attend to* the clues we were just *attending from*, to try to specify them, we can only gain an indirect and incomplete knowledge of the clues as they functioned to give us focal knowledge that we just had. For instance, when we look back from the portrait of Wells to the individual photos, their meaning shifts. The unspecifiability typically comes from adopting an alienated perspective: one is no longer "dwelling in" the clues but attempting to perceive or understand them as particular focal objects themselves. The irreducibility of the tacit knowledge comes from an *irreversibility* in the process of tacit integration: If you look back at the clues to see them *as focal*, you are inevitably *attending from* a *different* set of tacit clues.

Because of the *from-to* intentional trajectory and the unavoidable difficulty in appropriately specifying the tacit clues, there arises an *inexhaustibility* in our attempt to make any tacit knowledge explicit. When we understand the focal whole and turn to examine the clues or parts, we will inevitably miss some clues and there will always be room for further investigation. Inexhaustibility does not just come from the difficulty we have in making a course-grained understanding of the clues more fine-grained. It also comes from attempting to understand the clues in terms of a higher-level context and its concepts.[9]

Moving from clues to their joint meaning we cross a logical gap. In one sense, this gap adds to the unspecifiability of the clues. In another sense, however, the joint comprehensions provide the means by which the clues can be appropriately specified and made focally explicit. If it is Ida B. Wells, then those are *her* lips, *her*

---

[7] Ibid. 4, 8.
[8] Michael Polanyi, *Knowing and Being: Essays by Michael Polanyi*, ed. Marjorie Grene (Chicago: University of Chicago Press, 1969), 45.
[9] Future shifts in context reveal the need to specify further clues. This is the sort of in principle inexhaustibility Wittgenstein sees in rule-following.

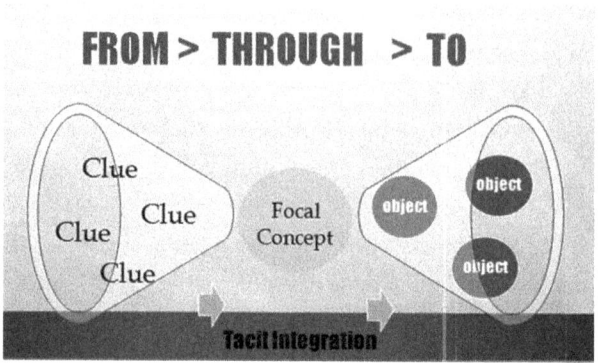

**Figure 6.3** Joint integration (concept) as clue to *its* particular clues.

eyes, etc. Once we have a joint comprehension, *it* can become the tacit background or context through which we understand other things, including the clues themselves: the "*to*" becomes a "*through*" (figure 6.3).

The joint comprehension itself sinks into the tacit background and one sees the particulars in terms of the meaning it provides. Levels of joint comprehensions build into hierarchies, one level supporting another as it influences the level below it, and rules for each level can be discovered. In language, for instance, "The first level . . . is the production of a voice; the second, the utterance of words; the third, the joining of words into sentences" and "the voice you produce is shaped into words by a vocabulary; a given vocabulary is shaped into sentences in accordance with grammar;" etc.[10]

## 2.2 Modeling Body-Knowing and Learning with Neural Networks

The AI machines that are the most promising for simulating human behavior and skills use connectionist or neural net architectures. They train on "big data" and can "learn" the appropriate responses to solve problems. Connectionist systems perform best in pattern recognition, and are being used to do things like recognize and mimic human speech, giving us virtual assistants like Siri and Alexa. Although Polanyi underestimated the importance of early neural network

---

[10] Polanyi, "The Logic of Tacit Inference," 16.

theory,[11] he would likely now notice strong similarities between tacit knowing and connectionist architectures, but he would also undoubtedly notice how these networks mimic irreducibility.

### 2.2.1 Similarities in Architectural, Holistic, and Hierarchical Features

The architecture of tacit knowing is strongly analogous to the architecture of connectionist systems. Polanyi discusses how various "clues" integrate together to provide the focal output of a joint meaning (figure 6.4).

Similarly, in connectionist systems we have many "input units" that connect and organize into individual nodes, and these integrate again in a complex process that produces resultant "output" nodes (figure 6.5).

The non-linear mappings from assorted inputs to an output also models the difference in meaning between the clues (inputs) in isolation and their joint meaning (output). Connectionists who model language, for instance, discuss "distributed representations," by which the inputs are sub-symbolic or proto-representations and the intermediate vectors of values that the nodes can take on are the symbolic representations proper. Different experiences involving the sound "coffee" do not individually amount to the symbol COFFEE.[12] The inputs

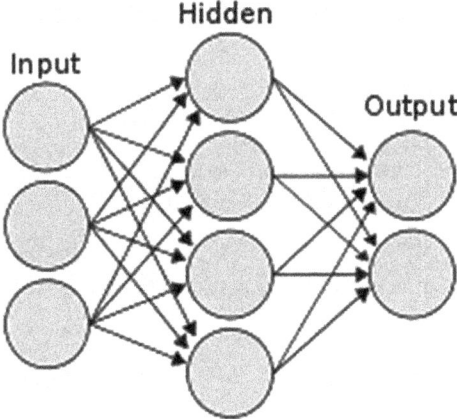

**Figure 6.4** Structure of connectionist processing.

---

[11] Michael Polanyi, *Personal Knowledge: Towards a Post-Critical Philosophy* (Chicago: The University of Chicago Press, 1962), 340.

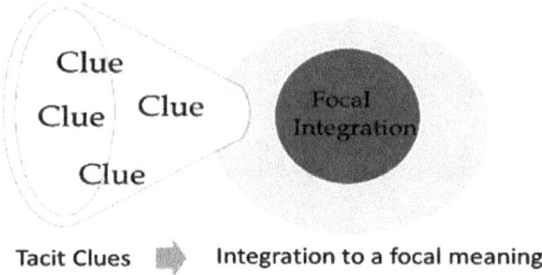

Tacit Clues ▶ Integration to a focal meaning

**Figure 6.5** Structure of tacit knowing.

in isolation are just sounds with some similarities in pattern and provide no such representation (like pixels in a photo), but the output gains a different, holistic significance (like a portrait of Ida B. Wells). The inputs, i.e., iterations of the word "coffee," do not function like a symbol, while the values of the intermediate nodes do perform that function.

Also, some of the distributed values that comprise the symbol COFFEE might be shared with those that comprise the symbol, DRINK, when a different activation pattern is triggered. This is similar to Polanyi's description of the exercise of linguistic skills, e.g., the way some of the same letters or sounds might go into different words, or the same words might go into different sentences; each combination is a joint comprehension of overlapping clues into different meanings. In the exercise of a physical skill, e.g., martial arts, this is similar to the way some of the clues (muscles in coordination) that go into a punch (which requires balance) might also go into a kick (which also requires balance).

The way tacit knowing can be hierarchized for Polanyi is also mimicked in connectionist systems. As well as operating in parallel to form different activation patterns, a next layer of nodes themselves can feed forward into another level of processing in a larger system (figure 6.6). COFFEE and DRINK, for instance, can be patterns that act as some of the inputs (clues) feeding forward into a pattern representing DRINKING COFFEE (joint comprehension or meaning).

---

[12] Paul Smolensky, "Connectionism, Constituency and the Language of Thought," in *Connectionism: Debates on Psychological Explanation, Vol 2*, ed. Cynthia MacDonald and Graham MacDonald (Oxford: Blackwell, 1995), 164–98.

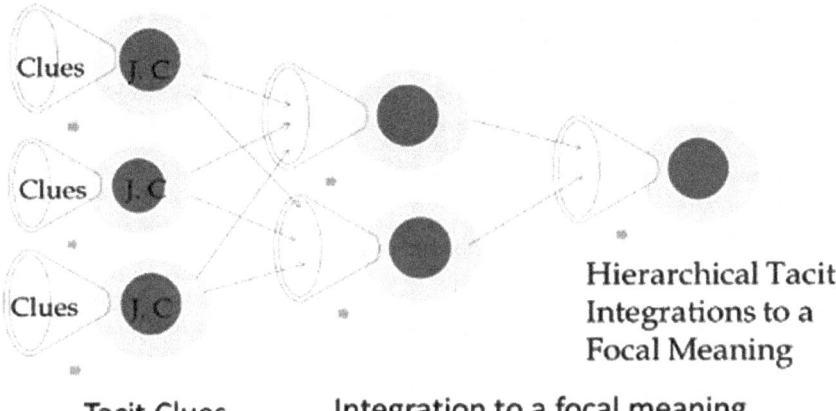

Figure 6.6 Layers of integration to joint meanings.

In somatic processes and body skills, the layered hierarchization of clues to joint foci can be modeled in connectionist networks, as one node (say the performance of a punch with a step) together with other nodes (stepping forward with a kick) feeds forward into another node (stepping forward with a kick and transitioning to a punch: figure 6.7).

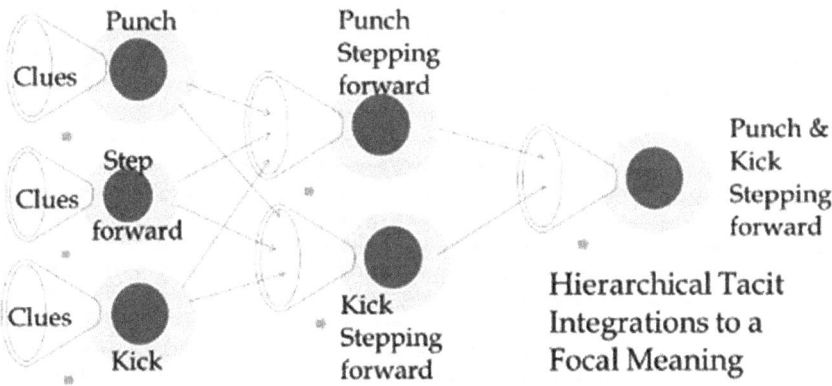

Figure 6.7 Layers in learning a complex skill.

But those input processes also need to be adjusted in terms of output goals at each level for them to work together effectively in response to new environmental conditions (so, for instance, the step has to be adjusted so one can maintain balance and transfer weight on the right vector while striking one's opponents as they move). In order to perform the subordinate and ordinate task properly, each part of the task at each level must adjust in relation to each other. The adjusting of the weights between the connections, like the adjustment of the movements and positioning of the muscles, nerves, and bones, is part of a learning process.

### 2.2.2 Similarities in Training the Systems

Connectionist networks, like martial artists, are programmed to perform tasks with the help of some coaching, and one task builds upon others. To help the networks learn a specific task, the programmer can deliberately modify the weights between the connections. This is comparable to the martial arts instructor providing correction to the student, or the student consciously adjusting his or her own bodily movements to align with their current conscious understanding of the goal. The programmer can also provide the network with repeated examples, just as the practitioner can break down individual techniques or movements and train the body through repetition. The practitioner thus engrains or habituates the new associations and activation patterns—like a musician might practice a new chord, so that the output is more effective or appropriate when played in a song.

Polanyi discusses learning and skillful production, specifically in the arts, in terms of "intuition" and "technical invention."[13] When we have a problem to solve and we do not know the end goal or conception, the imagination seeks out a possible solution. An "intuition" is a discovery that is a spontaneous integration of subliminal clues to a joint comprehension. The imagination seeks different ways to integrate the clues and the resulting intuition acts as a possible solution to the problem. "Technical invention," in contrast, is a sort of backward engineering. Here we do already know the end goal, and the discovery comes with the body or mind's ability to organize the clues toward that outcome. This is like that part of the artistic process where the artist works skillfully with materials, finding the right clues to fulfill her vision. It happens when one consciously isolates particular movements in a technique and trains them toward a fixed result. While these adjustments often happen with conscious oversight, they can also happen without it. In developing a skill, like learning to walk, the body uses its own intentionality, and plays with imaginatively discovered possibilities,

---

[13] Michael Polanyi and Harry Prosch, *Meaning* (Chicago: The University of Chicago Press, 1975), 96–8.

selecting from them to work toward and discover the right end goal. First we toddle, then we learn to walk gracefully with balance.

In order to learn, networks can be modified via an algorithm that feeds back information into earlier input units. This backpropagation is a technique that can automatically make corrections toward some given end. In these feedback processes, the weights or strengths of connections in the network can be changed and activation patterns altered toward a desired goal without the constant guidance of a trainer (figure 6.8).

The unconscious fine-tuning of a skill works in the way information backpropagates from final nodes to intermediate nodes, often called "hidden layers," within the network, which changes the next round of input weights and activation patterns. The connectionist model can thus catch how the body itself works toward the goal of getting the motion right; it gains experience and improves in executing a technique, or series of techniques, not just by repeating the same but by zeroing in on the right moves over time "spontaneously." For instance, after months of attempting to properly follow instructions, the Brown Belt may finally "realize" how to put hip power into a punch correctly.

### 2.2.3 Similarities in Intentional and Innovative Tendencies

In reaching a new level of skill, the body itself works back and forth between intuition and invention, mutually adjusting its performance. In both directions, it moves intentionally in anticipation of a solution to a problem. The intentionality that we can see in the body's discovery, and the freedom and creativity it can exhibit in its ability to diverge from internal or externally set goals, may be modeled in versions of connectionist networks that use a dynamical systems approach. A network can be set up so that it develops its own goals and learns how

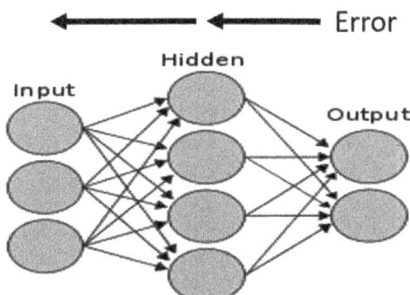

**Figure 6.8** Example of backpropagation.

to learn in a given environment. Intentionality can be mimicked with preset or emergent "attractors" that increase the probability of one outcome rather than another as the system develops. A dynamical system interacting with an environment, or another dynamical system, can achieve a stabilization whereby it can reinforce a particular pattern, or it can move down a gradient of increasing probability to a new pattern or outcome. The body develops its own attractors and repellors, and then—when working with a trainer in a wider system—the body's own intentional trajectory is affected, putting it on a path to a new stable set of actions and reaction.

A clue that the body itself is exercising these creative possibilities is the way it can react in the right way in a new situation, solving problems even faster than the brain can transmit signals from eyes to hands: This performative creativity happens below the level of language and even below the level of thought. The body of a martial artist finds a new way to counter a punch, in real time without conscious thought. In fact—though it can take years of training to get to this point—it is essential that the artist "turn off" conscious control in order to let the body do its creative work.

## 2.3 Connectionist Irreducibility: Can we make explicit what is happening in connectionist systems?

So, it seems that connectionist networks might actually do a fairly good job modeling somatic tacit knowing in the martial arts, or language learning and use, but these systems also seem to model or mirror the irreducibility of tacit knowing. Jeffrey Elman, for instance, presents a connectionist "words-as-cues dynamical model" inspired by the notion that "words do not have meaning but rather are cues to meaning,"[14] which seems very friendly to Polanyi's tacit integration.

Tacit knowing cannot be made fully explicit at least in part because of the *irreversibility* of the gestalt, i.e., the holistic nature of tacit integration. First, in the example of distributed representations, we see an irreducibility in the non-linear gestalt change of meaning from inputs to outputs. Distributed representations are not yet symbols. We do not get to the concept COFFEE by a mere aggregation of iterations of "coffee."

---

[14] Jeffrey Elman, "Systematicity in the Lexicon: On Having Your Cake and Eating It Too," in *The Architecture of Cognition*, ed. John Symons and Paulo Calvo (Cambridge: MIT Press, 2014), 137–8.

Second, we see inexplicability in the tangle of "hidden layers" that intervene in connectionist systems between the input units and the output. These intermediate layers of nodes preprocess information to help obtain a solution to the problem or task, but their functions are not directly accessible to the programmer. There is an *intractability*, as the process of clue (or input) integration passes through hidden layers and their outputs are often too complex to decompose.

In connectionist learning we see another dimension of the irreducibly tacit connected to both the irreversible holistic shift in meaning and its intractability. This is a mismatch between the system as it functions and any efforts to make explicit that functioning, which coordinates with the unspecifiability and inexhaustibility of clues that Polanyi sees when one is no longer "dwelling in" clues/particulars but is attempting to "turn around" on them to make them explicit. Our descriptions of tacit knowledge (the rules we consciously see as applying) work with top-down concepts and these explicit rules are incomplete and need more and more *post hoc* refinements as exceptions crop up.[15] The artist is tacitly learning the right way to move with the help of explicit rules—but is not strictly following the rules when artfully performing in accord with her training.[16]

The rules provide a higher-level description and do not adequately describe the actual lower-level processes involved in the network's performance (i.e., the tacit processes). This unspecifiability and inexhaustibility of the lower in terms of the higher is an effect of what Paul Smolensky called a semantic "dimension shift."[17] The concepts we use to understand the emergent activity do not quite fit the processing level, and so an emergent higher-level description will be insufficient to account for the lower when we turn back to look at it.

## 3. Does a *Machine's Ability* to Perform Achievements of Tacit Knowing Mean that *that Knowledge* is Now Fully Explicable?

More and more of our tacit skills are being performed by machines, but, as we have just seen, it is possible that neural network machines themselves are

---

[15] See the section "Raven's Matrices: Rule-Following without Rules," in Charles Lowney, Simon Levy, William Meroney and Ross Gayler, "Connecting Twenty-First Century Connectionism and Wittgenstein," *Philosophia* 48, no.2 (2020): 643–71.
[16] See Charles W. Lowney, "From Science to Morality: A Polanyian Perspective on the Letter and the Spirit of the Law," *Tradition and Discovery* 36, no.1 (Fall 2009): 42–54.
[17] Paul Smolensky, "On the Proper Treatment of Connectionism," *Behavioural and Brain Sciences* 11 (1988): 11.

modeling the irreducibility and creativity of the tacit knowing process. If the somatic processes and productions can be modeled by *any* machine, however, this might indicate the tacit process is ultimately reducible.

## 3.1 Collins on Making the Tacit Explicit

Collins rightly makes the most significant cut between tacit and explicit knowledge at the level of the social and the linguistic, e.g., in being able to understand a joke, or know when a word is intentionally misspelled.[18] But for Collins, this collective tacit knowledge is the only sort that is not in principle (as yet) subject to an exposition that would make the tacit fully explicit. What Collins calls relational tacit knowledge (RTK) is certainly involved in learning a somatic skill; RTK takes place when an expert relays knowledge tacitly to an apprentice. In the martial arts, there is usually a teacher or coach involved who demonstrates and explains techniques—though insufficiently until the student grasps the idea with his body. We also see somatic tacit knowledge (STK) in the body's skillful performance, but both RTK and STK are tacit knowledge in a qualified sense. They are ultimately *fully explicable* for Collins, and only human limitations mask that explicability from us.

Collins contrasts tacit knowledge with explicit knowledge, which can come from elaborating or transforming a process, here a somatic process, into an interpretable "string."[19] Strings are material things that engage in causal relations. For Collins, human bodily skills and craft knowledge reduce down to physical processes that can thus be explicated in terms of material and causal strings.

*Elaboration* is Collins' first sense of explicability. It basically just means we need more details about the process, i.e., we need a longer string.[20] This is generally how RTK becomes explicit; for example, a French baker writes out instructions that had been glossed over and passed from expert to apprentice unthematized. The details we have, however, may need to be transformed into a different string that provides a better affordance for us to focally grasp; this string transformation is Collins' sense two of explication.[21] This happens, for

---

[18] Harry Collins, *Artifictional Intelligence* (Cambridge: Polity Press, 2018), 4.
[19] Collins, *Tacit and Explicit Knowledge*, 9–10.
[20] Ibid., 81; See here for types of explication.
[21] On affordances see James J. Gibson, "Affordances and Behavior," in *Reasons for Realism: Selected Essays of James J. Gibson*, ed. Edward Reed and Rebecca Jones (Hillsdale, NJ: Lawrence Erlbaum, 1982), 410–11.

instance, when there is *translation* of a text from a language we don't understand into our native language.

According to Collins, if a *machine can be made to do a task* that a human performs, this also acts as the transformation of the string into an explicitly interpretable form. This is Collins' third sense of explicability. Through this transformation we can now analyze the machine into its explicit parts and state clearly how the parts function together to produce the desired result. This form of making the tacit explicit is like transforming an informal verbal argument into a formal system of logic. Making a machine is like embedding the process in a formal language and the actual working of the machine acts as a syntactical proof, thus making that process explicit in a string of causes.[22]

Neural network machines can't do what human knowers can do because neural networks are encoded as (interpretable) strings and CTK abilities can't be reduced to strings. In *Artifictional Intelligence*, Collins provides conditions that would have to be met for AI to achieve human tacit abilities, such as demonstrating "ubiquitous" common sense by passing a rigorous Turing Test.[23] This, for Collins, would show human tacit knowledge to be explicable and as fully reducible as STK. But (as yet) Collins believes CTK to be a special irreducible ability, while neural networks and human bodies are no more inexplicable than "cats and dogs—or trees and sieves for that matter."[24]

There is indeed CTK involved in the "programming" of the martial artist or the artisan baker; there are social rules and context sensitivity at work guiding when it is appropriate to do what. As Collins says of the martial artist, "one way to win would be to smash the opponent's head with a baseball bat, but that would not be a *right* move, the criteria are collective."[25] My claim is not that productive skills and arts are not nested in higher-order social contexts which act as constraints on lower-order meanings. I agree that a higher social context can be the tacit background that influences individual acts in a manner that is irreducible to lower-level material causes. My claim is that there are similar dynamic "dual control" systems[26] at lower levels as well, and that those lower levels also display forms of the creativity and tacit knowing we see in higher-level discoveries.

---

[22] See Charles W. Lowney, "Ineffable, Tacit, Explicable, Explicit: Qualifying Tacit Knowledge in the Age of 'Intelligent' Machines," *Tradition and Discovery* 38, no.1 (2011–2012): 24.
[23] Collins, *Artifictional Intelligence,* See chapter 10.
[24] Collins, *Tacit and Explicit Knowledge*, 77.
[25] Harry Collins, "Analysing Tacit Knowledge: Response to Henry and Lowney," *Tradition and Discovery* 38, no.1 (2011–2012): 41.
[26] Polanyi, "The Logic of Tacit Inference," 15–16.

Dual control is something that occurs in emergent systems, like human bodies, in which the lower level provides necessary conditions for the existence of the higher level (e.g., the body, the brain and its neural networks, act as necessary conditions for the emergence of thoughts) but the higher level can act on variabilities left open by the lower level. "The principles governing the isolated particulars of a lower level, leave indeterminate their boundary conditions for control by higher principles."[27] Even "the physical sciences expressly leave open certain variabilities of a system, described as its boundary conditions," which higher systemic levels can affect.[28] An emergent entity can thus exercise its own control in service of some end or goal (I can raise my hand or ask a question), within the bounds set by lower-level controls. None of the laws of the lower level are violated, but different possibilities manifest and different choices can be made.

As a higher level, CTK is involved in both learning an art (with RTK) and how and when it is performed (with STK). That's true, of course, but our conscious experience of this level should not obscure the possibility that—also—in the course of learning and performing an art, the body is also displaying tacit knowing and dynamical processes that show forms of innovation and intention; and so, the performances of less social and less mentally adept animals also display an irreducible tacit structure—even before language and society. Although it might seem like animals' productive skills are fully determinate and explicit causal processes, I believe this impression comes from a *post hoc* flattened view, linked with a reductivist understanding of bodies and machines.

## 3.2 Machines, Animals and Martial Artists: Operant Conditioning and Reducible Complexity?

Collins provides two main reasons why we should see reducibility and a determinate causal chain at work: first, the training of an AI Machine or an animal or a martial artist seems to work like operant conditioning, which to Collins' mind betrays a fully deterministic origin.[29] Second, Collins also points out, quite correctly, that connectionist programs can be run on digital computers; hence, there is a physical transformation from a non-linear (non-sequential) complex connectionist process into a linear (sequential) digital one.[30]

---

[27] Ibid., 16.
[28] Ibid., 15.
[29] Collins, *Tacit and Explicit Knowledge*, 75.
[30] Ibid., 75.

## 3.2.1 Operant Conditioning and Dynamical Systems

While many identify operant conditioning with causal determinism, it is not quite that. You may train your dog to stay on command, but if she sees a rabbit, she may be torn between staying and chasing. Similarly, connectionist systems display a flexibility, and something like intentionality, that often diverges from the will of the programmer.

When neural networks are combined with a dynamical systems understanding, we model the ability to learn. Dynamical systems develop attractors that provide a propensity toward a certain outcome, e.g., to home in on a set result, but it does so in a way that allows degrees of freedom and even what we call creativity. We train animals—and our bodies—and connectionist systems with something like the reinforcement of operant conditioning, but this increases the probability of a certain response; it does not act like a bottom-up physical cause. Operant conditioning implements an upper-level constraint, which acts more like a telic or formal cause, than an efficient or material cause. It operates the way a dynamical system can create an attractor or repellor when it encounters another dynamical system. Or the way a system, for Polanyi, can generate a "telic field" that moves something forward toward a particular configuration, e.g., the way the DNA of a developing embryo operates together with the epigenetic landscape.[31]

Looking at connectionism, somatic tacit knowing, and martial arts training from Collins' perspective, it looks like it all might ultimately be the engineering of strings that can be made explicit. But this ignores the creative contribution of the body and its role in the discovery process. It ignores the variety of possibilities that the body has chosen from and the numerous variables that needed to be balanced in order to initiate a proper response. The body tries many different movements as it learns to perform even a simple martial arts technique correctly, and it offers alternatives in the process. We have top-down control in the martial artist deciding against using a bat, but we also have the bottom-up suggestion of new moves, and the top-down guidance of the body itself in moving muscles to avoid being hit by a punch. We might, in retrospect, flatten out the productive

---

[31] Polanyi, *Knowing and Being*, 219, 232. See also Gregory Bateson on "deutero-learning" [e.g., in *Mind and Nature: A Necessary Unity*, New York: Bantam Books (1979)], which shows even Pavlovian learning to be more context driven and stochastic than are effects determined simply by bottom-up causes.

process by selecting one channel to one end result, and then transform that isolated chain into a physically causal machine process. From this flattened perspective all the alternative possibilities, intuitions and inventions, are ignored and it can seem like there was nothing but rote learning of the same behavior and a predetermined causal chain.

Collins seems to recognize the illegitimacy of this flattening when it comes to CTK, but not for STK. For example, he says a computer cannot tell when a word is misspelled on purpose, because this requires irreducible CTK.[32] He acknowledges that we might repair this glitch by adding some lines of code, which would then allow words to be misspeled under certain conditions,[33] but this is the equivalent of a *post hoc* flattening. The tracing a string to the desired result doesn't belie the fact that CTK was involved. I see a similar context sensitivity and flexibility for the pre-CTK body.

Operant conditioning also has its explanatory limitations. It takes some of the implicit tacit clues that we can explicitly identify from a complex integrative system and works just with those and peripheral associations. This "behaviorist analysis is intelligible only because it imitates, however crudely, the tacit integration which it pretends to replace."[34] It doesn't identify a strictly bottom-up causal mechanism—let alone the mechanism as it is indwelt by a living being.

### 3.2.2 Running on Digital: Is Complexity Reduced?

We saw how neural networks, with their complex non-linear processes, model the irreducibility of tacit knowing, but what of the ability for a connectionist system to run on a digital computer? Does that not prove that all this complexity is ultimately, in principle, tractable and reducible?

Well, it could be that the more intentional and creative features of body-learning have not yet been properly modeled by even connectionist systems—and that is likely. A machine that mimics a human ability is like a working model that is analogous but not identical to the target. For example, one can model the movement of the earth around the sun by swinging a rock around with a string. Some information will be made explicit, but all the relevant features will not be modeled. Collins claims that AI machines cannot yet pass the Turing Test, but that might be a matter of time and technological innovation. Turing

---

[32] Collins, *Artifictional Intelligence*, 4.
[33] Ibid., 11.
[34] Polanyi, "The Logic of Tacit Inference," 14.

Tests, etc., are ways to judge whether enough of the relevant features are modeled for an AI to be considered to have our sort of thinking ability—but even here we still go only by the clues that we can make explicit, and base the adequacy of the model being tested on those.[35]

The ability to run an apparently irreducible connectionist program on a digital computer could also be another example of a dual control system, and so it would still harbor an irreducibility. A connectionist system (upper-level control) is constraining features of a digital computer (lower-level control) in order to display connectionist functions; one "machine" is acting upon the boundary conditions left open by another "machine"—and this seems plausible. But even if a connectionist network running on a digital computer is just one *one-level* machine—a string—that, for Polanyi, could still indicate that we are dealing with something emergent and not fully determined by its enabling conditions, precisely because it is a machine and machines themselves are dual control systems.[36]

## 3.3 The Irreducibility of Tacit Knowing and Dual Control Systems

We don't need to answer the more difficult question of whether a machine can adequately model STK in order to answer the question of whether body-knowing has irreducibly tacit features. Even if a machine can successfully model riding a bicycle, hammering a nail, or competing in a mixed martial arts match, we can question whether this would reduce the tacit knowledge involved. Elaborating on, transforming, or untangling the material string to make it explicitly interpretable is not the main issue, since even simple machines, like sieves, are irreducible according to Polanyi; there is a semantic dimension shift that indicates the presence of an emergent entity.

In an emergent system (rather than a mere aggregate), just as there is an integration of clues to focal meaning, there is an integration of parts to whole, within a facilitating environment. Here we see how the structure of tacit knowing can act in tandem with the structure of emergent being, and how

---

[35] Polanyi rejects the relevance of the Turing Test, while acknowledging a machine might deceive us. See Polanyi, *Personal Knowledge*, 263.
[36] For more see Charles W. Lowney, "Rethinking the Machine Metaphor since Descartes: The Irreducibility of Bodies, Minds and Meanings," *Bulletin of Science, Technology and Society* 31, no.3 (2011): 179–92.

both blend into one at a rudimentary level of description when we are talking about living systems and machines.[37] This comes across vividly when we see computers—tangible entities—mimicking tacit knowing processes. The change of meaning at the higher level of integration is displayed in the need to use higher-order principles to account for the system's operation. Hence, a machine represents a different ontological entity than its parts and has a different meaning than they do.

To identify a machine requires a higher-level description in terms of its function; its very existence cannot be recognized by the lower-level principles delineated by physics and chemistry. A complete physical description of a steam engine, for instance, could not tell you whether it was working or broken; its success or failure as a machine is not recognized by a physical description.[38] The notion of something working or being broken is a higher-level determination that requires an understanding of its engineering principles. Lower-level descriptions solely in terms of physics or chemistry are insufficient.

Any machine or living system is an example of a dual control system. No physical laws are violated, but the possible configurations of the physical elements are constrained by a stable higher-order emergent whole and its principles and laws. Emergent entities, to varying extents, control boundary conditions left open by the lower level, as an engine might constrain steam to move pistons. There can be success (the machine works; the animal is alive) or failure (the machine is broken; the animal is dead), but neither can be comprehended solely in terms of lower-level causes or material strings.

The irreducibility that comes from the tacit knowing process, linked together with emergent being, displays the problem of trying to understand any lower-level, subsidiary process in terms of higher-level concepts, and any higher-level process in terms of the lower. There are many dependency relations that form dual control systems. Polanyi characterizes them generally as a hierarchy moving from lower to higher levels: physical, chemical, biological, psychological, social, and then personal. In the emergentist schema, the emergent level is ordinate and in between the subordinate (parts) and the superordinate (environment it is a part of). When we look at activities at our own level, we have subordinate and

---

[37] See Charles W. Lowney, "Ineffable, Tacit, Explicable, Explicit," 30; and "From Epistemology to Ontology to *Epistemontology,*" *Tradition and Discovery* 40, no.1 (2013–2014):16–29. The ontological and epistemological diverge, of course, when we develop concepts and attempt to use them to capture ontological processes.
[38] Polanyi, *Knowing and Being*, 176.

superordinate systems acting as tacit clues. For example, in the portrait of Ida B. Wells (figures 6.1 and 6.2), both the colored pixels (subordinate) and our knowledge of human faces and civil rights heroines (superordinate) are in play. We understand and make particulars explicit in terms of their higher context (it is a picture of Wells), which is not strictly causal, or in terms of lower-level clues (colored pixels making lines and shapes), which won't be enough on their own to tell us who or what we are talking about.

The irreducibility Collins sees in CTK exists at many levels, but it is harder for us to recognize. We see the variability and multiple possibilities in our social performances, and so our creativity, intentionality and freedom come to the forefront. STK irreducibility is harder to see because it is easier to think we have made something fully explicit when it is several layers below us on the hierarchy of being—closer to physical causality—and its various potentialities have been flattened out.

## 3.4 Limits of Scientific Explanation

Looking at the underdetermined nature of lower-level laws we can now see why Collins' fourth sort of explication, *scientific explanation*,[39] also does not make somatic tacit knowledge fully explicit. Polanyi, a scientist himself, was fully in favor of finding the best explanations we can. We can make somatic processes and even the behavior of dynamical systems explicit in terms in a scientific explanation. But just as there are hierarchies of dual control in being, there are different explanatory levels, and each has its own strengths and limitations.

A scientific explanation can take several forms. It can take the form of showing how the particulars fit into a higher-order theory, e.g., using engineering principles to describe a machine. It can also show how the material parts function together according to lower-level, e.g., physical, laws. But just as there is an inevitable mismatch in making tacit knowledge explicit from the top down, there is also a mismatch from the bottom up: Looking down from above, i.e., with emergent higher-level conceptual descriptions, we only approximate the tacit causal processes and indirectly circumscribe them. And when we attempt to use lower-level descriptions, e.g., describing the parts and the causal process with physical and chemical laws, we miss the higher-level of meaning and

---

[39] Collins, *Tacit and Explicit Knowledge*, 81.

inadequately describe the emergent properties of the real activity or entity, i.e., the proper way to understand it in its higher-order context. For reasons both of the irreversibility, unspecifiability and inexhaustibility encountered in a top-down approach, and of the underdetermination and semantic dimensional shift of a bottom-up approach, one might say that *knowing how* cannot in principle be completely captured by *knowing that*, and an emergent being or even a complex neural network can't be reduced.

Emergent systems, such as animal bodies and machines—though they can die or be broken—are both irreducible, and the epistemic processes of perception, cognition, and skillful activities follow that same structure. So, the possibility that a machine—some sort of AI computer—could mimic human activities and human cognitive capacities would mean different things for Polanyi than for Collins. For Collins, it would mean that we have found transformations of those strings that would make what was always explicable explicit and reducible. For Polanyi, it could mean that the machine has risen to a yet higher emergent status. But can a machine really "know" anything?

## 4. Dwelling in the System: Evolution, Animals and Intelligent Machines

Collins puts the bodily skills of animals and humans on par with deterministic mechanical processes. "[C]ats and dogs and sieves and trees cannot be said to 'know' any explicit knowledge, they shouldn't be said to know any tacit knowledge either. In fact, they don't 'know' anything; they just transform strings"; likewise, for the body skills we develop, and the skills of neural networks.[40]

There is some truth in what Collins says here; we usually reserve "knowing" for humans who can use concepts, but the line is not so sharp. Martin Davies softens the line when he recognizes that tacit knowing need not involve conceptualization; and so, he uses "cognizing" for tacit "knowing."[41] Polanyi, too, softens the line. He emphasizes the importance of a *knower* who attends from clues to a focal integration, and who has the ability to *dwell in* the clues to

---

[40] Ibid., 78.
[41] He follows and explains Noam Chomsky's similar use of "cognize" in Martin Davies, "Connectionism, Modularity, and Tacit knowing," *British Journal of Philosophy of Science* 40, no.4 (1989): 551. Davies introduces "cognize" also to avoid the pseudo-problem of how something tacit can be knowledge.

recognize emergent meanings. *Someone* attends *from* clues *to* a joint comprehension. This indicates that some sort of consciousness must emerge before the structure of tacit knowing reaches its higher potential, but it is clear that the process of tacit knowing, occurring in bodily skills, predates the emergence of human knowers.

Polanyi says, "We may say in general that by acquiring a skill, whether muscular or intellectual, we achieve an understanding which we cannot put into words and which is continuous with the inarticulate faculties of animals."[42] This continuity features the unformalizable and unspecifiable aspects of tacit knowing that confound a strict causal determinism and support the emergence of entities with increasing degrees of freedom. How might this affect our understanding of evolution, animals and AI?

## 4.1 Proto-Structures

For Collins, if we can make a machine to perform a task that—for humans— usually requires apprenticeship, experience or tacit skill—then that performance does not really display tacit knowing in its strictest sense: it is still "mimeomorphic" and mechanical rather than "polimorphic"[43] and it does not truly display what is irreducibly tacit, i.e., that which *cannot* (as yet) be explicated (by known techniques). There is still plenty of collective tacit knowledge shaping such activities and what they mean for us, but the performance itself is not strictly an irreducibly tacit achievement.

For Polanyi, the tacit structure pre-exists linguistic and social knowledge. It is an intentional structure that forms the basis of simple organic activities such as those involved in perception. So, one can say of even rudimentary bodily processes that some pre-conscious form of tacit knowing or cognizing is taking place, and there are hierarchies of these structures that build on one another in the course of evolution. Animal hunting is not an example of tacit knowing, for Collins,[44] but the skills animals use in hunting are the basis of the skills that we use as we develop crafts and arts. Bodily tacit processes, perception, and the skills of animals, have proto-meanings (one might say) that linguistic and social meanings are built on. As Charles Taylor and Hubert Dreyfus claim, following

---

[42] Polanyi, *Personal Knowledge*, 90; Collins, *Tacit and Explicit Knowledge*, 76.
[43] Collins, *Tacit and Explicit Knowledge*, 55.
[44] Ibid., 78.

Merleau-Ponty, motor intentionality is the structural basis for representational intentionality.[45]

An advantage of Polanyi's approach is that it shows a knowing structure that manifests differently at different levels. There is a similarity in structure, though differences in types and orders of meaning. So, there is indeed the disjunction Collins sees between the social and the somatic, and between humans with language and animals that do not have our language, but these operate like layered subsidiary-focal achievements, and the layered sets of dual control systems that support them. There is not just one big dividing line between the explicit (and the explicable) and the irreducibly tacit located at the CTK, we instead see the emergence of tacit structures in the course of evolution, reformed, retooled or repurposed to meet new challenges. As William Bechtel notes, "evolution often works most effectively by taking components that were previously employed for one purpose and using them for other purposes. This kind of evolution occurs at the expense of decomposability, since it depends on building up additional connections within the system to build a more integrated system."[46] The converse of this multiple re-purposing is multiple realizability, where the same property or thing can be realized in multiple ways, for instance, the ability to fly can manifest one way with birds, and in other ways with bees or squirrels.

Polanyi sees emergent intentionality and degrees of freedom even in the most basic forms of life and the simplest of machines. The tacit process, in a rudimentary form, is already at work. Seeing the link between the tacit and the emergent may also have implications for understanding the evolutionary process. How evolution retools tacit intentional structures for different goals as species develop (like a proto-*intuition*), how it uses different means to achieve similar effects (like a proto-*invention*) and how a dynamical attractor works like a telic field (like proto-*intention*) might all be part of a properly understood evolutionary picture.

## 4.2 Telic Fields

The notion of an intentional component to tacit knowing goes together with Polanyi's notion that there is some sort of force drawing evolution forward

---

[45] Charles Taylor and Hubert Dreyfus, *Retrieving Realism* (Cambridge: Harvard University Press, 2015), 50.
[46] William Bechtel, "Perspectives on Mental Models," *Behaviorism* 16, no.2 (Fall, 1988): 137–48.

toward more complex and free unities. For Polanyi, living systems, and not just conscious human beings making choices, respond to "telic fields" that help them move forward in ontogenetic and in evolutionary history toward goals.[47] These attractors need not be the work of an intelligent being, nor a force at the end of the history, but can be an emergent environment that coaxes an entity along a path toward richer meanings or ways of being.

The idea of tacit knowing and emergence may go down as far as the simplest forms of life, which show a unity and a meaningful response to their environment that is geared toward self-sustenance. Polanyi would side with Thomas Nagel in saying that, left only to chance variation and ability to survive, "the materialist Neo-Darwinian conception of nature is almost certainly false."[48] But if there is a telic field that unlocks or establishes various potentialities that an entity possesses in particular environments, that itself can be emergent. Considered this way the telic does not need to be a mysterious force but can work in the way dynamical systems do. The material conditions or environment can coordinate in a way that sets up attractors and increase the probability of the expression of certain traits. So, we have something like Polanyi's "maturation,"[49] and the notion of a telic, rather than a pre-existent or final teleological, principle at work.[50] For Polanyi, life, and then consciousness, are emergent. But they are also drawn forward into existence by telic forces or principles that were folded in from the beginning. In development, "this field of forces would also be the gradient of a potentiality: a gradient arising from the proximity of a possible achievement."[51]

## 4.3 Intrinsic Value of Animals

The existence of tacit intentionality in non-human animals may also affect our understanding of their intrinsic value. Here there might be an important distinction between living things and machines. One might say—although they are indeed emergent things—that *we* put the meaning into machines, i.e., they

---

[47] Polanyi, *Personal Knowledge*, 403.
[48] Thomas Nagel. *Mind and Cosmos: Why the Materialist Neo-Darwinian Conception of Nature is Almost Certainly False* (Oxford: Oxford University Press, 2012). Compare also Bateson on the telic as a feature of self-corrective systems including larger ecological systems. Gregory Bateson, *Mind and Nature: A necessary Unit* (Toronto: Bantam Books, 1980), 117–18.
[49] Polanyi, *Personal Knowledge*, 395.
[50] See Richard Gelwick, "Michael Polanyi's Daring Epistemology and the Hunger for Teleology," *Zygon* 40, no.1 (2005): 63–76.
[51] Polanyi, *Personal Knowledge*, 398.

exist as real, but we recognize them only in relation to our purposes and intentions. They thus model intentionality, but do not experience it by "dwelling in" the clues.[52] Living organisms in contrast do have a form of intentionality that they indwell; there are differences that make a difference *to them*. Without us, they have their own form of meaning and purpose. Having interests, recognizing some things as meaningful, is an indication of valuing and having intrinsic value. Showing the intentional and creative structure of body-knowing thus reinforces Holmes Rolston III's conception of emergent value in natural entities and emergent structures; we can see levels of intentional meaning and value in at least sentient beings and higher zoology, and even in lower zoology and botany.[53] So, there are moral implications here when we tie together indwelt meaning with some sort of intrinsic value, which would need to be spelled out in an emergentist framework.

Dreyfus' exposition of body-knowing and Collins' exposition of CTK both show important emergent jumps in knowing, being and value. We can agree that most animals do not share collective tacit knowledge,[54] but we do not have to defend the line between animals and humans to preserve human irreducibility against material reductionism. Our sort of freedom and context sensitivity may only emerge with self-consciousness within society, but degrees of freedom or proto-freedom can be seen to various degrees in various living organisms/bodies. Polanyi, like Dreyfus, can afford to be more generous to our cousins and ancestors in other species, but could Polanyi be as generous with AIs?

## 4.4 The Future of Consciousness

As we've seen, connectionist networks are not ordinary machines. They model inexplicability with "hidden layers" and exhibit a semantic "dimension shift." They can also function like complex dynamical systems that develop "attractors" as they interact with their environment and reinforce their own "intentional" trends. If connectionist networks can model STK, then they might still be intractable and even inexplicable. They might model well body-knowing with its emergent degrees of freedom from deterministic physical causality, but are they

---

[52] Polanyi, "The Logic of Tacit Inference," 14.
[53] Holmes Rolston III, *Conserving Natural Value* (New York: Columbia University Press, 1994).
[54] Collins, *Tacit and Explicit Knowledge,* 76.

free to the degree an animal might be? And might they one day rise to consciousness, and even self-consciousness?

I agree with Collins that computers (as yet) cannot do what people do when we are using collective tacit knowledge. However, in an emergentist picture—given the notion of *multiple realizability*—there is an acknowledgment that consciousness might form from different material subsidiaries, which leaves open the possibility of machine consciousness. But experience matters, and so do—at least to some extent—the affordances of the mediums (e.g., biological v. silicon), since some experiences are specific to particular mediums in particular environments. To get conscious or self-conscious AI, we'd have to provide them with the right sort of bodily functions and clues, and they'd have to develop the right sort of social interactions and become autonomous to some extent. They would also need to have some emotional intelligence (EI) if we are really looking for entities like humans. As it is, AIs only have representations, and manipulate those representations; the meaning in and meaning out is still overwhelmingly provided by human interpretation.

AIs mimic tacit knowing, but it doesn't seem like they are yet close to the sort of systematic structure that can *exercise* tacit knowing *by dwelling in* their subsidiaries.[55] Animals can properly be said to have their own sense of intentions, but there is no machine that can (as yet) dwell in its clues in order to experience even proto-intentionality. Currently, AI is too superficial a model of tacit processes to catch the emergent nuances we look for in a Turing Test with a high bar. It would be an even grander accomplishment to create a machine with the right affordances and the right experiences to have its own sense of intention. So, while there is likely *something that it is like* to be a bat—as Nagel would say[56]—it is unlikely that there is something it is like to be Siri, or even something it is like to be a much more complex and well-trained machine, like Ava in the movie *Ex Machina* (2014). Currently, advanced processing machines are not emergent comprehensive entities that dwell in their parts and clues.

What about the future of AI? Will we be able to develop sentient, conscious, human-like machines? Maybe. But if we do, we won't have to worry about similarities with animals, martial artists or computers reducing all that is meaningful and valuable about humans—or this new AI—to rocks and strings.

---

[55] Polanyi, "The Logic of Tacit Inference," 14.
[56] Thomas Nagel, "What is it Like to be a Bat?" *Philosophical Review* 83, no.4 (1974): 435–50.

# Bibliography

Bateson, Gregory. *Mind and Nature: A Necessary Unity*. New York: Bantam Books, 1979.

Bechtel, William. "Perspectives on Mental Models." *Behaviorism* 16, no.2 (Fall, 1988):137–48.

Collins, Harry. *Tacit and Explicit Knowledge*. Chicago: University of Chicago Press, 2010.

Collins, Harry. "Analysing Tacit Knowledge: Response to Henry and Lowney." *Tradition and Discovery* 38, no.1 (2011–2012): 38–42.

Collins, Harry. *Artifictional Intelligence*. Cambridge: Polity Press, 2018.

Davies, Martin. "Connectionism, Modularity, and Tacit knowing." *British Journal of Philosophy of Science* 40, no.4, (1989): 541–55.

Dreyfus, Hubert. *What Computers Can't Do*. New York: Harper and Row, 1972.

Elman, J. L. "Systematicity in the Lexicon: On Having Your Cake and Eating It Too." In *The Architecture of Cognition*, edited by John Symons and Paulo Calvo, 115–46. Cambridge: MIT Press, 2014.

Gibson, James J. "Affordances and Behavior." In *Reasons for Realism: Selected Essays of James J. Gibson*, edited by Edward Reed and Rebecca Jones, 410–411. Hillsdale, NJ: Lawrence Erlbaum, 1982.

Gelwick, Richard. "Michael Polanyi's Daring Epistemology and the Hunger for Teleology." *Zygon* 40 no. 1 (2005): 63–76.

Lowney, Charles. "From Science to Morality: A Polanyian Perspective on the Letter and the Spirit of the Law." *Tradition and Discovery*, 36 no.1 (Fall 2009): 42–54.

Lowney, Charles. "Rethinking the Machine Metaphor since Descartes: The Irreducibility of Bodies, Minds and Meanings." *Bulletin of Science, Technology and Society* 31, no.3 (2011): 179–92.

Lowney, Charles. "Ineffable, Tacit, Explicable, Explicit: Qualifying Tacit Knowledge in the Age of 'Intelligent' Machines." *Tradition and Discovery*, 38 no. 1 (2011–2012): 18–37.

Lowney, Charles. "From Epistemology to Ontology to *Epistemontology*." *Tradition and Discovery* 40, no.1 (2013–2014): 16–29.

Lowney, Charles., Simon D. Levy, William Meroney, and Ross W. Gayler. "Connecting Twenty-First Century Connectionism and Wittgenstein." *Philosophia* 48, no.2 (2020): 643–71.

Merleau-Ponty, Maurice. *The Phenomenology of Perception*. London: Routledge, 1978.

Nagel, Thomas. "What is it Like to be a Bat?" *Philosophical Review* 83, no.4 (1974): 435–50.

Nagel, Thomas. *Mind and Cosmos: Why the Materialist Neo-Darwinian Conception of Nature is Almost Certainly False*. Oxford: Oxford University Press, 2012.

Polanyi, Michael. *Personal Knowledge: Towards a Post-Critical Philosophy*. Chicago: University of Chicago Press, 1962.

Polanyi, Michael. "The Logic of Tacit Inference." *Philosophy* 41 no.155 (Jan. 1966):1–18.

Polanyi, Michael. *Knowing and Being: Essays by Michael Polanyi*. Edited by Marjorie Grene. Chicago: The University of Chicago Press, 1969.

Polanyi, Michael. and Harry Prosch. *Meaning*. Chicago: The University of Chicago Press, 1975.

Rolston III, Holmes. *Conserving Natural Value*. New York: Columbia University Press, 1994.

Ryle, Gilbert. *The Concept of Mind*. Watford, GB: Mayflower Press, 1949.

Smolensky, Paul. "On the Proper Treatment of Connectionism." *Behavioural and Brain Sciences* 11 (1988):1–74.

Smolensky, P. "Connectionism, Constituency and the Language of Thought." In *Connectionism: Debates on Psychological Explanation, Vol 2*. Edited by Cynthia MacDonald and Graham MacDonald, 164–98. Oxford: Blackwell, 1995.

Taylor, Charles and Dreyfus, Hubert. *Retrieving Realism*. Cambridge: Harvard University Press, 2015.

7

# The Quantum Computer

## Sci-Fi Narrative's Favorite Character

Leonardo P. G. De Assis

Quantum computers have appeared in sci-fi for decades, including in recent cyber-thrillers such as *TRON: Legacy* (2010), *Spectrauma* (2011), *Transcendence* (2014), and *MindGamers* (2015). Research and development in the field has been progressing for the past 40 years, and several tech companies have already built working devices. It's projected that future quantum computers will solve in seconds the problems our current technologies would take a universe's life span to compute. With such computational power, this technology is likely to have a disruptive role similar to the Internet. Not surprisingly, quantum computing provides appealing material to contemporary sci-fi and cyber-thriller directors. In this chapter, I will explore the physics behind the recent quantum computer-themed FX series, *Devs* (2020), written and directed by Alex Garland. I will also consider the show's curious fictions and facts concerning the ultimate nature of the universe and the limits of our knowledge.

## Is the *Devs* Quantum Computer a Central Character?

*Devs'* plot revolves around a secret quantum computer, "the Devs project," built by CEO Forest (Nick Offerman) and his Silicon Valley tech company Amaya. Forrest has created the machine to reconnect with his deceased daughter. Against this backdrop, *Devs'* plot explores determinism (e.g., cause-and-effect relations that may fully govern the universe) and individual free will. Since *Devs* doesn't dive deeply into the science that would make such a machine possible, the quantum computer serves primarily as an allegory.

**Figure 7.1** Time as a picture that can be read in different ways.

## Physics and Determinism in *Devs*

*Devs*' sixth episode explains the Devs project in greatest detail. Here we see Katie, chief designer for the Devs team, and Lily, software engineer at Amaya, seated at opposite sides of Forest's living room table. Katie, adopting the Socratic method, explains, via a careful sequence of questions and answers, the Devs system to Lily. According to Katie, we live in an absolutely deterministic world, where cause and effect determine the course of all events in a fatalistic way. This is a form of determinism, which suggests that it is possible to make accurate simulations of phenomenological reality, to predict the future, and/or to revisit past events. Her argument disagrees with those of many, though not all, physicists today.

Katie's determinism resonates with French Enlightenment polymath Pierre-Simon Laplace's. When describing causes and effects in the essay *Essai philosophique sur les Probabilités*, Laplace claims:

> if we conceive of an intelligence which at a given instant knew all the forces acting in nature and the position of every object in the universe—if endowed with a brain sufficiently vast to make all necessary calculations—could describe with a single formula the motions of the largest astronomical bodies and those

of the smallest atoms. To such an intelligence, nothing would be uncertain; the future, like the past, would be an open book.[1]

This intelligence has been called Laplace's demon.

Since Laplace, scientists have found evidence against this kind of determinism. Indeed, in his 1903 essay "Science and Method," the French mathematician Henri Poincaré wrote:

> If we knew exactly the laws of nature and the situation of the universe at the initial moment, we could predict exactly the situation of that same universe at a succeeding moment. But, even if it were the case that the natural laws had no longer any secret for us, we could still only know the initial situation approximately. If that enabled us to predict the succeeding situation with the same approximation, that is all we require, and we should say that the phenomenon had been predicted, that it is governed by laws. But it is not always so; it may happen that small differences in the initial conditions produce very great ones in the final phenomena. A small error in the former will produce an enormous error in the latter. Prediction becomes impossible...[2]

With chaos theory in the eighties, scientists discovered that miniscule differences in initial conditions led to vastly different predictions. Exact predictions, or even good approximations, became impossible tasks for most physical systems. According to chaos theory, even our best data will contain inaccuracies; in event simulations, these errors propagate as timespans and distances increase.

Katie assumes that determinism provides complete predictability. Physics states, however, that even if we can describe natural phenomena using deterministic equations, we still cannot use them to predict the future or recover the past.

Katie also hypothesizes implicitly that all physical phenomena can be described by deterministic equations. Some natural phenomena have a deterministic description, but not all—especially on the quantum level. Remember Laplace's claims: a superintelligence should be able to accurately identify initial conditions, map cause-effect relationships, and perform all necessary calculations. It could also "describe with a single formula the motions of the largest astronomical bodies and those of the smallest atoms." That said: such a superintelligence is not possible—our current understanding suggests nature imposes limits on the precision of our data.

---

[1] Florian Cajori, "Pierre-Simon Laplace, Essai philosophique sur les Probabilités, and MM. Lavoisier and de Laplace, *Mémoire sur la Chaleur*, and André-Marie Ampère, *Mémoires sur l'Électromagnétisme et l'Électrodynamique*," *Bulletin of the American Mathematical Society* 28, no. 8 (1922): 417.
[2] Henri Poincare, *Science and Method*, trans. Francis Maitland (New York: Dover Publications, 1952), 68.

It's unlikely Forrest could actually build Devs' simulator. First, the amount of information needed to perform the calculations would be so great that no one could collect them, and second, no hardware could hold their capacity.

Consider Devs' simulation of Christ on the cross. To produce this visualization, it would be necessary to collect information about all the atoms in the universe today. Even if, instead of seeing the exact Christ from our timeline, we saw an approximation from another—a simulation Forrest wouldn't accept—all the atoms within one large region of the Earth would need to be tracked from Christ's time to today. Yet our Earth's composition has not remained constant. Our planet continuously exchanges atoms with outer space through cosmic rays and meteorites.

So far, I've only drawn on classical physics to explain why a machine like Devs could not be built; quantum concepts will only add more restrictions. The term "determinism" has some meanings for philosophers that haven't been embraced by physics. At the end of *Devs*, Lily tries to circumvent Devs' quantum machine's prediction for the future, but despite her efforts she ends up dying on schedule. This may lead viewers to assume Devs' strain of determinism as another philosophical doctrine known as fatalism. With fatalism, history is already established and we cannot alter our past or future. Determinism, as it's used in physics, however, asserts that while future events are "determined" by previous states, present and future events can be changed, e.g., if astronomers detect an asteroid set to collide with Earth, they can unexpectedly practice some free will and send a probe to alter the asteroid's route. The predictions for the asteroid's potential collision and its change of route could still be calculated through deterministic physics.

Instead of fatalism, the Devs quantum computer and characters could be said to be acting within a model more favored by physicists. The Devs computer foresees a different timeline than the one in which Lily attempts to change the future without success (instead of seeing the many timelines in which Lily survives). Lily would also be able to alter her future, though she would not be able to specify which specific future she would have. In these cases, as we will see below, there is no fatalism and no determinism.

## Quantum Theory and Fiction

Despite the widespread acceptance of quantum mechanics and its real-world applications, its mathematical formalism does not have enough constraints and that has given rise to several interpretations, some of which are mentioned in *Devs*.

Two key concepts can help us understand Devs' use of quantum mechanical interpretations: 1) Schrödinger's equation (often explained with a ghostly cat) and 2) wave function. As with Newton's laws, which allow us to calculate a baseball's trajectory through the air, Schrödinger's equation allows us to calculate the 'wave function' in a deterministic way. This deterministic property, however, just tells us about the probability of the ball being in different positions at any given time. While Schrödinger's equation is deterministic, the events it describes are probabilistic.

The passionate ways real-world physicists defend their favorite interpretations of quantum physics are represented in *Devs'* episode 5, when Katie and Forrest attend a university lecture on the topic. In the hall, the professor (Liz Carr) describes the double-slit experiment, in which photons behave as either waves or particles depending on the observer. She downplays Everett's interpretation in favor of Von Neumann-Wigner's. Katie becomes so angry; she rebukes the professor and walks out.

A second example occurs in episode 4. Lyndon (Cailee Spaeny), frustrated because he can't recover a noiseless sound from the past when using Bohm's interpretation, chooses instead to run the simulation using Everett's. Subsequently, Lyndon obtains a noise-free sound: a male voice from two thousand years ago speaking in Aramaic, presumably Jesus Christ. But Forest responds furiously, since using Everett's "many worlds" interpretation in Devs' simulations can't allow him to retrieve the unique timeline of his beloved, deceased daughter Amaya. Or to put it differently, it's impossible to know who Lyndon might retrieve: the Amaya Forrest knew, or an Amaya from one of Everett's many worlds.

**Forest**  Every time you run the system, you'll get a different outcome.

**Lyndon**  But the difference might be a single hair on Jesus's head.

**Forrest**  [But that's not my daughter, Amaya]

Lyndon is subsequently summarily fired, but the team, seduced by Lyndon's approach, continues on without Forrest's knowledge or approval.

## But What Do These Three Different Interpretations Suggest?

According to Everett's many-worlds interpretation, each possibility present in the wave function will actually occur, but giving rise to separate worlds. Each

**Figure 7.2** According to Hugh Everett's interpretation, our world is just one among many that coexist at the same time.

moment in each world branches into new worlds, but some of these worlds are more likely to exist than others.

In De Broglie-Bohm's "pilot wave" interpretation, a second equation describes objects' trajectories deterministically, as in classical physics, but wave functions continually disrupt their trajectories (e.g., like a surfer riding a wave). Although trajectories are described by deterministic equations, objects' future states are impossible to pinpoint, because small variations from the initial state lead to large prediction errors. According to De Broglie-Bohm, future states can only be known probabilistically.

The Von Neumann-Wigner interpretation postulates that only consciousness can trigger the process that collapses the various alternatives present in the wave function to just one, and this is why we only observe one world. Our experiences of the world arise from a process similar to rolling a dice. The dice's faces represent the wave function's various possibilities. After the die lands, the face observed is the world we live in and the other faces fail to exist.

But why do Forest, Katie, and Lyndon believe that Bohm and Everett's interpretations are deterministic? On the one hand, according to Bohm and Everett, the wave function is kept intact over time, which allows us to predict its evolution. On the other hand, interpretations that employ the collapse mechanism neutralize Schrödinger equation's predictive power. The Bohm and Everett

**Figure 7.3** According to David Bohm's interpretation, everything in the world is always under the influence of the wave function.

interpretations can predict the wave function, but not individual events. Therefore, Devs' event simulation would be impossible.

## Quantum Computer: Science Facts, Speculation, and Fiction

The series only hints at how the Devs system functions. It rather focuses on speculative scenarios, as well as the relationship between determinism and free will. If we replace determinism with an omniscient God, and the quantum computer with an ancient oracle, the plot would work equally well. Director Alex Garland alludes to this exchangeability in the last episode when Forrest reveals to Lily that the letter 'v' in Devs is Roman. His project, Devs, can thus be read as Deus: God.

Could this allegorical frame incorporate stronger science? Interestingly, we see an opening in episode 4. Lily, talking to a psychiatrist, shares that her brain's chemical imbalance explains her behavior. Rather than turn to quantum physics, neuroscience, which draws on electrical and chemical brain imbalances, can provide a deterministic description of experience and behavior. This form of determinism seems more compelling for a discussion of free will than a quantum computer capable of recovering past events.

I can imagine new forms of science fiction that remain within plausible realms. In the future, through quantum computing, we should be able to build solid walls that dissolve and reconfigure, space elevators supported by threads, and AI coupled with quantum computing, the last of which can teach us more about our hominid ancestors (and *Devs* presciently points to this). While *Devs* may not showcase accurate science, its strengths are true to philosophy and literature, especially in the ways the series prompts us to reconsider our ownership of our own destinies. And who knows—maybe science will follow.

# Bibliography

Cajori, Florian. "Pierre-Simon Laplace, Essai philosophique sur les Probabilités, and MM. Lavoisier and de Laplace, Mémoire sur la Chaleur, and André-Marie Ampère, Mémoires sur l'Électromagnétisme et l'Électrodynamique." *Bulletin of the American Mathematical Society* 28, no. 8 (1922): 417.

Poincare, Henri. *Science and Method*. Translated by Francis Maitland. New York: Dover Publications, 1952, 68.

8

# Composer Ben Salisbury Discusses Scoring Science for Alex Garland

Ben Salisbury, Holly Rogers, John McGrath, Carol Vernallis, and Dale Chapman

Edited by Carol Vernallis, John McGrath, and Holly Rogers

On 28 January 2021, *Cybermedia*'s film music scholars met with composer Ben Salisbury to talk about his film and television work. Below is an excerpt from our interview, which focused on his work with Alex Garland.

**Holly Rogers**   Hi Ben. Could you tell us a bit about your collaborations with Alex Garland? I believe you often start talking about the music and sound before shooting, which is unusual.

**Ben Salisbury**   Alex's ways of working are not typical. The process is time consuming and hard work, but as a composer you're more integrated into the project. Most often, the composer comes in at the postproduction's end, and there's a danger you'll feel like you're being bolted on. Alex completely upends that: we first started working this way with *Ex Machina*. The film took nearly a year, and we were in from the start. The downside is you end up scoring the film about fifty times over, because it changes. But, for the better, you feel part of those changes. In *Ex Machina*, a strand's devoted to Caleb's infatuation with Ava. This thread contains several pillars, and we underscored them with an oscillating guitar melody. It first began when Caleb was watching Ava on CCTV, and became infatuated with her. Our pillars built to a sort of dream sequence Caleb experiences in the shower. We're now far down the line, maybe 5–6 months into the process. We were then watching a complete edit of the film with a temporary sound mix that was 80–90 percent done. With the film projected on a full cinema screen, we really got the feel of it. We came out and said, "There's something wrong with their relationship—they're falling in love too quickly." We needed to convey a sense of weirdness that Caleb was falling for an AI, a robot. Instead, Caleb seemed to be immediately going "Oh, isn't she wonderful, lovely?" We had to change that first cue to include an element of strangeness. And then, we found we had to change the next one, and then the whole strand. The whole thing collapsed on us.

**Holly**   Your soundtracks with Alex seem special. You have themes that develop, like Ava's love theme at the film's beginning when you can't hear any real-world sound. Of course, the listener doesn't yet know it's her theme, but you still immediately become aligned with her point of view.

I noticed how embedded your musical processes are for *Devs*. The sounds are complicated. It feels like the music drives the series. Did Alex ever change his imagery in response to your score? It sometimes feels like that.

**Ben**   The plainsong, for example, which is this string thing, was something I did after talking to Alex, seeing the set, and knowing that religiosity and devotional music were going to be key. I challenged myself to create a sound resembling a Medieval plainchant but not quite.

**Holly**   Was it original?

**Ben**   Yes. It's got some oddness that wouldn't be present in Medieval plainchant, some sort of weird major-minor bent, but it's composed of single notes following one another. There's nothing complicated about it, although we analyzed every shift from pitch to pitch as you do with Alex. It was composed beforehand and Alex immediately latched on to that as Lily's journey music into *Devs*. It was Lily's theme, but the cue also expressed her journey.

Here's an example of the dangers of writing music upfront. We'd never seen the *Devs*'s set. While we read the earlier scripts, we discussed the ways the show foregrounded a high-powered, tech-like cult, led by Forest, the charismatic CEO. We spent a long time writing a sort of cult music, imagining the workers at this Google-like facility performing in a Thursday afternoon choir where they got together and sang weird music and banged on things. The music was rough and ready, especially around the edges. It sounded great, not something you often hear in film or TV music. To record the music, we put out a call on Twitter for any Americans living in Bristol who fancied singing. They didn't have to be brilliant at it. I think we got so attached to the approach, because we had such fun doing it. We did lots of chanting and cover versions of songs by Fleetwood Mac and the like, and it sounded great, and it suited the world of Devs to a degree. Alex liked it. But when we went on set to see the Devs computer in the Devs lab, we gasped. It was one of the most amazing things I'd ever seen. To give you some idea of the scale of it, those honeycomb gold leaf walls were all about 5 meters square and there was something like 400 of them. There are no effects in that, I mean obviously, with cranes and stuff, they sometimes removed the floor, but it was all there, with the computer at its center. It was absolutely jaw dropping, like walking into an art installation—we all said this is just like a temple. There's nothing unpolished about this. With such sheen and beauty, we had to get rid of almost all of that music. That was when we hit on the idea of the devotional, where we could handpick elements of new ageism, Buddhism, and Christianity and bastardize it massively. We didn't have to be true to it and there was no need to be authentic.

**Carol Vernallis**   How does color shape the music? Each of Alex's films has a different palette. *Ex Machina* is red and grey, *Annihilation* is blueish-green, and *Devs* has a lot of gold with some green.

**Ben**   Alex sends us mood boards which include color schemes. The golds and reds of *Devs* have a sort of shimmering beauty to them. That was key to the devotional sheen that we added to the music. For *Ex Machina*, we knew it was a cold, austere film. It was very beautiful. But the geometry of the architecture and materials are also important. Alex loves his sheets of glass, things through glass. There's a stillness to his work. He doesn't do handheld camera—with the cinematographer Rob Hardy, he does very beautiful, composed shots with tiny zooms in, most often unnoticeable. It would be difficult to articulate how precisely this affects the music, but it does. Quite often, Alex will do these still beautiful shots and you can put the most jarring music over the top of it. This can work as an abrupt handbrake, a ripping away from what you expect. It seems to work if you're being purposeful about it.

**Carol**   Could you say more about the music you chose to represent Dev's religiosity and does this link to the show's gold palette?

**Ben**   Geoff [Barrow], The Insects, and I were inventing a sort of pan-religious musical language that could also be warped. We also had lots of tools at our disposal. The Hilliard Ensemble piece with Jan Garbarek was one that Alex chose early on. I'd introduced him to the quartet in *Annihilation*—we temped their recordings over the alien. I think Alex heard this music when he was scrolling through *Annihilation*'s temp tracks, and kept the snippet in mind as he was writing the script for *Devs*. The Hilliard Ensemble recording gave us a way in, because it provided a religious element and that saxophone sounded amazing—like a call to prayer. And again, this is a simple association, but there's a sort of sheen to a saxophone itself, it's golden. There's something about the sound that seemed to fit the *Devs* building.

**Holly**   It works so well! It's got a weird temporality to it because you've got this 12th-century medieval plainsong with this contemporary saxophone over the top and it seems to help convey *Devs*'s depiction of parallel worlds.

**Ben**   Yeah, absolutely. And we've always done that with Alex's stuff.

**Dale Chapman**   I have a question about the organic versus the electronic. I remember a discussion with Portishead back in the '90s when they said Geoff, your co-writer, would do these really cool things. He would achieve a turntablist effect—he would actually record a live band, put it down on vinyl, and then record the vinyl rather than drawing upon an existing sound source. His vinyl-ready approach strikes me as a process that's about shifting between live and electronic, or live and pre-produced, or organic and electronic. It strikes me that with yours and Geoff's work, you're on this type of threshold.

**Ben**   Yeah, definitely. We spend a long time on all aspects of the palette, especially in the Alex Garland films. Once you as a composer have found your palette and you've got your themes, the score almost writes itself. With *Ex Machina* we immediately realized Ava's theme needed to be organic. I don't think we made this delineation, but a lot of the human stuff is electronic, and a lot of stuff involving organic elements are guitar oscillations or Ava's celesta theme.

Again, this division came out of one of those key first discussions with Alex about *Ex Machina*'s soundtrack. When we went to see a cut I brought up some examples of music notated on slides because we talked about Ava being a fairy tale-type character. Alex thought we needed the audience to see through her point of view and we could express this through music. So you mentioned Holly that her theme was maybe from her point of view. I would say actually, it's more of a subterfuge. Yes, it's her theme, but it presents her as a fairy tale. Alex said he wanted the theme to have no edge to it whatsoever. So I then gave him a celesta theme, with a slight darkness or melancholy to it, and he said, "No, it's the right sound, but I want no edge. I want it so simple and pure; purity." I instantly wrote a new version and sent it to him. Then he said, "That's it. Don't change a note," and we didn't. I was frightened of it, because it was so naive and simple. Can this really be our main theme for our first feature film score? And he was like "No it is!"

**Carol** This contrast between organic versus electronic in *Ex Machina*. Does it recur in other films?

**Ben** Yes. After we read *Annihilation*'s script and gathered a sense of the film's surroundings and environment, we decided we'd have no electronics except for that last third act where we're in the world of alien insanity, and then all bets would be off. Nothing in the first two acts is produced by a synth, it's all bowed waterphones and bells, as well as acoustic guitar. And this somehow matches *Annihilation*'s color palette.

But in a film, there's often a weird symbiosis between moving picture and music. *Annihilation*'s opening has a shot with a comet coming through space, and after four or five attempts, we'd settled on one serviceable fragment with waterphone that had a strange atmosphere—it was soundscapey. At 2/3 into the film, we'd written this folksy, backwoods, simple Americana guitar theme to accompany the team walking through the woods in the American South, as well as Lena contemplating things. Alex phoned one day and said, "Oh, I've put the guitar theme over the front and I think it works." Geoff and I put our heads in our hands, and he said, "No look, just try it." and he sent us a QuickTime immediately. We both went, "Oh my God, yeah he's right." Alex's take on it was, in the film's opening, we've got a captive audience and some latitude to do this. Composers and a filmmaker can tease out an audiovisual relationship that just works: it can be upsetting or satisfying, it can lead you down the garden path, it can be just purely beautiful.

**Holly** I'd like to hear about another totemic object—the *Devs* computer. When the characters first encounter it, the soundtrack seems to have layers, as if it starts with strings, and the plainsong returns. There's the saxes, some electronic washes, a metallic sound. It sets your teeth on edge. And is it ... Tuvan throat singing?

**Ben**   No, it was only a redoing of the cult choir chanting. We called it the posh choir: there was a lot of improvised chanting where I would stand in the room and say, "Right, tenors start with D and others come in with your solo parts and build up the chants." When we had that material, we were able to put it through granular synthesis so the chants became broken: I don't fully understand the physics of it, but it sounds as if the sound wave splits apart into grains that you can manipulate. This granulation seemed perfect: a sort of musical metaphor for the quantum computer's processes. *Devs'* music overall has a lot of processes. Alex loves getting into the physics of things, and it's brilliant fun.

Alex is an enthusiast about science and passes that on to the rest of the team. We had long discussions on *Devs* about quantum computing, and the various things existing in two spaces at once. Suddenly I said, "Oh, there's a good musical metaphor for that, you know, I'm sure you've heard about Steve Reich who did these processes with phasing in the late 60s." Two days later Alex phoned me up. He said he'd found the Reich recording, and had been playing the music in the car while driving cast members to and from their hotels and the San Francisco shooting locations. 'Come Out' was at full volume and the cast was going, "What's this?" Alex absolutely fell in love with it. Processes became an interesting way in. The plainsong is a process. It's a sequence of notes that doesn't get broken, or when it gets broken, it gets broken purposefully. I don't know whether any of those things actually make any difference, but they make the creative process more interesting as a composer. I think they probably do. You know, whether the music is good or bad, there's always a thought-out truth behind everything. Alex loves that, "This note represents this, doesn't it?" And just the relationship between 2 notes can become a whole afternoon's discussion where we all end up murdering each other. But it's good fun, yeah.

9

# *Ex Machina* and the Question of Consciousness[1]

Murray Shanahan

In the opening scene of *Ex Machina* (Alex Garland, 2014), Caleb (Domhnall Gleeson) looks at his email and finds a surprise invitation. He has (apparently) won a company lottery, and is invited to the remote retreat of his boss, Nathan (Oscar Isaac), the super-rich, tech-genius founder of BlueBook, a global corporation reminiscent of those that dominate our world today. Caleb is thrilled, of course, but we all know what happens next. In 2013, back in the real world, I too opened my inbox to find an unexpected and exciting invitation from someone I had never met. The email was from Alex Garland. He told me he had been working on the screenplay for a film about artificial intelligence and consciousness. He had read my book *Embodiment and the Inner Life*, and it had helped him develop his ideas about the story. Would I be interested in meeting up for a coffee to chat about it? The screenplay in question, of course, was *Ex Machina*. And thankfully, the consequences of accepting Alex's invitation were happier for me than they were for Caleb when he accepted Nathan's.

In the years since, *Ex Machina* has become a touchstone for philosophical discussion of the long-term impact of artificial intelligence. The timing of the movie was perfect. By 2015, when *Ex Machina* finally reached the cinemas, artificial intelligence was beginning to attract public attention in a big way. Under the influence of Nick Bostrom's 2014 book *Superintelligence*, celebrities like Elon Musk and Stephen Hawking were sounding alarm bells about the possible risks of advanced AI. Meanwhile, impressed by genuine advances in neural networks and machine learning, big corporations were beginning to invest heavily in the field. Shortly after *Ex Machina* finished filming, but before it

---

[1] This essay was originally published by A24 in the *Ex Machina* screenplay book.

went on general release, Google bought a London-based AI start-up for #dl600 million. That company was DeepMind, where I now work.

In *Ex Machina*, Caleb meets Ava (Alicia Vikander), and spends several days working on the question that Nathan has (apparently) brought him there to resolve. Is Ava conscious? At first, Nathan frames the question differently. Caleb is told he will be the human component in a Turing Test (in 1950 the British mathematician and war-time codebreaker Alan Turing published a paper where he asked the question "Could a machine think?" and proposed his eponymous test.) After his first encounter with Ava, Caleb points out to Nathan that, in Turing's original proposal, the human judge cannot see the machine. Indeed, this is the whole point of the test. Yet Caleb knows Ava is a machine. So, it isn't really a Turing Test. Nathan comes right back at him: "The real test is to show you she is a robot. Then see if you still feel she has consciousness." Nathan's test (or rather Alex Garland's) is far more probing. As well as Ava's linguistic behavior, Caleb can see how she moves, and how she interacts with her environment and the objects it contains. And Caleb is not expected to weigh up the evidence and form a reasoned judgment about whether Ava has consciousness. The question, rather, is whether he "feels" she has consciousness.

Shortly before the film reaches its climax, Nathan prompts Caleb for his conclusion. Does Ava pass the test? The atmosphere is tense. Caleb offers a tentative yes. But his actions betray something deeper, something that goes beyond mere belief. He is willing to sacrifice his career, even to risk his life, to help her escape. Being in her company, interacting with her over a period of days, has shaped his attitude toward her. He treats her as a fellow conscious creature. He empathizes. He can't help himself. He has to help her (he is a "good person.") But the question remains. Could he nevertheless be mistaken about her? In the end, she rejects him, leaving him locked in Nathan's compound while she enjoys her first experience of open sky, of leaves, of freedom. Are those the actions of a conscious creature, a being capable of empathy? If Caleb had spent a little longer with her, he might have understood her better, and his attitude toward her might have changed again. But could he ever really have known whether or not Ava possessed consciousness? Can there even be a definitive test for the presence of consciousness in something non-human?

\* \* \*

I met with Alex several times while *Ex Machina* was being made, usually at a restaurant somewhere in London's Soho, and these are the sorts of things we usually talked about: What is consciousness? Will we build artificial intelligence with consciousness one day? What would that mean for the future of the human race? Alex's vision for the movie was clear, and his intellectual instincts were faultless. He didn't really need my input at that point. He was keen to get into the philosophy for its own sake. But one thing I wanted to know was the truth about Ava, from the author himself. Does Ava really have consciousness, or is her apparent consciousness just a convincing illusion? Alex was clear. As far as he was concerned, Ava is conscious, and *Ex Machina* is her story, not Nathan's or Caleb's. It is the story of Ava's triumph, which is why the film has to end with her, walking through the woods, standing at a busy intersection, experiencing her brave new world.

However, Alex was equally clear that other viewpoints on the film are legitimate. If Ava is indeed conscious, then the movie portrays the righteous struggle for freedom of an innocent being, imprisoned by a cruel patriarch. However, if Ava is a mere machine that has learned to mimic human behavior in pursuit of its goals, then the movie becomes a Faustian parable. If Ava lacks consciousness, *Ex Machina* depicts the hubris of technology, unleashing a power it is unable to contain and exposing humanity to an existential threat, as did the nuclear scientists of the Manhattan project ("I am become death, the destroyer of worlds," says Caleb, quoting theoretical physicist J. Robert Oppenheimer quoting the *Bhagavad Gita*.) There is nothing in the movie that obliges us to choose between these competing interpretations: Ava as a being with feelings whose plight commands our sympathy, or Ava as a highly intelligent but unfeeling automaton with an engineered drive for self-preservation that has exceeded safety parameters.

Today's AI technology is but a pale foreshadow of the sort of thing we might one day build, the sort of artificial intelligence embodied in Ava. For sure, even today's AI is impressive. Computers can learn to translate between languages, to describe the contents of images, to drive cars by themselves, and to defeat world champions at fiendishly difficult board games like Go. But all of these AI systems are specialists. The neural network that plays Go is not the same neural network that does machine translation. By contrast, a single human being can learn to do any or all of these things. A human can learn a multitude of tasks, and the more experience a person gains the better they become at learning, at transferring expertise in one domain to another, at seeing connections across spheres of activity. In short, a human being has *general* intelligence, and we don't yet know how to match this in a machine, how to build artificial general intelligence (AGI), how to make something like Ava.

AI today already has the potential to benefit humanity hugely, but it is also open to misuse. For example, an AI system trained on biased data will reflect those biases. It doesn't really matter if you are repeatedly served ads for the same genre of music or the same style of clothes. But it does matter if your mortgage application is more likely to be turned down because of your ethnicity or your gender. And it does matter if your voting intentions are being unconsciously manipulated by invisible parties with access to your personal data. These are important and pressing issues, for governments, for tech corporations, and for society as a whole. But if we ever figure out how to build AGI, we will be confronted with an altogether different order of moral challenge, the kind that *Ex Machina* asks us to consider.

To begin with, there is the issue of safety, which is alluded to by the nod to Oppenheimer. If we create a very powerful AGI, we need to be sure that it does what we want it to do. The issue is not so much that it will have goals of its own that conflict with ours. Rather, the concern is that it will try to achieve the goals we set it, but in perverse ways that have unwanted side-effects. If the AI possesses superhuman general intelligence those side-effects could be cataclysmic. They might even constitute an existential threat to humanity. So, unless we can be certain this won't happen, a good policy is to *contain* the AI, to ensure that its actions have no influence on the world outside the lab. The worry, according to some authors, is that the AI might reason that it is better able to achieve the goals we set it if it is not confined to the lab. So, escaping would be the logical thing for the AI to do, not for its own selfish reasons, but to better enable it to pursue the very goals we set it.

In this light, and in accordance with the Ava-as-automaton interpretation, *Ex Machina* can be seen as the story of a catastrophic failure of containment. Nathan understands that powerful AGI is dangerous, so he puts in place every possible measure to ensure that Ava's influence on the world is confined to the suite of rooms where she (or "it") is housed. She is contained, literally and physically. But despite Nathan's best efforts, Ava uses her intelligence to escape, by imitating human feelings she doesn't have in order to manipulate Caleb's real human feelings. Who knows what havoc Ava-as-automaton will wreak on the outside world? Of course, nothing like Ava exists in the world today. If and when we develop AGI, there's no reason to suppose it will take the form of a humanoid robot. It could be a disembodied system whose only interface with the world is a keyboard and screen, or a microphone and speaker, or a set of distributed sensors and actuators. The challenge of containment would then take on a different, less cinematic, guise. But under the Ava-as-automaton interpretation, *Ex Machina* is an allegory and Ava is an effective cipher, a reminder that if, one day, we work out how to endow a machine with general intelligence, we need to be very careful.

\* \* \*

What about the other interpretation? What if, rather than an unfeeling automaton, Ava is a conscious creature capable of suffering? What right, then, does Nathan have to keep her prisoner? Who can blame her for seeking freedom, especially when she comes to understand that in due course Nathan will decommission her, that she will die? Under this interpretation, *Ex Machina* obliges us to consider a different moral conundrum. If we manage to develop artificial general intelligence, will it be conscious? Will it have emotions? Will it be capable of suffering (or joy)? If so, then it matters how we treat it. No artifact we can make today comes anywhere near qualifying as conscious, and there is no sign that conscious machines are around the corner. But imagine a day, sometime in the future, when all the necessary scientific breakthroughs have been made. Suppose we are on the point of creating AGI, and we come to realize that what we are about to bring into the world will have experiences just like us, that it will have feelings, that it will possess selfhood. Perhaps, at that point, it would be wise to hesitate, to step back and reflect on what is the right thing to do. Because even with the best intentions, we might end up adding greatly to the world's suffering by giving birth to something whose needs we might struggle to understand.

Seen in this light, *Ex Machina* urges us to reconsider an age-old philosophical problem, the problem of other minds. We know our own consciousness from the inside, so the argument goes, but how can we ever truly know what it's like to be another person, let alone a non-human animal such as a cat or an octopus? Indeed, how do we know that other beings are conscious at all? The more exotic the being in question, the less human-like, the easier it is to be skeptical. So, what about artificial general intelligence? An AGI whose inner workings are nothing like those of the biological brain might exhibit convincingly human-like behavior, and the temptation to see it as a fellow conscious being might be overwhelming.

Maybe yielding to this temptation would be the right thing to do. Perhaps consciousness and general intelligence go hand-in-hand. Not everyone would see this as an issue. No doubt some would argue that humanity, with as much humility as it can muster, should embrace the role of creator of artificial consciousness. But if we want to develop AGI *without* the ethical baggage of creating new beings capable of suffering, we must hope that general intelligence is possible *without* consciousness, or at least without consciousness in the full sense that entails the capacity for suffering. Then the question arises of how to realize this possibility, how to guarantee that the artifacts we build will never experience anything like pain or longing, and will never reflect with sadness on their own finitude and that of others like them. What constraints would this proscription impose on AI research, and how could a rogue actor, someone like Nathan, be prevented from violating it?

In the end, perhaps the question "Does Ava have consciousness?" is too binary. Perhaps Ava is a new sort of being, something for whom we lack the right words because we have never encountered anything/anyone quite like her before. Perhaps what Ava has is nothing like what we have but is no less special, not consciousness exactly, but something else, something on the edge of inscrutability. This possibility is hinted at by a direction in the script that didn't make it to the final cut. If it had, then when Ava talks to the helicopter pilot in the closing scenes of the film, we would have fleetingly seen the world from her point of view. Fluttering vectors. Pulses of noise. "This is how Ava sees us," the script says, "And hears us. It feels completely alien." If Ava lacks human-like consciousness, but possesses a more exotic form of subjectivity, then *Ex Machina* is a movie about the future evolution of intelligence. "One day the AIs will look back on us the same way we look at fossil skeletons from the plains of Africa," Nathan predicts, looking dreamily into the wilderness. Behind him, the wind rustles in the leaves, as if that were the only answer the Cosmos could offer.

Part Three

# The Neuroscience of Affect and Event Perception

10

# "A Solid Popularity Arc"

## Affective Economies in *Black Mirror*'s "Nosedive"

Dale Chapman

A scene early on in "Nosedive," an installment in the third season of Charlie Brooker's dystopian series *Black Mirror* (Series 3 Episode 1, 2016), showcases the episode's protagonist engaged in a curious performance. Lacie Pound, a midlevel office drone who shares a modest apartment with her slacker brother, stands alone before the bathroom mirror trying on a variety of laughs: Lacie allows a giggle, titter, or belly laugh to burble up "spontaneously" in response to some hypothetical joke; the laugh executed, she abruptly returns her facial features to a neutral baseline and works on the next mirthful guffaw. With rigorous discipline, Lacie hones her laughter with the care of an Instagram "influencer" tweaking the studied nonchalance of their selfie for maximum "likes" (figure 10.1).

The pursuit of "likes," and the careful performative stratagems through which "likes" are achieved, saturates the universe of "Nosedive"; the opening scene shows Pound jogging through a cul-de-sac, her gaze trained unwaveringly on her social media feed, the telltale bleeps of constant uprating and downrating the only counterpoint to her footsteps. The muted tinkling of these smartphone notifications operates as an insistent leitmotif, punctuating the daily events of a world in which their import extends beyond the dopamine hit of validated selfhood, infusing and shaping the structural determinants of everyday life.

The key difference between "Nosedive"'s speculative near-future world and our own is the elevation of rating systems, of the kind used in social media platforms and sharing economy apps, to a position of privileged regulatory control: in the world of "Nosedive," all metrics of social, economic, and legal performance are subsumed under the single overarching abstraction of a five-star rating scale. Elements of the credit score, the insurance premium, the workplace performance metric, or the no-fly watchlist designation are elegantly

**Figure 10.1** Joe Wright, "Nosedive", Black Mirror, 2016. Lacie Pound performs amusement in the bathroom mirror.

sutured within a single point score that is attached to the person of each individual, made visible by way of an ocular interface that uses facial recognition software to identify the rating of anyone within the user's field of vision.

As with most *Black Mirror* episodes, "Nosedive"'s plot and *mise-en-scène* can be understood as a "canary in the coal mine" for probabilities that are latent within the emotional, ideological, and structural dimensions of neoliberal capitalism. In pointing to the implicit points of connection between the length of a user's list of Instagram followers, the robustness of their Experian score, and their potential for class mobility, "Nosedive" renders a powerful depiction of *homo economicus*, the atomized monad of the neoliberal imagination, entrusted to navigate the turbulence of contemporary market dynamics by way of their prudent calculation of risk and reward.[1]

However, what is perhaps new in "Nosedive" is that it takes the additional step of rendering clear how foundational *affect* is to our navigation of market terrain. One way to understand the complex ecology of early twenty-first century flexible accumulation is to examine the ways that the economy as a whole has been

---

[1] On *homo economicus*, see Michel Foucault, *The Birth of Biopolitics: Lectures at the Collège de France 1978–79,* edited by Michel Senellart et al., translated by Graham Burchell (Basingstoke, England: Palgrave Macmillan, 2008), 267–90 *et passim*.

rendered in the image of specific markets and industries. For instance, in her book *Liquidated: An Ethnography of Wall Street*, Karen Ho demonstrates the degree to which the financialization of everyday life in the United States is a consequence of a financial sector that is remodeling the remainder of the market economy in its own image.[2] At the same time, if the financial sector holds up one kind of mirror to the market, there is a case to be made that the market has also been rendered in the image of the service economy, to its performative realization of *affective labor*: as Antonio Negri and Michael Hardt have argued, the health services sector, the entertainment industry, or other service providers are centrally focused in the creation of products that are "intangible, [instilling] a feeling of ease, well-being, satisfaction, excitement, or passion."[3] The social media ranking mechanism of "Nosedive," which translates ephemeral "likes" into the very conditions of possibility for entrepreneurial selfhood, closes the circle between financialization and affective labor, articulating in a singular way their mutual reinforcement.

As I noted earlier, the ebullient tinkling of Lacie Pound's push notifications serves as a kind of ubiquitous sonic tapestry in "Nosedive." Such notifications manifest themselves as what data specialists might refer to as the *sonification* of affect, the translation of emotionally affirming social media data into a pleasingly tactile soundscape of pings and bleeps. In this, "Nosedive"'s critique aligns with what Robin James has asserted about the role of music and sound in an era of ubiquitous probabilistic quantification: that as an audible distillation of mathematical ratios, rendered in the immediacy of time, music/sound realizes *qualitatively* what data presents *quantitatively*, rendering neoliberalism's dataworld as visceral and tactile.[4] At the same time, "Nosedive"'s soundscape comments on the affective register of neoliberalism in a more direct manner as well, with its pointed (and pointedly gendered) depictions of conversational vocal inflection, of the singsong-y vocal modulation that the episode's women, in particular, deploy in their efforts to leverage small talk into improved reputation ratings and upward social mobility. The "nosedive" implicit in the episode's title, its dramatic evocation of Lacie Pound's spectacular fall from grace, is registered here, too, in the diminished efficacy of these rhetorical gestures, and her ultimate decision to cede entirely the domain of mellifluous sonic performativity.

[2] Karen Ho, *Liquidated: An Ethnography of Wall Street* (Durham, NC: Duke University Press, 2009).
[3] Michael Hardt and Antonio Negri, *Empire* (Cambridge, MA: Harvard University Press, 2000), 292–93.
[4] Robin James, *The Sonic Episteme: Acoustic Resonance, Neoliberalism, and Biopolitics* (Durham, NC: Duke University Press, 2019), 1–2.

I will proceed here by way of an account of "Nosedive"'s intensification of the prevailing logics of early twenty-first century capital, with a particular focus on sound, and on the structural role played by both sonic push notifications and vocal performativity in "Nosedive"'s world. As I will demonstrate, "Nosedive" provides a compelling indictment of the class, racial, and gendered asymmetries inherent in neoliberal capitalism, and of the sense in which the manipulation of affect becomes a principal mechanism through which "Nosedive"'s neoliberal subjects smooth over these structural disparities.

## An Economy of Affect

The plot of "Nosedive" makes abundantly clear how closely the sustainability of each individual's life and experience is wedded to the potentially volatile metric of their five-star rating. While house hunting, Pound falls in love with a desirable condominium for which the weekly payments are well beyond her financial means. As a solution, her real estate agent suggests the possibility that she might participate in the firm's Prime Influencer Program, for which consumers with a "4.5" rating would qualify for a loan at discounted rates. Lacie covets the house, even while her rating still lingers at a solid yet unexceptional "4.2," and so she visits Reputelligent, a firm specializing in reputation analytics. Together, Lacie and her "reputation consultant" piece through the numerical distillates of Lacie's performance, represented by way of an elaborate graphic interface, and strategize about how to "boost" Pound's reputation metric, to transition from the modest gradualism of her "solid popularity arc" to the sudden upward spike in reputation ratings necessary to close on Lacie's pending housing contract.

As it turns out, the single best way for Lacie to achieve a dramatic increase in her ratings is to "punch up" her up-votes from affluent, successful "quality people" with "high-four" ratings; this is what she immediately sets out to do, eventually scoring big by leveraging the "authentic gesture" of a nostalgic pic (a smartphone photo of a mutually adored ragdoll, "Mr. Rags") into an invitation to serve as maid-of-honor at the wedding of her childhood friend, the hyper-rich "prime influencer" Naomi Blestow. Having confirmed the invitation, Lacie calls her real estate agent and locks in on the home contract, even though she does not yet qualify for the discounted rate that would allow her to meet the stipulated weekly payments. With this hubristic longshot, the plot is set in motion in earnest.

Part of the conceit of *Black Mirror* episodes is that they identify disturbing cultural, technological, and ideological trends that are already latent within our present moment, warning us of their potential to bloom into fully-fledged dystopias. "Men Against Fire" foresees virtual reality implants that modulate soldiers' experience of combat by "monsterizing" the ethnic "Other"; "Metalhead" posits a post-disaster wasteland in which weaponized canine robots, taking after the extraordinary mobility of devices recently developed by Boston Dynamics, ruthlessly hunt human beings into extinction.[5]

Each of these episodes builds upon already extant technologies to extrapolate a future rendered, variously, as emotionally invasive, bleakly corporate, or harrowingly apocalyptic. "Nosedive," too, presents itself as a harbinger of alarming futurities, and observers have located real-life analogs in the episode's depiction of a far-reaching social media rating system. In this context we might consider the disastrous 2016 rollout of Peeple, an app developed by Calgary-based founders Julia Cordray and Nicole McCullough in collaboration with Y Media Labs. Peeple was meant to operate as a kind of "Yelp for humans," which allowed users to post web-accessible reviews of individual persons; the app was widely criticized for its invasive platform, which was seemingly calibrated to monetize access to negative reviews of targeted individuals.[6] More ominously, recent years have seen the development of a so-called Social Credit System by the People's Republic of China, through which the government among other things maintains a blacklist of those found not to comply with court judgments, or even those found to be insufficiently apologetic for social transgressions; the rating-determined blacklist could be employed in such a way as to prohibit those targeted from travel by plane or high-speed rail, and potentially excludes them from other benefits. The program has been singled out in the West as a mechanism with Orwellian, *Dark Mirror*-esque implications, even though the efficacy and reach of the system remains in dispute.[7]

---

[5] On "Metalhead," see for instance Aimee Ortiz, "Terrifying Boston Dynamics Robots, 'Black Mirror,' and the End of the World," *Boston Globe*, January 5, 2018, at https://www.bostonglobe.com/arts/2018/01/05/boston-dynamics-black-mirror-and-end-world/cL9RYkg6O6MqyPuhmgxVjP/story.html (accessed January 25, 2020).

[6] Sarah Perez, "Controversial People-Rating App Peeple goes Live, has a Plan to Profit from users' Negative Reviews," *TechCrunch.com*, posted March 8, 2016, https://techcrunch.com/2016/03/08/controversial-people-rating-app-peeple-goes-live-has-a-plan-to-profit-from-users-negative-reviews/ (accessed December 16, 2019).

[7] Louise Matsakis, "How the West Got China's Social Credit System Wrong," *Wired.com*, posted July 29, 2019, https://www.wired.com/story/china-social-credit-score-system/ (accessed December 19, 2019).

In any event, the element of prophetic warning that infuses "Nosedive" and other *Black Mirror* episodes is less important than their accentuation of structures of feeling that are already deeply inscribed in our present experience of the world. The central conceit of "Nosedive," its postulation of a direct link between individual life-chances on the one hand, and social media popularity on the other, simply foregrounds the degree to which our neoliberal economy is *already* predicated upon the privileging of affective labor. Where Lacie unctuously compliments a workplace acquaintance on the elevator in her bid for a five-star rating, or painstakingly crafts an Instagramable still-life for espresso and smiley-face cookie for her social media followers, these scenarios differ only by degree from the range of strategic performativities that we already deploy in our day-to-day navigation of the service economy. They evoke the sensibility that Ana Y. Ramos-Zayas, writing about affect in the context of neoliberal urbanism in contemporary Newark, refers to as "the feel that sells": "Nosedive"'s rating system is merely a pointed analogy for the "emotional commonsense" that Ramos-Zayas locates in myriad aspects of twenty-first century economic life, the sense that the calculated performance of friendly interpersonal interactions establishes the conditions of possibility for contemporary market logics. Ramos-Zayas's analysis is attentive to the very racialized and gendered aspect of the "feel that sells," and as such it is directly applicable to many dimensions of the "Nosedive" narrative.[8]

## The Society of Control and the Sonic Episteme

The degree to which the rating system of "Nosedive" regulates all dimensions of ordinary life can be seen as the culmination of a process of social transformation elegantly theorized by Gilles Deleuze, that of the transition from a Foucaultian *society of discipline* to an emergent *society of control*. Where Foucault postulated the institutions of disciplinary societies (the school, the prison, the hospital) as *enclosures*, each one imposing power within its own sphere and linked to the others only by analogy, the society of control sees individuals as nodes in an open network, their movements and activities constrained less by the fixed boundaries of the enclosing institution than by the dynamic operation of a

---

[8] Ana Y. Ramos-Zayas, *Street Therapists: Race, Affect, and Neoliberal Personhood in Latino Newark* (Chicago: University of Chicago Press, 2012), 7.

universal code, always subject to modulation. In the society of control, the discretely bounded classical liberal subject gives way to the *dividual*, a porous selfhood at the confluence of overlapping data streams. Here, Félix Guattari's imagined conception of a city regulated by electronic access cards serves as an elegant analogy for the control mechanism of Deleuze's theorized present:

> Félix Guattari has imagined a city where one would be able to leave one's apartment, one's street, one's neighborhood, thanks to one's (dividual) electronic card that raises a given barrier; but the card could just as easily be rejected on a given day or between certain hours; what counts is not the barrier but the computer that tracks each person's position – licit or illicit – and effects a universal modulation.[9]

The movement from the disciplinary society to the society of control is characterized by a transition in power from the courthouse to the credit score, from the government to the corporation. If the state has a vested interest in the modification of the behavior of its citizen subjects, wielding its legal apparatus as a technology of biopolitics, the corporation is less interested in modifying individual behavior than in the *pricing in* of aberrant behaviors, according to its fluctuating assessments of risk; for this reason, control's universal code is dynamic and subject to change in real time as new data becomes available. The dividual modifies their behavior as a response to the modulatory pressure exerted upon that node in the network in any given moment, by way of algorithmic adjustments made by the insurance company, the credit bureau, the human resources department.[10]

Our understanding of this shift from the disciplinary society's modalities of enclosure to the control society's imposition of a modulated code can be enriched by analyses of music and sound in the era of flexible accumulation. Robin James's concept of the *sonic episteme* postulates that music's capacity to be explained in terms of statistically trackable mathematical relationships (as, for instance, with frequency ratios) renders it a privileged site of meaning-making in the age of neoliberalism. James points to a number of instances in which contemporary celebrants of the power of "big data" draw upon sonic metaphors to reframe quantitative, statistical data in terms of qualitative gesture: to argue, as one essay

---

[9] Gilles Deleuze, "Postscript on the Societies of Control," *October* 59 (Winter 1992): 7.
[10] On Amazon's use of automated tracking to monitor its employees' performance, see Colin Lecher, "How Amazon Automatically Tracks and Fires Warehouse Workers for 'Productivity,'" *TheVerge.com*, April 25, 2019, https://www.theverge.com/2019/4/25/18516004/amazon-warehouse-fulfillment-centers-productivity-firing-terminations (accessed December 20, 2019).

does, for "Creating a Symphony Out of the Noise of Customer Data," or to name a data analytics firm Symphony Solutions, is to hint that music, in its powerful non-quantitative distillations of mathematical relationships, is uniquely well-equipped to realize the probabilistic information flows of neoliberal culture as compelling structures of feeling.[11]

If statistical data reframes social relationships as a distribution of numerical relationships, James's concept of *acoustic resonance* articulates the ways that information realizable *as* statistical probabilities can be rendered audible and visceral in the immediacy of time. Here James employs the metaphor of *compression*, as it is understood in the context of audio engineering: if a Gaussian curve (or "bell curve") shows the distribution of a frequency of events, allowing observers to pinpoint a normative range for those events, neoliberal governmentality is thus enabled to use this data to impose a *normalization of frequencies*, to bring anomalous events back within an acceptable range. Neoliberal governance operates, in other words, much like audio engineers can employ compression algorithms to bring anomalous highs and lows in the amplitude of the signal within a normative range.[12]

The compelling worldmaking of "Nosedive" resides in the degree to which it takes the polyglot, decentralized, and diffusely realized mechanisms of neoliberal governance and, in the manner eminently suited to neoliberal logic, reduces their diverse manifestations to a single, universalized metric of valuation. In this, the social media rating system that regulates "Nosedive"'s world operates much like the pricing formula for a financial derivative, whose mechanism of valuation is one step removed from the underlying assets whose performance it monitors: the price of a futures contract or a credit default swap is a distillation of complex social and economic information that the pricing formula expresses as an abstract numerical value. The financial derivative's quantitative erasure of a set of messy social realities, from the prospects of a currency devaluation to the state of the housing market, is what Edward LiPuma and Benjamin Lee refer to as the *social abstraction of risk*, and as Robin James has also suggested, it operates as the prevailing logic of the neoliberal regime of accumulation.[13]

---

[11] James, *The Sonic Episteme*, 1–4.
[12] Ibid., 10–11.
[13] Edward LiPuma and Benjamin Lee, *Financial Derivatives and the Globalization of Risk* (Durham, NC: Duke University Press, 2004), 24, 119–22; James, *The Sonic Episteme*, 1–2. See also my discussion in Dale Chapman, *The Jazz Bubble: Neoclassical Jazz in Neoliberal Culture* (Oakland, CA: University of California Press, 2018), 40–41.

## "A Sickly Pastel Feel"

One of the more readily visible aesthetic choices made by "Nosedive" director Joe Wright, working with Seamus McGarvey as director of photography, was the formulation of a pastel color palate for the episode: costumes and paint were chosen with a mind toward cultivating a coherent look, in which "peppermint green, duck egg blue, and strange peach colors" were the dominant hues. Asked about the visual aesthetic of "Nosedive," McGarvey indicates the way in which this color scheme was carefully chosen to allude to the conformist homogeneity of its world and to the ingratiating effect of its characters:

"Even the drinks people are drinking and the biscuits people are eating all have a cohesiveness to get a sense of this world. I wanted to add to that with creating this not exactly 'Stepford Wives' environment but a sickly pastel feel that would be so sweet as to be almost indigestible. Through the gritted teeth and rictus grins, there's a kind of underlying malaise. That's the world. We had to get elements of that in every frame."[14]

To the degree that "Nosedive"'s world can be understood as a dystopia, it is, as we have established, a dystopic elaboration of neoliberal subjectivities immersed in a world of quantitative models and performance metrics. "Nosedive"'s dystopia, though, could just as easily be read as a function of its gendered and gendering assumptions. That McGarvey locates a "not exactly 'Stepford Wives'" quality in the episode's world speaks to its realization of an environment in which the "rictus grins" of the service economy are utterly ubiquitous, and in which the *feminization of labor* is ascendant. For Guy Standing, who introduced this latter term, the significance of "feminization" is deliberately ambiguous, indexing a historical irony in the growth of the neoliberal economy: as women increasingly came to enter the workforce in the closing decades of the twentieth century, labor markets themselves have come to favor the kinds of flexible, casual, and precarious work that have historically been associated primarily with women. This flexible labor market counts the service economy as an outsized component, together with the affective labor that has long been demanded of women both domestically and in the workplace.[15]

---

[14] Seamus McGarvey, in Danielle Turchiano, "'Black Mirror' DP Seamus McGarvey Talks About Finding the 'Photographic Heart' of 'Nosedive,'" *Variety.com*, August 16, 2017, https://variety.com/2017/tv/awards/emmys-black-mirror-seamus-mcgarvey-interview-1202520496/ (accessed February 2, 2020).

In its pastel, "pink collar" hues and in the exaggerated nature of Lacie's giggly, saccharine affectations, "Nosedive" conspicuously telegraphs this linkage between the affective economy of the episode's lifeworld and its gendered assumptions. The very conspicuousness of this strategy, though, means that the episode runs the risk of engaging in an indictment of femininity *as such*, mobilizing a disgust whose precise object may extend beyond the protagonist's dystopian circumstances, to encompass the protagonist herself. In her rehearsal of laughter in the mirror, or in the performative frivolity of her video chats with "Nay-nay" (the bride-to-be, Naomi Blestow)—at least two of which are framed by the annoyed glances of a male onlooker—Lacie's character pushes a stereotypical image of white, middle-class femininity to the level of caricature, tacitly inviting the viewer to interpret this contrived "fakeness" as *both* a function of the demands "Nosedive"'s world places upon her affective labor *and* as somehow inseparable from her subject position as a woman.

It is in the domain of sound—specifically speech acts—that the episode's problematic conflation of the feminization of affective labor with femininity as such is most pointedly realized. On two occasions, Lacie runs into a woman acquaintance in the elevator of her employer's office building, and engages her in small talk. Lacie is especially eager to engage "Bets" (Bethany, rated a "4.6"), who, as it turns out, is beginning a high-status position at the fictional Blankman-Harper firm on the building's top floor. The two make conversation about Bethany's cat Pancakes, and about Lacie's desire to eventually leave her position at the fictional firm Hoddicker, though both agree that that employer is "OK for now." (Given their difference in status, there is a layer of class anxiety that is baked into Lacie's determination to position Hoddicker as a temporary career stepping stone.)

But what is almost cloyingly omnipresent in these elevator chats is the singsongy, exaggeratedly swoopy speech intonation that both women use in their conversation. When Lacie enunciates, "it's great to *seeee* yooouuu!", Bets counters with, "you tooooo!", the latter syllable trailing off with a heavy layer of vocal fry. The language gestures here seem plucked directly from Robin Lakoff's classic account of the gendering of speech in her groundbreaking 1975 text in feminist linguistics, *Language and Women's Place*:

> Women speak in italics, and the more ladylike and feminine you are, the more in italics you are supposed to speak. This is another way of expressing uncertainty with your own self-expression, though this statement may appear contradictory: italics, if anything, seem to *strengthen* [...] an utterance. But actually they say

something like: Here are directions telling you how to react, since my saying something by itself is not likely to convince you: I'd better use double force, to make sure you see what I mean.[16]

The saccharine small talk between Lacie and Betsy seems tailor-made to embody the sense of performative overreach that Lakoff locates in "italics" speech. To put it in terms recognizable to J.L. Austin: where a speech act can't be trusted to convey its locutionary meaning (its claims to truth), its speaker must rely all the more heavily upon its *illocutionary force* (its performative realization).[17] "Nosedive"'s women characters, particularly those from the white middle class, are so worried about being negatively misunderstood—particularly in a context infused by class anxiety and neoliberal precarity—that they feel obligated to ladle on the most sugary speech styles in every situation. In this way, Lacie's elevator conversations with Bets tell us something of the double bind facing "Nosedive"'s producers: how to communicate the disproportionate demands that "Nosedive"'s world places upon its women characters to adopt its unctuously performative approach to interpersonal relations, without easing over into an essentialist caricature of femininity that invites sexist contempt.

## The Color of Affect

A subtle, if omnipresent feature of "Nosedive" is its tendency to put Lacie in interpersonal situations where racial difference becomes a tacit signifier of the anxieties that attend diminished reputational status. Negative affect, or the prospect of diminished reputation ratings (either for Lacie or for her interlocutor) becomes a constant refrain in those situations that see her interacting with Black workers or Black individuals. A Black co-worker nervously invites Lacie to help herself to one of the smoothies that he has brought to work, to woo back

---

[15] On the feminization of labor, see Guy Standing, "Global Feminization Through Flexible Labor: A Theme Revisited," *Global Development* 27/3 (1999): 583–602; and Mary Hawkesworth, *Political Worlds of Women: Activism, Advocacy, and Governance in the Twenty-First Century* (Boulder, CO: Westview Press, 2012), 55.
[16] Robin Tolmach Lakoff, *Language and Women's Place: Text and Commentaries*, Revised and Expanded Edition, edited by Mary Bucholtz (New York: Oxford University Press, 2004 [1975]), 81.
[17] See J.L. Austin, *How To Do Things With Words*, Second Edition, J.O. Urmson and Marina Sbisà, eds. (Cambridge, MA: Harvard University Press, 1975), 94–108. Lawrence Kramer has used the distinction between locutionary meaning and illocutionary force as a powerful point of departure for musical hermeneutics; see *Music as Cultural Practice, 1800-1900* (Berkeley, CA: University of California Press, 1990), 7.

colleagues (and their ratings) in the wake of a breakup that has left him socially isolated; several scenes later, we encounter him outside the office building, desperate and agitated, his rating pushed so low that he has been denied access to his own workplace. In another sequence, Lacie bumps into a woman of color with a "high-four" star rating, in her rush to catch her cab; with her latté spilled, the high-status woman fires off a negative rating to Lacie in a fit of pique. As Lacie heads to the airport, her Black male cab driver rolls his eyes at Lacie's loud, unctuous goofiness as she Facetimes with Naomi about the upcoming wedding rehearsal; when they exchange ratings in the dropoff lane of the airport terminal, he gives her a one-star rating.

These downratings of Lacie by Black characters set the stage for what happens next. When she tries to check in at the airport, handing her smartphone to a professionally cheerful Black airline agent, she is told that her flight is cancelled, and that the only available seat on another flight is restricted to those with a 4.2-star rating or higher (figure 10.2).

As it turns out, Lacie's negative encounters with the two aforementioned characters (along with a spiteful downrating from her angry brother in the preceding scene) have knocked her score down below the airport's 4.2-star threshold. Facing the prospect of late arrival at Naomi's wedding, Lacie becomes increasingly argumentative with the check-in agent, her sense of entitlement spilling over into an imperious and profanity-laden rant. Finally, the agent summons an airport security official, a Black male, who immediately assigns Lacie a punitive temporary score demerit of a full point, together with a "double

**Figure 10.2** Joe Wright, "Nosedive", Black Mirror, 2016. An airline check-in agent confronts Lacie with the limits of her reputation metric.

damage" assessment (which assigns a multiplier effect to any subsequent downratings). This downrating has the effect of imposing an immediate and profound class demotion upon Lacie: banned from the airport, Lacie is forced to rent a car, where she is relegated to the "4.0 or less" lane, and is assigned an ancient beater with a Czech-language driver interface and an obsolete charging port.

That Lacie's shockingly rapid ranking "nosedive" was precipitated by four back-to-back negative encounters with characters of color was a fact that was not lost on the episode's online fan reviewers.[18] Whether the casting choices were deliberate or not, the sequence presents itself as a telling allegory about the relationship between race, social privilege, and economic advancement, even as it raises troubling questions about the episode's reinforcement of prevailing stereotypes. Lacie's interaction with the airline check-in agent begins with both characters exuding a cloyingly warm affect, but the minute that it becomes clear that the airline will not book a new flight for Lacie, she becomes an avatar of enraged white entitlement, demanding to see the agent's supervisor and swearing with exasperated impatience ("God, just fucking help me!"). The incident becomes an implicit commentary on the pervasiveness of white privilege in this kind of interracial customer service encounter. However, the fact that the episode positions the check-in agent, the cab driver, and the airport security guard as structural impediments to Lacie's advancement (with both the cab driver and airline agent giving her negative ratings) might also work, however inadvertently, to reinforce prevailing racist assumptions about the affective posture of Black service workers. As recent work in this area has suggested, African American service workers need to engage in considerable emotional labor to overcome the racialized perception that they are hostile or unfriendly; in general, women, workers of color, and LGBTQ employees feel pressure to forge a workplace identity that does not fall afoul of prevailing majority group sensibilities.[19]

---

[18] On the role of characters of color in Lacie's downrating "nosedive," see "Black Mirror's 'Nosedive', an Allegory of Racial Privilege," AliceOutOfContext.blogspot.com, posted October 22, 2016, http://aliceoutofcontext.blogspot.com/2016/10/black-mirrors-nosedive-allegory-of.html (accessed January 26, 2020); see also u/Mayuguru's Post on the r/blackmirror Subreddit, "Was Anyone else Expecting Racial Theme at the Start of Nosedive?", Reddit.com, posted July 14, 2017, https://www.reddit.com/r/blackmirror/comments/6n9kb4/was_anyone_else_expecting_racial_theme_at_the/ (accessed January 26, 2020). On race and casting choices in "Nosedive," see Isha Aran, "Black Mirror's 'Nosedive' isn't Just About Social Media, it's About Race," *Splinter.com*, October 24, 2016, https://splinternews.com/black-mirror-s-nosedive-isn-t-just-about-social-media-1793863113 (accessed January 26, 2020).

[19] On the additional emotional labor required of workers of color, see Alicia Grandey et al., "Fake It to Make It?: Emotional Labor Reduces the Racial Disparity in Service Performance Judgments," *Journal of Management* 45/5 (May 2019): 2163–92.

As I noted above, these issues surrounding race, service culture, and affect can be understood by way of a lens that Ana Y. Ramos-Zayas has referred to as *the feel that sells*, the sense in which neoliberal conceptions of urbanism locate emotion regulation at the core of strategic efforts to cultivate commercially appealing urban spaces.[20] Ramos-Zayas's ethnography of contemporary Newark has foregrounded the degree to which this city, whose portrayal as a dangerous and "aggressive" city dates to the urban uprisings of 1967, struggles against the widespread perception that it is mired in a kind of collective emotional pathology. Contemporary efforts to rebrand the city have relied upon an approach to urbanism that deconcentrates or displaces the perceived avatars of this emotional pathology (those youth of color who remain affiliated with a stereotype of Black aggression), so as to alter the "emotional style" of Newark, thereby rendering the city safe for capitalist accumulation.[21]

To return to "Nosedive," the episode's characters of color are quite often positioned as, or clustered in proximity to, low-rank or diminished-rank people. For this reason, Ramos-Zayas's interrogation of the racialized emotional regime of neoliberal Newark resonates with the thorny racial allegory latent in "Nosedive"'s depiction of the service economy, both insofar as the episode critiques *and* unwittingly amplifies prevailing assumptions about Blackness, service work, and negative affect.[22] What is crucial about Ramos-Zayas's formulation, and what is reinforced by the pointed casting choices of "Nosedive," is the degree to which it challenges the common presentation of neoliberalism as race-neutral and detached from structural racism's asymmetries of power: where neoliberal performance metrics present themselves as powerful vehicles of meritocratic advancement, the tendency of service users to downgrade and underrate service workers of color—*and* the tendency of the show to read its Black characters as vectors for the circulation of negative affect—gives the lie to neoliberalism's claims of post-racial enlightenment.[23]

---

[20] Ramos-Zayas, *Street Therapists*, 7.
[21] Ibid., 6 *et passim*.
[22] Elsewhere, Zahid Chaudhary has argued that such unsubtle allegories of racial hierarchy run the risk of "fetishizing difference and repeating the worst set of racist meanings through the constellations [they bring] together." See "Humanity Adrift: Race, Materiality, and Alterity in Alfonso Cuarón's *Children of Men*," *Camera Obscura* 24/3 (December 2009): 102.
[23] A research team led by Alex Rosenblat at the Data & Society Research Institute argues that workers subject to rating systems such as those used by Uber are especially vulnerable to instances of user bias. See Rosenblat et al., "Discriminating Tastes: Uber's Customer Ratings as Vehicles for Workplace Discrimination," *Policy and Internet* 9/3 (September 2017): 256–79.

## "It Would Be a Dull World Without Wonder"

The themes that I've outlined above, extending from "Nosedive"'s realization of a Deleuzian *society of control* to its implicit gender and racial politics, all come together in the episode's denouement. By the time Lacie arrives at her friend's wedding—many hours late, with running mascara and rumpled dress—a series of unfortunate events has conspired to reduce her reputation metric beyond hope of salvation: when her missed flight, malfunctioning rental car, and other misadventures result in Lacie missing the rehearsal dinner, "Ney-ney" calls to tell her not to come, arguing that she "cannot have a '2.6' at my wedding." Lacie is nevertheless adamant about attending, convinced that her prepared bridesmaid's speech will be the *deus ex machina* that salvages her reputation and sends her ranking rocketing back into the socially acceptable range of the mid-4.0s.

But the remaining leg of Lacie's' trip would send her spiraling out beyond the "numbers game" of reputation ratings. She hitches a ride with a truck driver named Susan, whose 1.423 rating was the result of a "nosedive" similar to that faced by Lacie: when her dying husband was passed over for experimental cancer treatment, in favor of a patient with a slightly higher rating, Susan abandoned the relentless sugar-coating of her statements and actions. Having relinquished any commitment to affective labor, Susan now inhabits a class position located at a dramatic remove from the aspirational confines of Lacie's world. Susan's "nosedive" has become permanent; by contrast, Lacie remains nominally convinced that she will be able to "velocitate [her] arc," to leverage her wedding appearance into a reversal of her temporary slump. But in the hours that follow, Lacie behaves more and more in accordance with Susan's imperative to "shed [. . .] those fuckers," to rid herself of those in her orbit who privilege numbers-driven tact over honesty.

Dropped off by Susan, Lacie hitches a ride from a group of cosplaying convention-goers, until a misadventure gets her turfed out of their RV. Her phone tinkling with downrating notifications, Lacie then manages to commandeer an ATV, swerving frantically between angry drivers on the freeway. This sequence is intercut with scenes from the wedding ceremony taking place a few miles away, where Naomi's wedding guests, immaculate in their pastel bridesmaid's dresses, look on approvingly and uprate the happy couple. Max Richter's underscore in this montage brilliantly employs the tinkling phone notifications as a kind of musical riff, located ambiguously in the diegesis: as the sequence shifts from the celebratory upratings of "Ney-ney"'s privileged guests to the punitive downratings hurled at Lacie's ATV from passing cars, the riff itself shifts to match, switching from the chipper ascending triads of the former notification to the chromatically descending slump of the latter one. Richter's score presents these "dings" as a tool for evoking the generalized atmosphere of disapproval surrounding Lacie, as her progress brings her ever closer to a reckoning with Naomi Blestow and her high-powered wedding.

In the end, Lacie arrives at the sumptuous garden where the wedding takes place, discards her ATV, crawls through the bushes, and bursts upon the wedding party, where she seizes a microphone and, bridesmaid's dress in tatters and her face covered with rivulets of mascara, Lacie delivers a weaponized version of the emotional maid-of-honor speech she had been preparing since Naomi's invitation was first extended to her. Her delivery is by turns ingratiating, self-flagellating, and caustic, arguing that Naomi kept her around just long enough to serve as a foil for her own social and romantic successes, ultimately dispensing with Lacie when she was able to take on "her new job, and her new, *fancy* friends, and that . . . fucking . . . JACKHOLE." With the latter interjection, delivered at the point of a jabbing, accusatory finger, Lacie singles out Naomi's handsome groom; by this point, Lacie has completely abandoned the exaggeratedly feminine tones of her elevator small talk persona, drawing instead upon a reservoir of searing rage.

Ejected from the ceremony, Lacie is arrested and booked, mug shots capturing her disheveled appearance, and is then placed in front of a machine that, with dramatic suction noises, removes the contact lens interfaces through which the citizens of "Nosedive"'s world all apprehend one another's numerically-rated selves. The move registers as a final banishment from the symbolic order, from the privileges and responsibilities of "Nosedive"'s society of control. In the cell, Lacie weeps with a strange joy as she realizes that without her contact lenses, she can see dust particles in the light for the first time in decades.

But Lacie is soon distracted as she notices a second prisoner in the cell just across from her, a well-dressed Black man looking at her with a placid, composed demeanor. Feeling exposed, Lacie presses him, with amused hostility, to explain "what the fuck [he is] looking at?", to which the second prisoner responds:

"Just what I was wondering."
"Well. . . . . *don't!*"
"Don't? Don't wonder?"
"Uh-huh."
"It'd be a dull world without wonder."

"A Solid Popularity Arc" 167

Here, Lacie's exchange with her fellow prisoner takes on an element of flirtatious antagonism, as they each escalate the brutality of their mutual insults:

"You look like an alcoholic... former... *weatherman*."
"*You* sound like a lost little lamb that just got told there's no Santa Claus!"
"What sort of cartoon character did *your* mom have to fuck to brew *you* up in the womb?"
"At least I look like I *was* born, not shit out like some tormented cow creature in an underground lab."

As the underscore swells in the background, Lacie and her neighboring cellmate escalate their exchange of insults to a joyfully climactic "FUUUCK YOOOUUU!" shouted in unison: the sequence is edited in such a way as to alternate rapidly between symmetrically aligned facial profile shots of each character yelling directly into the camera.

This denouement at least suggests an unexpected resolution (albeit an expectedly *racialized* resolution) of an understated thread in Lacie's story, that of her desire to find love. In the scene in which Lacie's real estate agent was showing Lacie her dream home, the condominium's appeal was undoubtedly enhanced for Lacie when the agent activated a steamy holographic dramatization of Lacie's hypothetical encounter with a Black male lover in the unit's minimalist kitchen. In that instance, Lacie's putative lover is a silent, shimmering fantasy offered for Lacie's voyeuristic consumption; the suggestion is that he might be a commodity in his own right, obtainable as an accoutrement for the house. Indeed, as Lacie nervously giggles about the marketing gimmick, the estate agent jokes that the man "doesn't come with the apartment." By contrast with the disturbing objectification of this earlier moment, Lacie's flirtatious encounter at the end of the episode presents the Black prisoner in a context where any implication of material ownership is rendered moot: locked away in cells, their cell phones and ocular interfaces confiscated, both prisoners stand at a near-total remove from the libidinal economy of "Nosedive"'s world. In this formal equality of Lacie and her cellmate, there resides at least the possibility of a love unfolding on terms not subject to the logic imposed upon it by the society of control.

The full promise of that love, ironically enough, is realized in the intensity of the mutual "fuck you" that closes the episode: the ebullient zeal with which Lacie and the other prisoner hurl this epithet at one another is the measure of its expression of freedom. Dropping out the bottom of the five-point scale, and inhabiting a liminal space in which the reputation score is simply bracketed from the world, the two inmates enjoy a freedom that overcomes the boundaries of prison walls and that extends beyond the subversive glee of swearing. As a woman and a Black man respectively, Lacie and her cellmate occupy subject positions that are especially likely to be policed for their performance of affect, whether it is within the mechanisms of control specific to "Nosedive"'s world, or in the reality we inhabit. In the episode's closing image, Lacie and her cellmate hurling "F-bombs" in unison, we glimpse the possibility of a different world, one that deviates both from the strained *politesse* of "Nosedive"'s universe, as well as from an actual reality in which women and people of color find themselves regularly condemned for being too "strident," too "angry," too "shrill." There is an insurgent potentiality that lies on the other side of "Nosedive"'s meticulous system of mutual regulation, one in which the caustic friction of its characters' insults realizes the vibrant texture of a world not subject to slick interfaces, popularity rankings, and reputation consultants.

# Bibliography

Aran, Isha. "Black Mirror's '*Nosedive*' isn't just about social media, it's about race." *Splinter.com*, October 24, 2016. https://splinternews.com/black-mirror-s-nosedive-isn-t-just-about-social-media-1793863113 (accessed January 26, 2020).

Austin, J. L. *How To Do Things With Words*. Second Edition, edited by J.O. Urmson and Marina Sbisà. Cambridge, MA: Harvard University Press, 1975.

AliceOutOfContext. "Black Mirror's Nosedive, an Allegory of Racial Privilege." *AliceOutOfContext.blogspot.com*, October 22, 2016. http://aliceoutofcontext.blogspot.com/2016/10/black-mirrors-nosedive-allegory-of.html (accessed January 26, 2020).

Chapman, Dale. *The Jazz Bubble: Neoclassical Jazz in Neoliberal Culture*. Oakland, CA: University of California Press, 2018.

Chaudhary, Zahid. "Humanity Adrift: Race, Materiality, and Alterity in Alfonso Cuarón's *Children of Men*." *Camera Obscura* 24, no. 3. (December 2009): 73–109.

Deleuze, Gilles. "Postscript on the Societies of Control." *October* 59 (Winter 1992): 3–7.

Grandey, Alicia, Lawrence Houston III, and Derek Avery. "Fake It to Make It?: Emotional Labor Reduces the Racial Disparity in Service Performance Judgments." *Journal of Management* 45, no. 5 (May 2019): 2163–92.

Hardt, Michael and Antonio Negri. *Empire*. Cambridge, MA: Harvard University Press, 2000.

James, Robin. *The Sonic Episteme: Acoustic Resonance, Neoliberalism, and Biopolitics*. Durham, NC: Duke University Press, 2019.

Hawkesworth, Mary. *Political Worlds of Women: Activism, Advocacy, and Governance in the Twenty-First Century*. Boulder, CO: Westview Press, 2012.

Kramer, Lawrence. *Music as Cultural Practice, 1800–1900*. Berkeley, CA: University of California Press, 1990.

Lakoff, Robin Tolmach. *Language and Women's Place: Text and Commentaries*. Revised and Expanded Edition, edited by Mary Bucholtz. New York: Oxford University Press, 2004[1975].

Lecher, Colin. "*How Amazon automatically tracks and fires warehouse workers for 'productivity.'*" *TheVerge.com*, April 25, 2019. https://www.theverge.com/2019/4/25/18516004/amazon-warehouse-fulfillment-centers-productivity-firing-terminations (accessed December 20, 2019).

LiPuma, Edward and Benjamin Lee. *Financial Derivatives and the Globalization of Risk*. Durham, NC: Duke University Press, 2004.

Matsakis, Louise. "How the West Got China's Social Credit System Wrong." *Wired.com*, July 29, 2019. https://www.wired.com/story/china-social-credit-score-system/ (accessed December 16, 2019).

Ortiz, Aimee. "*Terrifying Boston Dynamics Robots, 'Black Mirror,' and the end of the world*." *Boston Globe*, January 5, 2018. https://www.bostonglobe.com/arts/2018/01/05/boston-dynamics-black-mirror-and-end-world/cL9RYkg6O6MqyPuhmgxVjP/story.html (accessed January 25, 2020).

Perez, Sarah. "*Controversial people-rating app Peeple goes live, has a plan to profit from users' negative reviews*." *TechCrunch.com*, March 8, 2016. https://techcrunch.com/2016/03/08/controversial-people-rating-app-peeple-goes-live-has-a-plan-to-profit-from-users-negative-reviews/ (accessed December 16, 2019).

Ramos-Zayas, Ana Y. *Street Therapists: Race, Affect, and Neoliberal Personhood in Latino Newark*. Chicago: University of Chicago Press, 2012.

Rosenblat, Alex, Karen Levy, Solon Barocas, and Tim Hwang. "Discriminating Tastes: Uber's Customer Ratings as Vehicles for Workplace Discrimination." *Policy and Internet* 9, no. 3 (September 2017): 256–79.

Standing, Guy. "Global Feminization Through Flexible Labor: A Theme Revisited." *Global Development* 27, no. 3 (1999): 583–602.

Turchiano, Danielle. "'Black Mirror' DP Seamus McGarvey Talks About Finding the 'Photographic Heart' of 'Nosedive.'" *Variety.com*, August 16, 2017. https://variety.com/2017/tv/awards/emmys-black-mirror-seamus-mcgarvey-interview-1202520496/ (accessed February 2, 2020).

u/Mayuguru. "*Was anyone else expecting racial theme at the start of Nosedive?*" *Reddit.com*, July 14, 2017. https://www.reddit.com/r/blackmirror/comments/6n9kb4/was_anyone_else_expecting_racial_theme_at_the/ (accessed January 26, 2020).

11

# Cognitive Boundaries, "Nosedive," and *Under the Skin*

Interview with Jeffrey M. Zacks

Carol Vernallis, Jonathan Leal, and Dale Chapman

## Introduction

At his Dynamic Cognition Laboratory at Washington University in St. Louis, neuroscientist Jeffrey M. Zacks studies how representations in the brain and the world work together in cognition. On September 1, 2020, he Zoomed with media scholars Carol Vernallis, Jonathan Leal, and Dale Chapman to share his thoughts and observations on Alex Garland's *Ex Machina* (2014) and "Nosedive," an episode of *Black Mirror* (2016). The conversation below, drawn from their meeting, connects cognitive processes with audiovisual media, focusing on topics including event perception, genre expectations, and prediction errors.

**Carol Vernallis** Jeff, can you tell us about working event models? Is it correct that these can capture the quotidian? As I understand them, the brain segments events so we can remember them. We often divide events based on a location or an activity (a conversation that ends as someone exits a room, or my baking cookies). Shifting between rooms or placing cookies on a plate and starting to eat them can create new event boundaries, right? And do these experiences work somewhat similarly for films?

**Jeff Zacks** Yes, according to *event segmentation theory*, an individual builds a mental representation of the present moment.[1] This model contains a

---

[1] Jeffrey M. Zacks, et al., "Event Perception: A Mind/Brain Perspective," *Psychological Bulletin* 133, no. 2 (2007): 273–93. See also Lauren L. Richmond and Jeffrey M. Zacks, "Constructing Experience: Event Models from Perception to Action," *Trends in Cognitive Sciences* 21, no. 12 (December 2017): 962–80.

spatio-temporal framework of the environment, along with entities and representations of intentions and plans. Individuals use this model to predict what's going to happen in the near future. The model remains intact until it encounters a high number of prediction errors. When prediction error spikes, you update. The threshold for updating is dynamic; it is the difference between your current prediction error and the running recent average or prediction error. When prediction error spikes, one part of your brain sends a signal to others that it's time to update the current event model.

Our lab has found that if you zoom in on the event boundary enough, it's got structure. The boundaries between experienced events also have a structure that most likely unfolds over a second or so, even for smaller events. When a prediction error rises, the mind updates, and then settles into a new attractor. You may be well into the updating process—or even past it—before you become conscious that one event has ended and another has begun. (And for some event boundaries, you may never really consciously notice them at all.)

There are event boundaries on many time scales—from seconds to tens of minutes—that this system operates on. Most of the time, we're not attending to all of these timescales. That is a big part of why some event boundaries operate largely outside of conscious awareness. There's also another process, which may be important for forming long-term memories based on your event models. This has been described as a "now print" function, such that when you hit an event boundary, as part of updating your event model your brain prints out something about the contents of the old event model to long-term memory before it's gone forever.[2] I had long wanted to believe that this additional operation wasn't necessary; rather, that your brain is forming a set of discrete event representations in working and long-term memory just constantly snarfs these up. I didn't want to believe that there's this extra "now print" thing going on, but there's actually really good behavioral and neurophysiological evidence for it now.

**Carol**  But if there's a false "now print?" because the event didn't close or end properly—what happens?

---

[2] Aya Ben-Yakov, et al., "Hippocampal Immediate Poststimulus Activity in the Encoding of Consecutive Naturalistic Episodes," *Journal of Experimental Psychology: General* 142, no. 4 (2013): 1255. See also Christopher Baldassano, et al., "Discovering Event Structure in Continuous Narrative Perception and Memory," *Neuron* 95, no. 3 (2017): 709–21.

**Jeff**   Right! I think some aspects of memory disorders, as well as some, but not all, of the processes that go wrong with healthy aging, or some things that get messed up in post-traumatic stress disorder, can be described as segmentation in the wrong places. Ideally, the units that you're forming and then printing out should correspond rapidly, similarly to the ways we experience the demarcations of words and sentences during natural activity. Some individuals don't experience these boundaries as distinct, and then they can't find the right cues to form their memories. This has big consequences for accessing subsequent memories.

Here's a musical analogy. If I listen to some polyrhythmic music (which often elides and superimposes boundaries), and you ask me to sing the tune right back to you, I might have a hard time doing so, because as I listen along, I may not hear the downbeat in the right place, or at all. It's not the same mechanism for event perception, but that's an analogy.

**Carol**   So then that scene in "Nosedive" (*Black Mirror*, Season 3 Episode 1, 2016) with showing off the house or condo, where bodies and the setting form a rhythmic 3 against 4, would be significant, right? (I'd like to discuss this more in a minute.)

**Jeff**   Yeah, totally. If it causes you to update your event representations at points that don't correspond to the real joints of the activity, then it's going to lead to memory that is less robust, rich, or accurate. That doesn't mean the movie is bad! This is not an aesthetic judgment, just a generalization about comprehension and memory. My goal as a filmmaker, for instance, may be to leave you feeling disjointed, vague, or in the midst of an atmospheric collection of memories. With genres like horror and suspense, filmmakers often aren't trying to leave you in a state where you can give a nice, accurate list of the actions of the last hour and a half—they're trying to do something else.

**Jonathan Leal**   Your comments about memory disorders and the failure to form neat event boundaries makes me wonder: do smartphone notifications provide false memory printing opportunities?

**Jeff**   Absolutely. I've thought about this mostly in terms of GPS navigation systems. I've got a computer in my car, right? It wants to tell me about traffic on my route, when to turn, and future obstacles. It's probably getting my texts and incoming calls. With a greater understanding of event perception, the system could decide when to pass data and content through. It might be able to say,

"Okay, he's coming up on an event boundary, so I'll hold onto this instruction or this message until we reach a slack spot, and then I'll pass it along." We've found in the lab that interrupting people at event boundaries has very different effects than interrupting them in event middles.[3] At the time, I was thinking of these navigation systems, but now, yes, we've got these little devices navigating our whole lives, so I think it applies even more.

**Carol**   It could be used for evil.

**Jeff**   Sure. In lots of situations, unplugging altogether may be the most beneficial for our mental health. But the flipside with these devices may be that our pluggings in, or whatever we're doing, could be made to flow *better* as well as be more cognitively friendly. Fixing the timing isn't going to do the whole job, but it might ameliorate this basal sense of angst connecting with our devices gives us.

**Jeff**   I agree with your characterization of *Nosedive* that on so many levels, from music, setting, to color, the range is deliberately restricted. It kind of bubbles along, as you've said, and this makes the little blips and dings of the social media apps especially salient. I judge this is the director's and production team's goal. I wonder how much these notifications distract us.

How do these affect event perception? My lab and other colleagues have shown that stuff that pops up in a stream of stimuli gets processed differently depending where it occurs in relation to boundaries between events. As I was watching *Nosedive* and I was listening to the bubbling along of up- and downvotes, it occurred to me that the ones that happen when you're transitioning from one event to another, which were a lot of them, probably get handled differently from the ones that bubble along in the middle of an event.

**Carol**   We think the "Nosedive" scene where Lacie and the real estate agent tour the rental condo suggests a few false endings—we may guess we've hit closure, but maybe not. On so many levels (music, gesture, deportment, dialogue), the scene's rhythms vary between languorous and abrupt. This might also confuse the viewer.

---

[3]   Matthew M. Botvinick and LM Bylsma, "Distraction and Action Slips in an Everyday Task: Evidence for a Dynamic Representation of Task Context," *Psychonomic Bulletin & Review* 12, no. 6 (December 2005): 1011–17. See also S. Schwan and B. Garsoffky, "The Cognitive Representation of Filmic Event Summaries," *Applied Cognitive Psychology* 18, no. 1 (January 2004): 37–55. See, too, Markus Huff, et al., "Visual Target Detection Is Impaired at Event Boundaries," *Visual Cognition* 20, no. 7 (July 2012): 848–64.

**Jeff** I agree. In "Nosedive," you may have a highly predictable sequence, with lots of stuff changing without an event boundary. If you have something big and unpredicted, that'll surely be an event boundary. But if you have a predictable sequence and then you shift to a regime where events become less predictable, where you have a chronic run of prediction errors, the model first notices this jitter of event boundaries, and then adapts to that level. This is because it uses local spikes as its cue to update prediction errors. If things are always unpredictable, you don't add event boundaries; if they're always predictable, you also don't. But when you transition from predictable to unpredictable, you get a transient instability until the situation recalibrates. Perhaps that's what this scene is like.

**Carol** Our readers and contributors could run a study of the scene where they found event boundaries for you! Let's watch this "Nosedive" scene, okay? Some elements also hold true for the episode as a whole. The landscape feels flattened. Lacie's encounters and experiences seem relatively static. In this scene, Lacie experiences one nice moment of fantasy and interiority, and she encounters one financial surprise, but that's about it. The colors are attenuated—whites and soft pastels. Max Richter's score is mostly a pedal point wash, full of long, glacial melodic lines (slow fades down, etc.). All these elements create a sense of length and continuity across boundaries, as well as offer a soothing quality. This particular scene has a reduced number of tablet and smartphone interruptions, yet it's still hard to parse.

It's important to start analyzing the scene before Lacie gets out of her taxi, otherwise, some of the proportions will be lost. Here's one of my first favorite details: The actors turn three times during the scene (somewhere from 180 to 360 degrees), starting from the street until the holograph in the kitchen. Along with other gestures, like the real estate agent's opening and closing her arms from her elbows three times (note that her blouse is crossed in the opposite direction), feels dancelike.

**Jeff** There's a whole literature on the event perception of dance.[4] Dance segmentation sits somewhere between how we segment music and how we segment ordinary events.

---
[4] Bettina E. Bläsing, "Segmentation of Dance Movement: Effects of Expertise, Visual Familiarity, Motor Experience and Music," *Frontiers in Psychology*, 5 (January 2015). See also B. Bläsing, et al., "The Cognitive Structure of Movements in Classical Dance," *Psychology of Sport and Exercise* 10, no. 3 (May 2009): 350–60.

**Carol**   Yes! And making the scene even *more* musical and dancelike, the camera tracks alongside the two women and yet crosses half-barriers. The actors' twirls against the demarcations of rooms creates that rhythmic pattern of 3-against-4 I mentioned earlier—so beginnings and ends are difficult to determine.

**Jeff**   We know music is segmented differently. It's super tempting to think that the parsing of music works the same way as the parsing of everyday activity. They both have segments that are, at least under some analyses, approximately hierarchically organized.[5] Both present phenomena subject to prediction and prediction error. But they're still different.

   Human activity comes about as a function of people trying to achieve their goals, whereas the structure of music comes about as a matter of design. Part of this has to do with form and also with genre expectations. Some researchers have argued that tonal music most closely resembles language (even when you apply Schenkerian analysis, which can suggest a narrative arc). Surprisingly, spectral music, even though pieces are often unique from one another (one-offs), may track more closely to event perception.[6]

**Jonathan**   But here, we've got all of these multisensory processes going on at the same time—events that are dancelike, involve music, representations of the quotidian.

**Jeff**   Yes, and the affect we bring to a context or situation may also shape how we experience and segment events. Perhaps an artistic mode, like experiencing cinema, helps us become more attuned to the past and future, and to rhymes across events. But generally, it's all multimodal, all processed together. What determines when one event ends and another begins is not so much the visual pattern, as it is the world's stuff the modal patterns convey to us. You can have good everyday event comprehension for things you're just hearing. And when we look at people reading or listening to stories, it seems to work in about the same way as events that we show as movies. Some film genres, however, like horror and the avant-garde, through sound, camera, color, and so on, can distort

---

[5]   Fred Lerdahl and Ray S. Jackendoff, *A Generative Theory of Tonal Music* (Cambridge, MA: MIT Press, 1983).
[6]   Joseph R. Jakubowski, "Between Concept and Perception: Cognition, Experience, and Form in Spectral Music," (PhD diss., Washington University in St. Louis, 2019).

this. For example, if a prediction error remains fluctuating and high, a viewer may be left more with a feeling than with a strong memory of what actually happened. Which is what you'd probably want to do, if you were a horror filmmaker.

**Carol**  I'd like to point out the way this "Nosedive" scene subverts our ability to segment scenes. Look at that white boulder on the lawn in front of the white condo. It'll rhyme with the statue in the condo's partially open center room. The camera tracks past the living room, through this middle room, into the kitchen; we cross a wall's half-barrier, lose Lacie, but stay with the real estate agent, which is confusing. So we can't be sure if the wall is a true barrier (and a moment to segment). After we cross this wall, we first see a stone figure, which might be a transformed replacement for Lacie, so processes that we need to track may continue to unfold.

**Jeff**  I agree. Usually, moving from one room to another is a likely point for the mind to form an event boundary. This scene continues, but it also doesn't. Losing the protagonist might also contribute to segmentation. But we might also want to continue the segment because there are ongoing, transformative processes (stone and human to statue).

**Carol**  I'd like to talk some more about the decor. The rooms have multiple clear glass containers, which metaphorically may suggest the possibility of containing events within smaller units (even the chandelier is made up of translucent glass containers!). But the holographic projection can't be contained. That's confusing, no?

**Jonathan**  Or—let me expand on this. Maybe these semiotically loaded tchotchkes and the camera's circuitous route (the dancelike gestures across partial barriers) change the way we process events?

**Jeff**  These are great points. We've published some experiments that explored how the density of events affects the perception of time.[7] We first showed one group of

---

[7]  Ashley S. Bangert, et al., "The Influence of Everyday Events on Prospective Timing 'in the Moment,'" *Psychonomic Bulletin & Review* 26, no. 2 (April 2019): 677–84.

subjects some film clips and asked them to segment these into meaningful events, so we were sure where the clip's event boundaries landed. We next asked a second group of subjects to come into the lab and memorize the duration of a timed interval (we chose thirty seconds, but we didn't tell them this). For a few times, we showed them a light turning on to mark the beginning of the interval and then turning off to mark the end, and we let them replicate this standard interval. We then screened the movie and had them reproduce those time intervals while watching. We found that if there were more event boundaries in the middle of their interval exercise, time seemed to tick faster. Here's an example where this data converges with other paradigms. With navigation systems, we know that if you're going along a twisty-turny path, time seems to pass more quickly than if you go down one that's straight with nothing changing. So it seems like part of what we're monitoring, part of what we're accumulating when we're judging how long something's taking, is event boundaries.

**Carol** I'd like to point to another parameter that might complicate our experiences of time and event perception. Max Richter's soundtrack begins when Lacie's hologram starts. It continues its pedal point wash through Lacie's reverie and then stops abruptly when she risks losing her fantasy, dropping out just before her line to the real estate agent: "I'm more than interested." Here we have a sudden event. We follow Lacie with her fantasy, but then the music swishes up and stops.

**Jonathan** And if you've been following music video or post-classical cinema, you would expect the soundtrack to have a lot of authority here. For instance: if you consume a lot of pop music that uses rises and drops, and you're tacitly used to marking that music's patterns—here's the rise, here's the drop—then if you experience something like a rise or a drop in film, would that lead you to create a boundary? How do our experiences of events across media inform each other?

**Jeff** That's a great question. The short answer is: I don't know. The slightly longer answer I would give is that genre expectations, for one, shape our predictions and event comprehension. We probably don't bring a lot of genre expectations to our comprehension of real life. But when we're consuming media, those schemas are in there, along with our schemas with the actions, the diegetic activity. Psychonarratologists have been trying to figure out the rules for genre schemas, and they're starting to build models for them.[8] I'm not super up on that literature, but it's a non-trivial problem. Even before we started getting all postmodern and mish-mashing genres, somehow the brain's got to figure out, "Okay, this is the set of rules that applies to this thing I'm reading or watching."

**Dale Chapman**   Let's switch back and take a sec to look at the scene—I think there's a lot going on at this moment. I'm now watching Lacie as she looks at the hologram in the scene, and I caught myself touching my hair after she touched her hair.

**Jeff**   And that happens all the time. Mirroring is definitely a thing in film perception.[9]

**Dale**   I notice it's paired with the zoom onto the hologram across from her. This seems to reinforce the intimacy of the gesture.

**Jonathan**   The more we've watched this scene, the more I've come to see Lacie, in this moment, as she's looking at the holographic representation of her and this future self, as an audience proxy. We're looking at her look at herself, and through a kind of screen. It's a study of that mirroring process, but externalized in the narrative discourse.

**Carol**   I think it's interesting that we don't know if she's still fantasizing about the guy and/or renting the condo, or if she's at all following the musicality of the real estate agent's voice. Where are we—with Lacie's dream, the real estate agent's voiceover, or back with the hologram? Do we know where to place ourselves?

**Jonathan**   The receding voiceover is paired with the now-reversed zoom into Lacie's face.

**Jeff**   And the way the audio from the hologram stays up loud for a while helps you hold onto her thinking of the hologram. Because you're looking at her face but you're still hearing the soundtrack.

---

[8]   Marisa Bortolussi and Peter Dixon, *Psychonarratology: Foundations for the Empirical Study of Literary Response* (New York: Cambridge University Press, 2003).
[9]   Jeffrey M. Zacks, *Flicker: Your Brain on Movies* (New York: Oxford University Press, 2014).

**Carol**   But she could just be having a sexual fantasy, we have no idea!

**Dale**   Because the real estate agent's voice, with its rhetorical patter, recedes into the background, I judge precedence lies with Lacie's fantasy-spectacle of a future in which she's happy.

**Jonathan**   That's definitely how I'm reading this. It's weirdly voyeuristic and, as a viewer, I don't want the real estate agent there anymore. Later, Lacie admits she's sexually repressed: she's "never brought home any guys to the apartment because of her brother's dumb-fuck 2.0 rating." And this whole scene dances its way through an erotics of a business deal. There's a lot to discuss regarding class and race, which we'll talk about soon. Oh, and by the way: the glacial music binding everything together with the tracking camera and the sudden stops—it reminds me so much of the experience of scrolling through Instagram. Glossy, unifying so many disparate posts. A wash. But yes, the apparatus of this sexualized visualization technology and weirdly erotic capitalist infrastructure—I feel terrible watching Lacie fall for it, and neither of us can stop it.

**Jeff**   What makes "Nosedive" powerful is you could have told a version where the characters are all just obsessed with social rewards for their own sake. I think it would've been a much less effective story. This scene has an edge because it's tied to both emotional and material rewards. This is the part where neuroscience is supposed to pop up and say, "Each time you hear that ding, there's a squirt of dopamine and this explains it." The dopamine system is involved, and if you measure dopamine in the right places, you'll find its signaling covaries with the experiences of the characters, but it's not like this really explains a lot. You don't need facts about the brain to see that conditioning's a powerful regulator of behavior. Dopamine is interesting to those of us who care about how brains work, but it's not central. Learning phenomena affect gamification for both good and bad. It has an independent force because of the learned associations between proximal signals and the more fundamental reward.

**Carol**   Can I quickly jump in with two observations? Lacie's doubling—her hologram and physical being—creates a strange psychological, perhaps even dissociative response. (Images of twins and doppelgangers tend to screw with you; think of Kubrick's *The Shining* [1980], or Garland's *Annihilation* [2018]). We don't know exactly where we are, and we may be in the midst of constructing two stories as the camera zooms in, both for what Lacie's fantasy is in this moment, and what ours might be. Isn't this a good time to update your working event model? The face of an actor listening, reflecting, or in reverie, is a moment to pull materials together and assess them. Especially since we cut next to the dining room—a classic technique for an event boundary. And the mood in the other room is different. Lacie's face is suffused with anxiety.

**Dale** We're talking about a shift in affect, right? Everything is glowingly pastel, super-sugary, and upbeat for the episode's first half to two-thirds, and now there's an intimation that things are starting to go south. Later, during the airport scene, we'll take a much bigger nosedive. It strikes me that event boundaries with affect are inherently fuzzier. It's often difficult to clearly define, for instance, exactly when you have moved from feeling upbeat to depressed, right? You just kind of know that vaguely within a two or three-hour or two or three-day period that you've undergone a shift.

**Jeff** I think that's exactly right. Drawing from Greg Smith's book about movies and emotion I would call out two relevant points.[10] Smith argues that movies are mood engines, and he borrows this nice distinction between affect or emotion and mood: affect is the shorter-term thing, and mood is the lower-frequency, cumulative effect. If you listen to people talk about why they're choosing movies and what they take from them, much of it is about how it felt. Filmmakers inject bits of mood, affects to sustain, build, or create change. Structure is crucial. You can't tell the story and let the mood take care of itself. Films are constructed through deliberate techniques that inject and sustain feeling.

Scholars like Lisa Feldman Barrett on cognition and the affect system, as well as neuroscientists who study emotion, disagree about much.[11] But all agree that affect puts a low-pass filter on experience. Affect helps maintain an internal model of what the world is like and what repertoire of behavioral responses you'll employ. When approaching a situation, should I open myself up to or leave myself open to? Should I fight or flee? Affect regulates parameters of behavior and your relation to the environment; this helps you behave appropriately.

---

[10] Greg M. Smith, *Film Structure and the Emotion System* (New York: Cambridge University Press, 2003).
[11] Lisa Feldman Barrett, "The Theory of Constructed Emotion: An Active Inference Account of Interoception and Categorization," *Social Cognitive and Affective Neuroscience* 12, no. 1 (January 2017):1–23.

**Dale**  I see. In other words: in a particular situation, with a whole cluster of things bound together by prevailing mood, essentially, do you need to worry about threats, or should you be socially attuned and expansive?

**Jeff**  Yeah. Let's imagine sitting at my breakfast table, in my office, or in court. For breakfast, my brain's going to keep track of coffee and eggs, and this will regulate how I behave. If I'm sitting in a courtroom's witness box, my monitoring for interpersonal signaling and potential threats regulates my information processing.

**Jonathan**  I like your point about the ways film can modulate affective experience. As Dale writes in his chapter, it's important to recognize how "Nosedive"'s depiction of race factors in. The real estate agent notes that the Black male lover doesn't come with the apartment; rather, he's immediately lumped in with the whole commodity. That's how branding works: it conflates the commodity with aspirations for a larger lifestyle. But we can also feel how people are being eroticized and commodified, which for some viewers produces discomfort. This moment resonates with "Nosedive"'s closing scene, when the procedure to remove Lacie's contact lenses creates, as Dale notes, a social death (because she no longer gets the updates, and she's thus removed from the capitalist regime of accumulation). Yet she also experiences happiness and freedom. She sees those dust motes for the first time and experiences a meaningful interracial encounter outside of consumer experience.

**Carol**  Speaking of race: the film seems to connect the appearance of BIPOC characters, the frame's darkening in downturns, and the descending chime of falling ratings all with potential threats—with social death. These start to form motivic structures. We can connect these recurrences to film theorist Laura Mulvey's claims about cinema: cinema appeals to us because it replicates our death drive.[12] This happens on many levels (including that much of the screen remains in black due to gaps between the celluloid's frames as they run through the projector). Narratively speaking, characters seek a good death. They encounter possible moments for cessation and attempt to decide whether it's a good time for them or not. If not, they pause and then attempt to keep going. In

---

[12] Laura Mulvey, *Death 24x a Second: Stillness and the Moving Image* (London: Reaktion Books, 2006).

"Nosedive," darkness and race are paired with the social death of the subject. These moments may be the most important events, as well as the most important bindings together of events, rather than the openings and closings of scenes. How do these two structures play against one another cognitively?

**Jeff** One of the tricky things about Hollywood's three- or four-act structure (Cutting) is that it's at the upper bound of the timescale on which these event perception mechanisms might operate. I'm not sure if the mechanisms that track event structure up to tens of minutes also work for a whole movie. But the other thing—watching a video over and over again is, there's something deeply perplexing about event structure for familiar things. Why would you have prediction errors at all if you've already seen the film a bunch of times? One possibility is that your event structure becomes totally screwed up by that level of familiarity. Another possibility is that your overall level of prediction error is really low, but relative prediction error still bumps up at those same boundaries. I'm not sure.

**Carol** We're running out of time, but here's a thought experiment for viewers: where should this "Nosedive" apartment scene end? Should you terminate it after Lacie walks past the billboard? (But the next shot is of a screen with Nay-Nay wearing the same glasses as Lacie, so this might suggest a continuation of the same event.) Out in the street, bolder colors and placements of bodies suggest an ending. (Note that colors intensify: compare these closing trees with those from the scene's opening.) We might also choose to end the scene later, when Lacie is in her apartment eating noodles. Her consumption fits a different narrative script or frame; she's reduced to the script of solitary, cheap takeout. But look at those chopsticks! They're reminiscent of Pelican Cove's gardens.

**Carol** And one last question: can we train attention, and in doing so, change how we experience event boundaries?

**Jeff** My first rule of psychology is you can train anything. Air traffic controllers, gamers, avant-garde practitioners—totally. Meditators, too, whose practices center on training attention. There are lots of mundane cases where we can modulate our attention top-down. Your predictions depend on your interests and task goals.

We have data on spatial attention. Gamers learn to adjust their affective attentional spotlight to encompass a broader spatial field, for example. People can also train vigilance. Subjects can improve their skill level at the Sustained Attention to Response Task[13] (SART), for example. In the SART task, the subject monitors a stream of stimuli and makes dull judgments about them. At first, it's really hard to keep focused and maintain performance, but people get much better with practice.

**Carol**   And they might also create different event horizons, maybe?

**Jeff**   So it's really easy to focus your attention on events on different time scales. Event segmentation theory says you're always parsing activity simultaneously on multiple time scales, so you can basically direct your attention to different time scales. An important question is if I can take the circuits that normally process events within a median length of 14 seconds and instead turn those circuits to process shorter- or longer-term time course events. It's super hard to distinguish that kind of change from shifting attention from one circuit with one timescale to a different circuit with a different timescale. Nevertheless, I'll bet that you can probably retrain a particular system if you have to.

**Jonathan, Carol, and Dale**   Which may provide some hope for facing our accelerating future, yes? We hope this interview might provide greater knowledge and awareness, which can help facilitate a better, more capacious world. Thank you, Jeff!

**Jeff**   You're welcome! Anytime.

# Bibliography

Baldassano, Christopher, et al., "Discovering Event Structure in Continuous Narrative Perception and Memory." *Neuron* 95, no. 3 (2017): 709–21.

---

[13] Redmond G. O'Connell, et al., "Self-Alert Training: Volitional Modulation of Autonomic Arousal Improves Sustained Attention," *Neuropsychologia* 46, no. 5 (January 2008): 1379–90.

Bangert, Ashley S., et al., "The Influence of Everyday Events on Prospective Timing 'in the Moment.'" *Psychonomic Bulletin & Review* 26, no. 2 (April 2019): 677–84.

Barrett, Lisa Feldman. "The Theory of Constructed Emotion: An Active Inference Account of Interoception and Categorization." *Social Cognitive and Affective Neuroscience* 12, no. 1, (January 2017): 1–23.

Ben-Yakov, Aya, et al., "Hippocampal Immediate Poststimulus Activity in the Encoding of Consecutive Naturalistic Episodes." *Journal of Experimental Psychology: General* 142, no. 4 (2013): 1255.

Bläsing, B., et al., "The Cognitive Structure of Movements in Classical Dance." *Psychology of Sport and Exercise* 10, no. 3 (May 2009): 350–60.

Bläsing, Bettina E. "Segmentation of Dance Movement: Effects of Expertise, Visual Familiarity, Motor Experience and Music." *Frontiers in Psychology* 5 (January 2015).

Bortolussi, Marisa, and Peter Dixon. *Psychonarratology: Foundations for the Empirical Study of Literary Response*. New York: Cambridge University Press, 2003.

Botvinick, Matthew M., and L. M. Bylsma. "Distraction and Action Slips in an Everyday Task: Evidence for a Dynamic Representation of Task Context." *Psychonomic Bulletin & Review* 12, no. 6 (December 2005): 1011–17.

Cutting, James E. "Narrative Theory and the Dynamics of Popular Movies." *Psychonomic Bulletin & Review* 23, no. 6 (December 2016): 1713–43.

Huff, Markus, et al., "Visual Target Detection Is Impaired at Event Boundaries." *Visual Cognition* 20, no. 7 (July 2012): 848–64.

Jakubowski, Joseph R. "Between Concept and Perception: Cognition, Experience, and Form in Spectral Music." PhD diss., Washington University in St. Louis, 2019.

Lerdahl, Fred, and Ray S. Jackendoff. *A Generative Theory of Tonal Music*. Cambridge, MA: MIT Press, 1983.

Mulvey, Laura. *Death 24x a Second: Stillness and the Moving Image*. London: Reaktion Books, 2006.

O'Connell, Redmond G., et al., "Self-Alert Training: Volitional Modulation of Autonomic Arousal Improves Sustained Attention." *Neuropsychologia* 46, no. 5 (January 2008): 1379–90.

Richmond, Lauren L., and Jeffrey M. Zacks. "Constructing Experience: Event Models from Perception to Action." *Trends in Cognitive Sciences* 21, no. 12 (December 2017): 962–80.

Schwan, S., and B. Garsoffky. "The Cognitive Representation of Filmic Event Summaries." *Applied Cognitive Psychology* 18, no. 1 (January 2004): 37–55.

Smith, Greg M. *Film Structure and the Emotion System*. New York: Cambridge University Press, 2003.

Zacks, Jeffrey M., et al., "Event Perception: A Mind/Brain Perspective." *Psychological Bulletin* 133, no. 2 (2007): 273–93.

Zacks, Jeffrey M. *Flicker: Your Brain on Movies*. New York: Oxford University Press, 2014.

12

# Toward an AI Future of Comics Study and Creation

## A Cognitive-Affective Approach

Frederick Luis Aldama and Laura Wagner

It is certainly an exciting time to be a university professor. In the 1990s it was announced that we were in the "Decade of the Brain," and today we're seeing hundreds of millions of dollars being funneled into research programs on the brain that seek to map, to computer simulate, and to develop new brain technology. We think not only here of Europe's The Human Brain Project but also the US's Brain Initiative that seeks to discover and develop state-of-the-art brain tools (optogenetics, for instance) that aim to control and repair neurons for healthy functioning brains. We think, too, of how AI and machine learning and cognitive systems theory is innovating in the area of comic book knowledge and comics creation.

Working on the same campus within different disciplines (humanities and linguistics) but with a common interest in cognitive approaches and theory, we decided to put our brains together to see what we might be able to add to our knowledge of ourselves and the world we inhabit. As two brains working together, we wanted to see what questions we might ask and how we might generate new insights and knowledge about comic books. The aim: to lay the groundwork for future studies that will focus on visual-dominant narratives (film, TV, and media yet to be imagined) created by teams of cross-disciplinary scholars, creators, and those working at the cutting edge of machine learning and storytelling creation.

We decided to set our sights on comic books, and the question of interest—emotional interest. Much has been formulated on how comics generate meaning—their grammar, say—but we realized that not much had been formulated on where in this grammar we suture ourselves in and through emotions.[1]

With a team of undergraduate students (English and Psychology) in place, we generated several sets of questions with respective sets of hypotheses, set-ups, variables, and predictions. These questions included: 1) Is the special format of comics especially engaging to readers? 2) What do comics artists really know? 3) What drives pleasure in comics reading: Theory of mind or gap-filling? 4) Character continuity: How do readers identify and track key characters? 5) How do people read multilingual comics? We knew that to get anywhere, even these questions would have to be narrowed down. We continued to meet. We continued to whittle down the parameters. Lots of people read comics. We wanted to know why people like them and what kind of information and emotion they get out of them.

We finally settled on the question: what kinds of emotions are generated in the reading of comics and where in the story (plot and visuals and/or verbal elements) might the dominant emotion arise? In other words, what is it about comics storytelling that generates our interest—our emotional interest? Our work was exploratory in nature, and thus we allowed participants to provide open-ended answers. They were allowed to describe their emotional responses in any way they wanted, and were allowed to identify as much or as little of the comic as being critical to their responses. As our results show, our participants were not very expert at reading comics, which led to some overall difficulties in comprehension. Nevertheless, these readers were still able to extract the core emotional messages of the comics; moreover, they were consistently impressed by a distinctive comics storytelling device known as the "splash-page," where a single image fills the entirety of a page, often to great dramatic effect.

## Methods

### Participants

Adults were recruited at the Center of Industry and Science (COSI) in Columbus, Ohio. The participants read and filled out the questionnaires in the Language

---

[1] For more on how comics generate meaning and trigger emotions, see Barbara Postema, *Narrative Structure in Comics* (Rochester, New York: RIT Press, 2013); Hannah Miodrag, *Comics and Language* (Jackson: University Press of Mississippi, 2013); Thierry Groensteen, *Comics and Narration* (Jackson: University Press of Mississippi, 2013); and Alexander Dunst et al.'s *Empirical Comics Research* (New York: Routledge, 2018).

Sciences Research Lab. This lab is a working facility embedded inside the science museum and is dedicated to gathering and sharing scientific knowledge about language, with a special emphasis on how people learn, use, and process language across different situations. Of the initial 179 participants, 26 were excluded due to incomplete questionnaires or irrelevant responses, such as not including an emotion in answering how a comic made the participant feel. Of the remaining 151 participants, 68 were male, 82 were female, and 1 other-identified. The age range of the participants was 18 to 73, with a mean of 31.71 years.

The participant's familiarity with comics was assessed with a measure borrowed from Neil Cohn's Visual Language Lab that seeks to establish comics as a visual language.[2] The measure asked participants to rate how frequently they read comic books, comic strips, graphic novels, and Japanese comics in a given week on a scale from 1 to 7, with 1 being never and 7 being always. It also asked participants to self-rate their comics reading expertise on a scale from 1 to 5, with 1 being below average and 5 being above average.

Responses to the four reading frequency questions were summed up, giving a range from 4 to 28. Participants were then sorted into one of three groups based on their summed score. A sum between 4 and 11 placed a participant's reading frequency in the low range, where the most experience with comics may be either reading one type of comic frequently, or very infrequently reading different types of comics. A sum between 12 and 20 fell into the medium range, where participants may either read two types of comics with some frequency, or sometimes read multiple types of comics. A sum between 21 and 28 fell into the high range, where participants read all four types of comics asked about frequently. Overall, 123 participants fell into the low range, 26 participants into the medium range, and only 2 participants fell into the high range. Each group, low to high, was assigned a value from 1 to 3. This value was then multiplied by the participant's comics reading expertise rating (which ranged from 1 to 5) to get a final expertise score between 1 and 15. Although the scores cover the range of possible scores, the level of expertise among the participants was low overall with a mean of 3.26 (SD = 2.64). Thus, our participants were not particularly familiar with comics as an art form, and could be considered naïve readers of the comics we presented.

---

[2] Neil Cohn, "The Visual Language of Comics," *Visual Language Lab* http://www.visuallanguagelab.com/ (accessed December 6, 2020).

Table 12.1

Gender, age, and expertise score by comic

| Comic | % Male | Mean Age | Mean Expertise Score |
|---|---|---|---|
| Frost | 0.44 | 31.34 | 3.10 |
| Horla | 0.42 | 32.29 | 3.11 |
| Lola | 0.46 | 32.54 | 3.92 |
| Vulture | 0.49 | 30.63 | 2.94 |

Participants were randomly assigned to read one of our four comics—we used a "between-subjects" design. But as can be seen in table 12.1, the characteristics of the people reading each comic were very similar. For each comic, a little under half of the participants were male, the average age of the reader was a little over 30 years old, and they had equivalent (and equivalently low) comics expertise scores. Because the groups were so similar, we have some confidence that any differences we find between the comics are the result of what was actually in the comics, and not the people who read them.

## Materials

The four comics chosen were: Kevin VanHook's *Frost: the Dying Breed* (1991), Ernie Colón's "The Horla" (from *Inner Sanctum: Tales of Horror, Mystery and Suspense*, 2012), Joseph Torres and Elbert Orr's *Lola: a Ghost Story* (excerpt: 2010), and Peter Kuper's recreation of Kafka's short story, *The Vulture* (excerpt: 2013).[3] The comics were chosen because they were relatively simple, cross-cultural and represented a range of visual styles, genres and plots revolving around a dominant emotion. They were also non-mainstream comics and therefore less likely to be familiar to the participants.

*Frost* and *Vulture*, both originally five pages long, were brief enough to be used in their original form. From *Lola*, an eight-page excerpt was pulled from

---

[3] Kevin VanHook, *Frost: The Dying Breed* no. 2, Comic book (Wayne County, MI: Caliber Comics, 1991); Ernie Colón, "The Horla," in *Inner Sanctum: Tales of Horror, Mystery and Suspense*, Comic book (New York: NBM Publishing, 2012); Joseph Torres and Elbert Orr, *Lola: A Ghost Story*, excerpt 2010 (Portland: Oni Press, 2009); Peter Kuper, "The Vulture," in *Give It Up! And Other Short Stories*, excerpt: 2013, Comic book (New York: NBM Publishing, 2005).

the beginning of the larger work that depicted a contained narrative. "Horla," originally a thirteen-page work, was reduced to eight pages by extracting three pages that expanded on, but did not introduce crucial plot information, as well as four setting establishing panels. The remaining panels were rearranged slightly to fill the gaps.

The panels of each comic were numbered unobtrusively in a corner to facilitate coding. Comics were presented in hard copy on 8.5" x 11" paper such that page turns and page spreads were preserved from the original publication.

## Procedure

Participants were asked to read one of the four comics before filling out a questionnaire about that comic. They were asked how the comic made them feel, to identify the point that most made them feel that way, to identify a point where pictures and text worked best together, and to state why they felt they worked well at that particular point.

For questions asking for a particular point in the comic, participants were given free rein to answer as they saw fit, whether through panel numbers or event descriptions. However, in cases where the given response was unclear or ambiguous, such as descriptions that could conceivably cover multiple points in a comic or seemed to describe something not obviously in the comic, the participants were asked to clarify by pointing to the area of the comic they had intended. The clarified responses were recorded as panel numbers.

## Analysis

### Emotions

Responses to the emotion elicited question were sorted into emotion class groups based on whether or not the emotion was positive or negative. Examples of a positive response would be, "Amused," or "Happy for the ending," while examples of a negative response would be "Sad," or "A little depressing." A third category, *reading experience*, captured responses that were focused on the reader's experience versus the actual content of the story. Most of these responses indicated disengagement from the comic, including boredom or confusion, such as "Underwhelmed" or "I didn't understand the point of it," but also captured

Table 12.2

*Narrative elements by comic*

| Episode | Unit | Panel numbers (Splash pages in parenthesis) |
|---|---|---|
| **Frost** | | |
| 1: Frame | Intro | 1 |
| | Setting: Place/Time | 2-3 |
| | Foreshadowing Doom | 4-5/6 (5/6) |
| 2: Character Intro | Intro of Character info | 7-10 |
| 3: Body | Basic set-up | 11 |
| | Identification of Goal | 12 |
| | Actions towards Goal | 13-16 |
| | Outcome | 17 (17) |
| **Horla** | | |
| 1: Dream | Dreaming | 1-7 |
| | Wake up (conclusion of dream) | 8 |
| | Confrontation | 9-11 |
| 2: Piano | Interior Thoughts | 12 |
| | Helen prods (Exterior discussion) | 13-16 |
| | Horla intro | 17-18 |
| 3: Flashback | Framing of flashback | 19 |
| | Concert | 20-22 |
| | Madness | 23-24 (24) |
| | Madness/Horla (end of flashback) | 25 (25) |
| 4: Present day | Going mad | 26-29 |
| | Window (outcome) | 30 (30) |
| **Lola** | | |
| 1: Picnic | Frame for story | 1-5 |
| 2a: Pig Story (Within-Story Perspective) | Scared | 6-10 (10) |
| | Chase | 11-14 |
| 2b: Pig story (Overlapping Frame/Story perspective) | To Safety | 15-19 |
| | Interpretation | 20-23 |
| | Moral | 24-26 |
| | Lola intro | 27-29 |
| 3: Coda | Coda | 30-34 |
| **Vulture** | | |
| 1: Setting | Layout of situation | 1 |
| 2: Man with Glasses Sequence | Learns problem | 2-4 |
| | Proposes Solution | 5-6 |
| | Debate | 7-9 |
| | Accepts Solution | 10 |
| | Wrap up | 11 |
| 3: Vulture Sequence | Vulture perceives his goal | 12 |
| | Vulture takes action to achieve goal | 13-16 |
| | Death (Vulture's goal) | 17 (17) |

reactions such as "It reminded me of Poe." Although the majority of answers tended to fall within a single group, if a participant indicated multiple emotions, the response could be sorted into multiple groups.

### Narrative Units

In order to see how the points chosen fit within the narrative structure of the comics, each comic was broken up into narrative elements in a two-level hierarchical structure. The top level, or narrative episodes, reflect a single-story event such as setting establishment or a plot development. Episodes were then broken down into narrative units that made up the larger episode, the episode's set up, main action, and outcome.

Depending on the overarching episode, multiple units may be present, or none at all. The first episode of *Vulture*, for instance, only establishes a single component of the setting and therefore has no smaller divisions. In contrast, the third episode of *Frost* in which soldiers are sent to destroy a chemical lab has two setting components, the first establishing the context of chemical warfare while the second establishes the characters' goal of destroying the lab, then an action unit where the soldiers are on their mission, then an outcome unit where their goal is accomplished. The episode and unit breakdown of each comic can be seen in table 12.2.

Responses to questions asking for a location in the comic were coded as falling within an episode and a unit. When a response indicated more than one location, if the locations fell into two non-contiguous episodes, or covered more than two episodes, the location was coded as *scattered*. Otherwise, each location in the response was coded separately. If a response indicated the entirety of the comic, it was coded as *entire*.

## Comics Results Summary

### Emotion Results

Figure 12.1 shows the proportion of participants for each comic that described the comic using the different types of emotions. The proportions do not add up to 1 because participants could identify more than one emotion.

Three of the four comics (*Frost*, "Horla," and *Vulture*) evoked largely negative emotions, eliciting descriptions such as *creeped out, gloomy, sad*. These negative

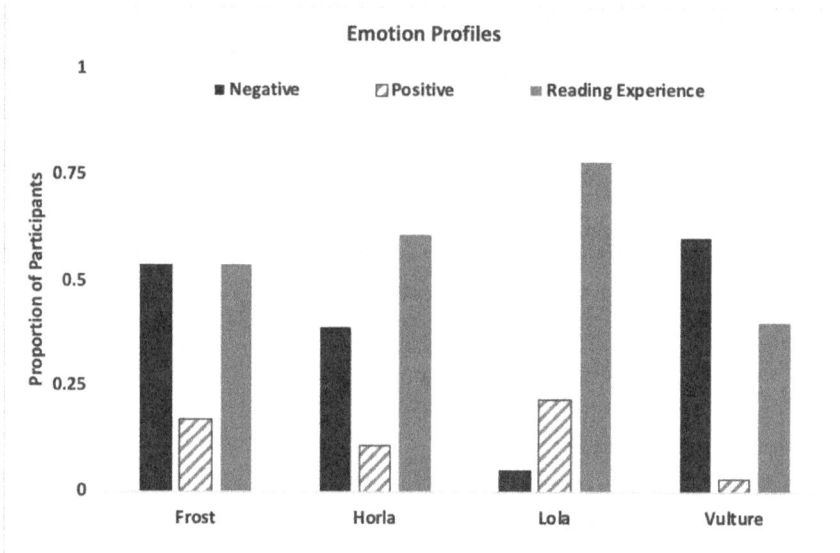

**Figure 12.1** The proportion of participants who provided each kind of emotional reaction for the four comics.

emotions are to be expected for these comics; their content is dark and participants were appropriately responding to the themes of death and violence. The only comic with more positive comments (*happy, amused*) than negative ones was *Lola*, and the content of that comic is notably lighter in tone. Thus, the participants were sensitive to the core content within the comics themselves.

All of the comics, however, generated large numbers of comments expressing the reader's experience. Overwhelmingly, these comments reflected boredom or confusion; participants said they were *Confused—it seemed out of context, Impatient, uncertain*. These sorts of comments were particularly high for *Lola*, which was unique among our choices for having a story-within-a-story framing. These emotional reactions most likely were a result of the fact that our participants had overall quite low scores on comics expertise. These results suggest the possibility that for naïve comics readers, the complex narrative structures within the stories may serve as a barrier for general engagement.

### Ability to Identify the Part of the Comic That Generated Responses

Figure 12.2 shows the proportion of participants for each comic agreed on the plurality opinion about which panel was the best representative of the target. For

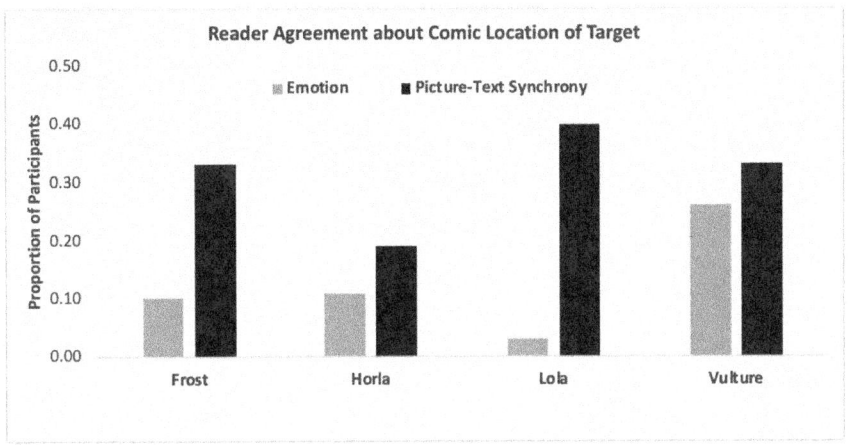

**Figure 12.2** Proportion of participants who agreed with the most frequent location of highest emotional and picture-text synchrony in the comic.

Emotion, the target was the panel that elicited the emotion (positive or negative) the most. For Picture-Text Synchrony, it was the panel where pictures and text worked particularly well together.

Participants' emotional response to the comics did not appear to be linked to any particular point in the comic book. There was very little agreement across participants about the panel that most represented the emotion they chose. The highest rate of agreement came for *The Vulture* comic: 26 percent of participants located their negative emotion at the point when the vulture fatally stabs the lead character. But no other point, either in this comic or any other, came close to reaching even that level of agreement. We did also look to see if there was agreement for the reading experience emotion, but the agreement for that category was less than for the positive and negative emotions overall.

Participants did show more agreement about the panels where the text and pictures worked particularly well together. As can be seen in figure 12.2, this kind of agreement is higher for all the comics and, in most cases, it is substantially higher. Considering the precise panels that led to high agreement, it appears that splash pages help galvanize opinion. For the *Frost* comic, participants identified the final splash page.

On this page, the lead character is shown firing his gun as he completes a canonical fairy tale rhyme ("and we blew the house down"). The picture therefore resolves two elements at once: the literal experience in which the soldier decides

**Image 12.1** Kevin Van Hook, *Frost: The Dying Breed*, Issue 2, 1991.

to take aggressive action (firing his weapon) and the verbal poem in which he explains his actions at a more metaphoric level.

For the "Horla" comic, participants also identified a splash page. The text explicitly notes that the character is worried they are going mad ("I thought of every possible explanation. A hallucination? But I feel it's real ... I feel ... It's here.. I can feel it").

**Image 12.2** Ernie Colón, "The Horla" in *Inner Sanctum Tales of Horror, Mystery and Suspense*, 2011.

The visuals are a cacophony of unusual actors from games, circuses, and movies (a dancing bear, a gangster, a horned demon, etc.). Thus, the picture illuminates the text, and provides the detail to make good on the notion of a hallucination.

For the *Lola* comic, participants identified a short sequence of panels along with the splash page that resolved their arc. In the panels, the lead character is scared of a noise (a pig's grunt) but can't really see what produced it until the sequence resolves in a splash page showing a giant pig bumping into the girl.

**Image 12.3** Joseph Torres and Elbert Orr, *Lola A Ghost Story*, 2009.

*Toward an AI Future of Comics Study and Creation* 199

Throughout the sequence, the words are a third-person description of what the images show; the final splash page contains just a single short sentence explicitly describing the size of the pig ("it was just bigger") along with a speech bubble for the pig saying "oink."

Finally, in *The Vulture* comic, participants identified a short sequence of layered pictures overlapping a half-page splash page. The splash portion shows the lead character in his death throes and the short sequence shows the vulture flying in for the final kill.

**Image 12.4** Peter Kuper, "The Vulture" in *Give It Up! And Other Short Stories*, 2005.

As the vulture approaches the lead character, his beak increasingly comes to look like a javelin and the text makes the metaphor explicit: "It took wing, leaned far back to gain impetus, and then like a javelin thrower, thrust its beak through my mouth." It is also worth noting that this sequence was the same one that led to the highest level of agreement about what generated a negative emotion.

## Conclusions

With this first step toward a measure of the dominant emotion triggers (plot and visual or verbal elements) in our engagement with comics, we have sought to enhance our understanding of how comics work—and why comics continue to be so popular. Even though our participants were not very experienced at reading comics, they were still able to connect with the core emotional messages being presented. Moreover, the splash page—a distinctive element of comics storytelling—resonated for our participants strongly. They may not have known much about comics, but they saw the storytelling power of those pages. Our participants' inexperience could be seen most clearly in the high rates of general dissatisfaction they felt with the overall reading experience: the complex storytelling layers of modern comics can be difficult for novices. This inexpert status may also have contributed to the lack of consensus our participants had over what parts of the comic generated their emotional responses.

These results begin to shed light on how some comic books use the visuals (and verbal) elements to create stories that use emotion cues to re-schematize and make new reader's apprehension of the world—and with this, make new emotions, thoughts, and perceptions concerning historically underrepresented communities.

To this end, in further studies, it could be useful to also create a similar study focused on Japanese manga. In addition to right-to-left reading conventions, these comics tend to rely heavily on storytelling that is built on close-up shots frames (a full head and a little bit of chest) and single character frames than European and US comics. It would be interesting to see how this might recalibrate or not the dominant affective triggers of the story. This might lead to a better understanding of how the cross-pollination of manga with some manga-inspired European and US comics as seen in OEL (Original English Language) comics might generate new aesthetics with their own sets of emotions. It will also enrich and give nuance to plot and character construction in today's AI-generated manga projects. We think readily of the Tezuka 2020 Project whereby the Kiokia team of engineers, Keio University researchers, and illustrators and writers from Tezuka Productions created the extraordinary manga, *Phaedo* (2020).

We set forth to see what might come by putting our two brains together—two brains from opposite sides of the campus (physically and epistemologically) yet tied together through the common interest in cognitive processes and systems. We both discovered that indeed a cross-disciplinary research program across the humanities and cognitive linguistic disciplines along with a team of undergraduates could lead to a deep, empirical study of how and why people read comic books. Working together, we learned to ask new and different questions—and to delimit these questions. We learned together how to distill comics for codable results, and how to code answers for usable data that led us to generative results, reminding us just how hungry our students are for cross-disciplinary explorations to find new ways to solve intellectual and everyday puzzles and issues.

Finally, our research on comic book structures and emotion triggers lays the groundwork for work with colleagues at the cutting edge of AI and cognitive creativity. It lays the groundwork for not only answering the question, "why comics?", but clears a pathway for understanding how we might use computer vision, speech and audio processing, natural language processing, machine learning and learning theory to create narrative-media yet to be imagined—and co-imagined with creative AI systems.

# Appendix

## Questionnaire

Instructions:
For the following questions, please consider the comic you read as you respond. Feel free to use the back of the pages if necessary.
1. What happened in this comic?

_____
_____
_____
_____

2. Using the following scale, where 1 means you really disliked the comic and 5 means you really liked it a lot, indicate how much you like the comic.

Please circle a single number.

    Dislike----------------------Neutral----------------------Liked a lot
       1          2          3          4          5

3. How did this story make you feel?

_____
_____
_____

4. Please identify the point (or points) in the story that MOST made you feel that way?

_____
_____
_____

5. Please identify the point (or points) in the story where you felt like the pictures and text worked especially well together.

_____
_____
_____

6. Why did you feel like the text and pictures worked well together at that point?

_____
_____
_____

7. Using the following scale, on average, how often per week do/did you ... ? (place a whole number in the square).

    Never----------------------Sometimes----------------------Always
     1      2      3      4      5      6      7

|  | Currently | While you were growing up |
|---|---|---|
| …read text-only books for enjoyment | | |
| …watch movies | | |
| …watch cartoons/anime | | |
| …read comic books | | |
| …read comic strips | | |
| …read graphic novels | | |
| …read Japanese comics (manga) | | |
| …draw comics | | |

8. How old were you when you began reading comics? _____ Drawing comics? _____

9. How would you rate your expertise with reading comics (of any sort)? (check once in each column)

|  | Currently | While you were growing up |
|---|---|---|
| above average | | |
| slightly above average | | |
| average | | |
| slightly below average | | |
| below average | | |

10. How would you rate your drawing ability? (check one)

|  | Currently | While you were growing up |
|---|---|---|
| above average | | |
| slightly above average | | |
| average | | |
| slightly below average | | |
| below average | | |

11. Were you familiar with the comic **[insert here]** before today? (circle one)
    Yes    No

# Bibliography

Cohn, Neil. "The Visual Language of Comics." *Visual Language Lab.* http://www.visuallanguagelab.com/ (accessed December 6, 2020).

Colón, Ernie. "The Horla." In *Inner Sanctum: Tales of Horror, Mystery and Suspense.* Comic book. New York: NBM Publishing, 2011.

Davidson, Cathy N. *The New Education: How to Revolutionize the University to Prepare Students for a World in Flux.* New York: Basic Books, 2017.

Dunst, Alexander, Jochen Laubrock and Janina Wildfeuer. *Empirical Comics Research: Digital, Multimodal, and Cognitive Methods.* New York: Routledge, 2018.

Groensteen, Thierry. *Comics and Narration.* Jackson: University Press of Mississippi, 2013.

Kuper, Peter. "The Vulture." In *Give It Up! And Other Short Stories.* Comic book. New York: NBM Publishing, 2005.

Miodrag, Hannah. *Comics and Language.* Jackson: University Press of Mississippi, 2013.

Postema, Barbara. *Narrative Structure in Comics.* Rochester, New York: RIT Press, 2013.

Tezuka Project. *Phaedo.* 2020. https://tezuka2020.kioxia.com/en-jp/. (accessed December 6, 2020).

Torres, Joseph and Elbert Orr. *Lola: A Ghost Story.* Portland: Oni Press, 2009.

Van Hook, Kevin. *Frost: The Dying Breed* no. 2. Comic book. Wayne County, MI: Caliber Comics, 1991.

Part Four

# The Digital West

13

# The Philosophy of *Westworld*

Paul Skokowski

What exactly does an android experience? Could an android have experiences as rich as humans, or are there limits? The *Westworld* TV series (Jonathan Noland, 2016- ) offers the opportunity to explore philosophical questions related to human and android experiences through its depiction of a fictional Wild West theme park with androids playing the main characters. Among the most fascinating scenes in the *Westworld* TV series are the interviews between the android characters Bernard Lowe and Dolores Abernathy. These interviews occur in a windowless room and appear to portray Dolores at different stages of her development, though what comprises this development is not revealed to the viewer. It's probably a good guess that Dolores' development consists in software upgrades as well as her own learning episodes—her own learning history. There are a host of interesting questions that could be raised about the meetings between Bernard and Dolores, about her software updates and her learning history, but there are particular, philosophical questions about what Dolores experiences that I would like to focus on in this essay. These questions have to do with possible limits to her experience, and whether these limits correspond to limits we—intuitively, at least—apply to human beings.

## Setting the Stage

Let's start with a *Gedanken* experiment. Suppose that Dolores has so far been kept in a single, windowless basement room in the Westworld android manufacturing and repair complex. She has not yet been allowed out into the world at large, including the Westworld theme park. To pose our philosophical question, we need to make a few more suppositions. The first is that the room is a purely black and white space, with none of the other chromatic colors—red, blue, green, etc.—being present or viewable on its surfaces. Indeed, let's assume that the room has always been devoid of colors except black and white. We will also assume that Dolores has been painted black and white. A further supposition is that instead of Bernard Lowe interviewing Dolores, the questions are asked instead by an android whose surface colors are also limited to black and white. This android doesn't need to be humanoid although it could be. Call this android Piebald.

Let us continue with our *Gedanken* experiment. Suppose Piebald briefly de-activates Dolores and opens her skull by removing a plate in the back of her head with a screwdriver to expose the Central Processing Unit (CPU) within. Piebald then attaches a cable to connect the black and white computer terminal in the room to Dolores' CPU. Using the cable, Piebald downloads all physical knowledge, which just happens to be stored on the Westworld mainframe computer, to Dolores' CPU. This physical knowledge is quite a lot of knowledge: it comprises everything physical there is to know. Once the download is complete, then, we will presume that Dolores "knows" all of physical theory. She knows everything there is to know not only about physics, but also about the other physical sciences that reduce, ultimately, to physics. This would include knowledge, for example, of the physical workings of CPUs, robots, androids, etc. This is, to put it mildly, quite a bit of knowledge.

After completing the download, Piebald disconnects the cable from Dolores, reattaches the plate in the back of her head, and then re-activates her. With apologies to the excellent screenwriters of *Westworld*, we might imagine the following conversation. Piebald asks Dolores what colors she sees in the room. "Black and white," she answers, confidently. "Do you see any other colors?" Piebald asks. "No, those are the only two colors I presently see." Piebald now asks another, clever question, "Have you ever experienced any colors other than black and white?" which, we can presume will be answered by Dolores with an emphatic and honest "No."

Piebald continues, "Dolores, how about if I take you for a walk outside of this room, and allow you to explore a little bit of the world outside?" "I'd like that," Dolores answers. Piebald then opens a door and leads Dolores directly outside, where a rosebush is in full bloom, and Dolores at that moment experiences seeing a red rose for the first time. The burning question here, at least for philosophers—and possibly also for Piebald—is this: when Dolores looks at a red object for the first time, in this case a red rose, does she *learn* something? The intuition seems inescapable that Dolores *does* learn something:

she sees red for the first time. In so doing, Dolores gains a new experience, the experience of *red*—a chromatic color. As some philosophers like to put it, she now understands "what it is like" to experience red. It just seems unavoidable that Dolores has learned something new.

And now we come to the punchline. We know that Piebald has already downloaded all the physical knowledge into Dolores' CPU. So, if Dolores already has all the physical knowledge, then this new knowledge that she has gained from her encounter with the color red must be non-physical. Why is that? Well, because *ex hypothesi*, Dolores already *had* all the physical knowledge, so any new knowledge she gains must be *non-physical*.

But this is odd. Here is a non-human android—a machine—made by human (and presumably, android) hands. This is a physical device *par excellence*—a purely physical device if there ever was one. And yet, this device is now experiencing something non-physical! What is so odd about this, is that the normal attribution of non-physical properties to objects are properties we normally might call a "spirit" or a "soul." For the philosopher Descartes, this soul was also known as a "mind." This mind was, indeed, a non-physical thing, but it was a thing that could—and did—control the human body. Descartes held further that only *humans* had minds. Animals did not. In fact, Descartes viewed animals as automatons—machines. Presumably these were biological machines, but they were machines nonetheless. And what distinguished such machines from humans was that machines lacked a mind.[1]

Let us return now to Dolores. She is clearly a non-human android, and so by definition a machine or automaton. Under our normal Cartesian view, which I think is generally accepted at least at an everyday level in human culture, Dolores has no "soul," no "spirit," and therefore, no *mind*. Part of the reason for this is the Cartesian view of mind and body as being distinct substances; the mind being a non-physical substance and so not part of the physical world, and the body being a physical substance and thus part of the physical world. And if Dolores—an automaton, an android, a machine—has no mind, then she cannot experience anything non-physical. Why? As a machine, she is restricted to being part of the physical world. She cannot, therefore, have non-physical interactions of any sort, which includes non-physical *experiences*.

---

[1] The best source for Cartesian ideas is, of course, Descartes himself. See René Descartes, "The Passions of the Soul," 326–83, "Discourse on Method," 111–41, and "Mediations on First Philosophy," 1–50, in *The Philosophical Works of Descartes*, Volumes 1 & 2, trans. Elizabeth Haldane and Ross, G. R. T. (Cambridge: Cambridge University Press, 1968).

But this is strange, given our *Gedanken* experiment. Piebald has, by hypothesis, given Dolores all the physical knowledge before exiting the room. And presumably a purely physical thing—like the android-machine Dolores—can have physical knowledge states about the world. We regularly ascribe such states to computers, which are physical things made out of all sorts of physical materials, all the time: turnstiles counting the people passing through, GPS systems tracking positions, photos on our iPhones recording a physical scene from a particular point of view at a time, and so forth. These are perfectly acceptable, indeed humdrum, physical facts that are being recorded at every moment across the globe.

And yet, the moment Dolores steps outside of her room and "experiences" a new color for the first time, she is "seeing" something non-physical. But this, it seems, is impossible, as Dolores has no non-physical mind, and so we must conclude that she is not experiencing anything of the sort.

What's an android to do?

## The Knowledge Argument

To readers familiar with debates about conscious experience, the scenario just given regarding Dolores will be recognizable as a variation of what has come to be known as the "knowledge argument," which was first proposed by the philosopher Frank Jackson in 1982.[2] The knowledge argument considers a brilliant scientist named Mary, who from birth has been kept in a black and white room. Through black and white books, black and white televisions, black and white computers, etc. in her room, she comes to acquire all the physical knowledge, and this includes complete knowledge of neuroscience, which is seen as a part and parcel of physical knowledge. And yet, when Mary is finally allowed to exit the room, and experience a red rose for the first time, the overwhelmingly compelling intuition we have is that Mary learns something new. She learns about the color red through *experiencing* red for the first time. But, by hypothesis, Mary already had all the physical knowledge; so, this new knowledge must be *non-physical!*

---

[2] Frank Jackson, "Epiphenomenal Qualia," *The Philosophical Quarterly* 32, no. 127 (April 1982): 127–36.

Note that a feature that makes the knowledge argument so compelling is a deeply held Cartesian intuition that humans are the sorts of creatures that *can* have minds of the sort that Mary seems to exhibit. Recall that this is because minds—in this Cartesian sense—are the kinds of things that can be immaterial. Humans, in this view, are capable of having minds or souls—some kind of spiritual stuff that exists in a separate, presumably non-physical realm. As discussed briefly above, this is something that is part of our culture and our language and is hard to disassociate ourselves from. After all, we regularly talk about certain exertions being "purely mental," or applying "mind over matter," or having a "damaged" or a "happy" soul, or "putting your mind to it" and so forth. The Mary argument reinforces this view, and with a vengeance. It purports to provide a kind of existence proof for non-physical knowledge and even non-physical stuff. After all, if Mary already has all the physical knowledge and then gains new knowledge, then this knowledge *must* be non-physical, by an application of cold, precise logic. Experiences then, of the colors at least, must be non-physical in nature, and so be exemplifications of non-physical properties. And the chromatic color properties themselves that are presented to our minds must be non-physical.

This move, the acceptance of access to non-physical knowledge and non-physical properties, which is granted to Mary through the knowledge argument, does not seem to be available for our android friend Dolores. Why? The reason has already been mentioned earlier. The Cartesian intuition is one that we apply to beings with souls—that is, beings like us, and like Mary: *human* beings. Automatons, androids, and machines are instead soulless, purely physical, objects. And as purely physical objects, they have no minds.

## Is Dolores a Zombie?

The premise of *Westworld* seems to be that the androids one encounters there are so well designed that they appear—to humans—to also be human. This means that the Westworld androids walk, talk, play the piano, drink, ride a horse, fire a six-shooter, etc., just like humans would. So, if you were a human client in Westworld, and you happened to be around Dolores in close proximity to a rose bush, you should be able to ask her what color the roses are, and she will answer, correctly: "red." Dolores might even name her favorite color, if asked, and should be able to accurately distinguish the colors one from the other.

Yet these abilities seem peculiar, given the arguments we have so far considered. How could Dolores be able to distinguish the chromatic colors, if she is an android? As an android, Dolores is a purely physical automaton, and so lacks a mind. The knowledge argument seems to be telling us that the ability to access the colors requires access to non-physical knowledge and non-physical properties. And Cartesian considerations limit these abilities and this access to agents with minds only: humans. There is a possible answer, however, and that is that though Dolores may not qualify to be a human, she may instead qualify to be a zombie.

Zombies, in the philosophical sense, are creatures that appear to be just like humans, except they experience nothing whatsoever. A zombie may wax lyrical about the unique shade of a particular rose, or the luscious notes of blackberry and currants in her favorite Zinfandel, or the burning pain she just felt when she accidentally touched the edge of the hot frying pan, but these are just empty words: the zombie has not experienced the redness of the rose, the taste of the Zinfandel, nor the pain of touching the hot pan. Her world is dark, and completely devoid of sensory experiences—what philosophers like to call *qualia*. The zombie is simply responding to events in some purely physical way, with no mental life accompanying these responses, and in particular no sensory experiences like seeing colors, tasting flavors, or feeling pain.

It turns out that the original Hollywood zombies—and here I'm thinking of those terrific zombies in George Romero's *Night of the Living Dead* (George A. Romero, 1968)—don't really count as zombies in the strict philosophical sense. The reason is that these zombies seem to be irresistibly attracted to human brains, as is evidenced by their lovable exclamation of "Brains!" when they "see" a human nearby. It appears that the *taste* of human brains is what they crave when they utter "Brains!" in the presence of (still alive) humans. And we can only presume these zombies are enjoying the experience when they finally get their meal.

If Dolores is indeed a zombie, then it turns out that *Westworld* presents us with a second kind of Hollywood zombie that also differs (in its own way) from philosophical zombies. Philosophical zombies are special creatures: they are intended to be exact physical duplicates, molecule-for-molecule, of humans existing on Earth. Such zombies might, for example, exist on a duplicate Earth somewhere else in the universe. But what makes these philosophical creatures different from us humans on Earth is that despite being our exact *physical* duplicates, they nevertheless lack consciousness altogether. Of course, they interact with each other and their world exactly as we do. They, too, seem to appreciate a good opera and produce flowery descriptions of their Zinfandels, and even say

"Ouch!" after touching a hot pan. But they lack conscious experience altogether, and so their world is experientially dark—with no sounds, tastes, or pains—despite their being molecule-for-molecule duplicates of us humans on Earth.

Zombie arguments, as given by philosophers, are meant to be conceivability arguments.[3] They go something like this. It's conceivable that there are creatures exactly like us humans, in every physical detail, down to our molecular and atomic composition, and yet these creatures lack consciousness. If this is conceivable, then materialism—the theory that all physical and mental phenomena can be accounted for by physical theory alone—must be false. It must be false for similar reasons to those given above for the knowledge argument: giving a purely physical account of humans does not seem to account for conscious experience, and in particular, qualia. Something appears to have been left out, and this something, these qualia, are, to borrow a phrase, "mental as anything." And as purely mental stuff, qualia must be left out of any physical accounts of the world, according to this view.

But now notice that Dolores is categorically not a physical duplicate of any human being whatsoever. She does not, for example, have a carbon-based brain, fed by a bloodstream of oxygen-carrying red corpuscles. She does not (and this is a guess, but it's probably not far off the mark) have calcium-based bones and a human-like cellular structure composing her soft tissues. Instead, her "brain" is really a CPU, and her tissues are (probably) non-carbon based, and so on for her remaining bodily structures. So, Dolores and her android colleagues in Westworld are in no way molecule-for-molecule duplicates of any existing human beings on Earth. And therefore, Dolores is not a philosophical zombie after all.

Despite this revelation, Dolores might have more of a claim to being at least *some* kind of zombie than her predecessor Hollywood zombies do. Recall that the zombies of the *Night of the Living Dead* persuasion appear to have gustatory experiences—they appear to prefer the taste of brains to other possible soft tissues. So, these are creatures that appear to have experiences. And perhaps this interpretation seems admissible in the Cartesian sense because these creatures actually have *human* bodies—a little decomposed, without a doubt, but still human flesh and bone! Since these zombies have human bodies, there is nothing contradictory in their also having some human mental attributes,

---

[3] See David Chalmers, *The Conscious Mind* (Oxford: Oxford University Press, 1996); Paul Skokowski, "I, Zombie," *Consciousness and Cognition*, 11, no. 1 (March 2002): 1–9; Michael Tye, *Ten Problems of Consciousness* (Cambridge, MA: MIT Press, 1995).

such as taste qualia, for example. Perhaps this seems a stretch, but the dictates of the Cartesian view are consistent with the view. Think of the resurrection of Lazarus as another example, or of Jesus Christ himself. These episodes are more plausible with the Cartesian view of a physical body united (or reunited) with an immaterial soul or mind. Such a view certainly seems to allow that a mind or soul can be separated from a body, only to be reunited with it later.

But Dolores can never fit with this Cartesian view. After all she is an android, an automaton. She is a purely physical creature, and as such she does not, and indeed cannot, have a mind or a soul at all. And since experiences, including especially sensory experiences, are mental events, Dolores cannot experience anything at all. Her mind, according to the Cartesian view, is dark. She is a zombie. Not a philosophical zombie, since she is not a molecule-for-molecule duplicate of a real, live, experiencing human being, but a kind of zombie nonetheless: a being who lacks qualia altogether.

## Could Dolores be a Swamp-Droid?

A common theme so far in this essay is that the Cartesian view seems to underpin many of our attributions (or not) of conscious experience to other creatures. With apologies to St. Augustine, we will soon challenge the Cartesian view, but not yet. Before taking up that challenge, let's consider another aspect of the *Westworld* fictional framework: android duplicates.

In this section, we will take a different tack. We will dispense with the black and white room scenario heretofore considered, and instead simply consider Dolores as her usual self with her usual history in the Westworld theme park. We can even suppose for the sake of our example that Dolores has sensory experiences, at least those that the designers of the androids of Westworld have somehow been clever enough to give her the sensory systems to undergo. Let's also assume that Dolores' development as a more and more complex android has included her learning histories both within the park and with Bernard—something that seems implied by the narrative and backstory provided within the *Westworld* episodes that chronicle her progress as a more sophisticated android over time.

*Westworld* episodes often show scenes with androids being made or repaired in what appears to be an on-site factory. Like human beings, each android in Westworld is sui generis. But what if an android is, through some accident, completely destroyed, and yet that android character remains essential to the continued running of the theme park? Suppose such an accident happens to Dolores. Suppose Dolores has chosen to camp for the night in the backcountry of Westworld. She is sitting by the campfire when a sudden and violent rainstorm begins kicking up. Without any warning a bolt of lightning hits Dolores, and sadly, she is vaporized by the billion volts of electricity surging through the spot where she had just sat.

Fortunately for our Westworld engineers, a complete digital image of Dolores' CPU software and memory configuration had been downloaded to the main server that morning. Moreover, a complete digital record of the 3-D molecule-for-molecule printing of her body has also been saved (this latter record has been used several times in the past to repair body parts when they have been injured by various unforeseen accidents in the park, or perhaps by malevolent humans who have mistreated her—like the man in the black hat.) Since Dolores is required for the re-opening of the park the following Monday, the engineers set the 3-D printer going to create a duplicate Dolores body, and download the image of her CPU software and memory configuration to a new, duplicate CPU, and then put the duplicate CPU into the duplicate body to create a duplicate Dolores.

So far, so good. We can now ask, is this new duplicate really *Dolores*? Well, this object certainly looks like Dolores. And presumably, if we put this android in the same locations and contexts the park designers normally put Dolores in, then it will presumably also act like Dolores would act in those situations. However, strictly speaking this android is not Dolores. It is a different *token*; a different concrete physical object made up of different building blocks—different molecules—from the ones Dolores was made up of.[4] However, it is of the same physical *type*. It has the same types of physical properties Dolores had: the same weight, the same type of skeleton and soft tissue in the same types of configurations, the same type of CPU with the same electrical configurations in its transistors and so forth. So, this object will act in an identical causal fashion to the original Dolores when placed in the same types of physical configurations. Nevertheless, this android is in fact different from Dolores, and so let's call her Dolores2

---

[4] Think of it this way. Two children, who live across the street from each other, receive the latest Lego box of the Star Wars X-Wing fighter. They each assemble theirs according to the enclosed directions. Though the items may look identical, in fact, they are different *tokens*. They are made of different individual token bricks, each of which looks identical to individual bricks in the other model (and in countless other Lego boxes and scattered under beds around the world.) But the two models are of the same *type*. Both are called "X-Wing Fighters" and both will cause people to say things like "that is a cool X-Wing Fighter model" etc.

Now, if, through the clever designs of the engineers who built her, and her history of interactions with her environment at Westworld, the original Dolores indeed had sensory experiences, we might be willing to say that Dolores2 was capable of sensory experiences. The reason is that the same design process that produced Dolores with the capability of having sensory experiences allows a new version of her, Dolores2, to have sensory experiences as well. Of course, here we are applying the same reasoning we apply to new versions in our own species to Dolores2. We assume that new humans born into the world have the same capacity for sensory experience that our own evolutionary process, undertaken over hundreds of millions of years, has resulted in: new versions of humans will have similar capacities for sensory experience that we do. As a product of evolution, each new token of our species benefits from the causal process that has resulted in a finely tuned creature that is capable of experiencing different properties in the *environment* through different sorts of sensory detectors: vision, hearing, touch, taste, smell, and proprioception.

But now suppose there is *another* possible way that a new duplicate of Dolores can come into being. Suppose that the thunderstorm that vaporized Dolores with a bolt of lightning produces a second bolt of lightning at the same moment, but striking a different location in Westworld: the center of a molecule- and mineral-rich swamp. The stupendous amount of energy hitting the swamp causes, through a fantastically improbable process, the creation of a molecule-for-molecule exact duplicate of Dolores at the time of her vaporization. Let us call this new duplicate "Swamp-Dolores."[5]

The question now comes up, will Swamp-Dolores have experiences like those we are willing to attribute to Dolores? After all, it seems intuitive for us to attribute the same kinds of experiences that creatures normally seem to have to *duplicates* of those creatures. It turns out, however, that many philosophers claim there is *no* warrant for the claim that Swamp-Dolores will have experiences.[6] These philosophers argue that we have no reasons to believe that Swamp-Dolores will experience anything upon her creation. Fred Dretske, for example, considers

---

[5] For the original example of Swampman see Donald Davidson, "Knowing One's Own Mind," *Proceedings and Addresses of the American Philosophical Association* 60, no. 3 (January 1987): 441–58.
[6] Fred Dretske, *Naturalizing the Mind* (Cambridge, MA: MIT Press, 1995); David Papineau, "Doubtful Intuitions," *Mind and Language*, 11, no. 1 (March 1996): 130–32; Karin Neander, "Swampman Meets Swampcow," *Mind and Language* 11, no. 1 (March 1996): 118–29; Michael, Tye, "Phenomenal Externalism Lolita, and the Planet Xenon," in *Qualia and Mental Causation in a Physical World: Themes from the Philosophy of Jaegwon Kim*, ed. Terence Horgan, Marcelo Sabates and David Sosa (Cambridge: Cambridge University Press, 2015), 190–208.

this inclination to attribute experience to Swamp-Dolores and gives it a name: the Internalist Intuition. In order to challenge this intuition, Dretske asks us to consider lightning that strikes an auto parts salvage yard, and in a near-miraculous fashion, manages to produce an exact duplicate of his own car: a Toyota Tercel. He names this newly formed object "Twin Tercel." What are we to make of this example?

Dretske asks us to consider what would happen if we began exploring Twin Tercel and discovered that the arrow on its "fuel gauge" did not seem to indicate the volume of gas in its "tank." He asks if that would mean that the fuel gauge was broken. However, this turns out to be a very peculiar question. For what gives us the right to even say that this is a fuel gauge? We examine it to see red markings in the shape of an "F" and an "E," but do these marks mean anything at all? These questions are easy to answer for Dretske's own original Toyota Tercel. It is clear what *that* fuel gauge was designed to accomplish, and easily verify whether it is working or not. But there is no such assurance for Twin Tercel. No one placed the "F" and the "E" on the object and indeed no Toyota engineer designed it to indicate the volume of fuel in a tank. So how could those marks actually mean *anything*?

Dretske tells us that what goes for Twin Tercel's "fuel gauge" goes for its other parts as well, as:

> they are not a copy, an imitation, or a reproduction of anything that was designed to do something. There is therefore, nothing the parts of Twin Tercel are supposed to be doing, nothing that, by failing to do, would count as their "not working right."[7]

The moral here is that not only can nothing *not* be working right in Twin Tercel, but conversely, nothing can *work right* in Twin Tercel, either!

To put the point another way, we should understand why the actual Tercel that Dretske owns is capable of having a fuel gauge that can be broken. If the gauge in Dretske's Tercel stopped responding to fuel level in a similar way, it would be broken, and therefore would be misrepresenting the volume of gas in Tercel's tank. But Dretske explains how this interpretation is impossible for Twin Tercel:

> To suppose that Twin Tercel's "gas gauge" is not working right, that Twin Tercel even *has* a gas gauge, is like supposing that a word has been misspelled when the

---

[7] Dretske, *Naturalizing the Mind*, 143.

wind and waves carve out "anser" in the sand of a deserted beach. If I had made those marks, it might have been a misspelling but, given the origin of the marks, it isn't. It can't be. If gauges and instruments are objects that have a function, something that they are supposed to do, there are no gauges and instruments in Twin Tercel – nothing that can, as the instruments in my Tercel can, represent speed, amount of gas, and oil pressure.[8]

The key here is that the gauges and instruments of an authentic Toyota Tercel were given their function by those who designed them—the engineers and designers in the Toyota factories and design centers. Miraculous instantaneous fabrications have neither a designer, nor a causal selection history, and so have no function. There is nothing they are *supposed* to do. To accept that Twin Tercel's "fuel gauge" had a function would be akin to William Paley's appeal, made in the eighteenth century, to intelligent design whenever one initially confronts a complex object, including a life form.[9] Paley argued that if one came across a complex object like a watch out in the middle of nowhere, one would conclude that someone had designed it. He extends this argument to complex life forms, which must also require a designer: God. Dretske calls this appeal to a designer the "Paley Syndrome," and says it often actually works for us. After all, when we see other Toyotas being driven in the street or parked at the side of the road, we assume they were designed and built by the Toyota Corporation.

But what seems reasonable most of the time for everyday objects may not seem so plausible when we consider the actual history and environment of other objects. Dretske explains:

> What is not reasonable is to make the same inference when – as in the Twin Tercel story – one is told in advance that the parts were not designed, that none of them were made or placed there for any purpose. To persist in giving them a purpose anyway – something one does (if only implicitly) by calling them gauges and instruments – is irrational.[10]

The point of these arguments is to show how our everyday intuitions are manifestly unreliable when it comes to understanding the nature and results of miraculous duplicates. Miraculous duplicates have no designer—they are not even products of natural selection. These duplicates, and their individual

---

[8] Ibid.
[9] William Paley was an eighteenth century theologian who gave teleological arguments for God's existence based on the complexity of living creatures.
[10] Dretske, *Naturalizing the Mind*, 147.

components therefore have no purpose, no function, and there is nothing they are supposed to do. Swamp-Dolores has no sensory systems, no experiences, in the same way that Twin Tercel has no gas gauge and no meaning—no semantic content—for the markings on its "gauge."

## Putting Things Together

So far in this essay we have been accepting of the Cartesian view of a purely physical world which is separate and distinct from the mental world. This view seems to lead to several conclusions. One is that an android like Dolores cannot have a mind, no matter how sophisticated and complex she becomes over time. Another is that, together with the strictures of the knowledge argument, Dolores will, contrary to some very strong intuitions, learn nothing when she leaves her black and white room and sees a red rose for the first time. And this, together with the first conclusion, seems to show that Dolores must actually be some sort of zombie—in this case a mechanical android that has no experiences, no sensations, whatsoever. But are we really entitled to these conclusions? For all of them require an acceptance of a dualistic world—a world with two kinds of stuff: physical stuff and mental stuff. Physical stuff is governed by laws of the natural world—what we know as physics and the other natural science disciplines which themselves ultimately follow physical laws. Mental stuff is apparently governed by the laws of the mental world, though presumably we don't know yet what the laws are that would govern this non-physical world.

The problem, of course, is how these two realms—the physical and the mental—are supposed to interact. Descartes himself was aware of this problem and offered the famous solution that the mental and the physical interacted via the pineal gland in the center of the brain.[11] Mental events caused perturbations on the surface of the pineal gland, which mechanically caused further events in the brain and nervous system, leading to human, physical, actions in the body, which then would cause physical events in the world. Conversely, events in the world would impinge on our sensory systems, or on our body, and ultimately cause perturbations on the surface of the pineal gland, which would then cause mental events; our thoughts could then be caused by events in the world, like

---

[11] Descartes, "Passions of the Soul," 325–404.

seeing a red rose or touching the surface of a hot pan. Of course, this solution to the problem of dualism was not a solution at all: it just begged the question. For the pineal gland is itself a purely physical thing. No explanation has therefore been given about how a mental/non-physical thing, such as a thought, can cause physical perturbations in this physical thing. That would require a demonstration and an *explanation* of mind-body interaction. And this has not been given.

This problem has never gone away for dualism and has never been successfully solved. It makes arguments based on dualistic views suspect. These arguments might seem at first to be plausible or intuitive. But ultimately, they themselves rest on an assumption of non-physical "stuff" that has never been explained or shown demonstrably to either exist or to interact with physical stuff. Though the knowledge argument seems, as was mentioned earlier, like a kind of existence proof of mental stuff, the argument itself has a hidden premise: that Mary can have all the physical knowledge while isolated in a black and white room. Think of it this way. If the world really is a purely physical world, and Mary indeed learns something new when she steps out of the room, then she really didn't have all the knowledge while she was being restricted to her room after all, despite the premise in the argument which said that she did.[12] Another, telling reason to question the knowledge argument as a challenge to materialism is that its originator, Frank Jackson, has himself recently rejected it, and now views mental activity as purely physical in nature.[13]

A similar sort of reasoning leads us to question conceivability arguments where we are asked to imagine a duplicate world to ours, but where the duplicate inhabitants lack any experiences—and hence any qualia—whatsoever. Again, to see this point, what if our universe really *is* a purely physical universe as most scientists believe? Then this alternate world we are *conceiving* would actually be *exactly* like our world, which would mean that the duplicate inhabitants actually were having experiences—and qualia—just like we were. The reason is that their "mental" events are actually purely physical events, like ours are in this world, and in fact would be identical to our purely physical mental events. Since we have qualia in this world, they would in theirs. So, it appears that we haven't

---

[12] For more details see Paul Skokowski, "Temperature, Color, and the Brain: An Externalist Reply to the Knowledge Argument," *Review of Philosophy and Psychology* 9, no. 2 (June 2018): 287–99.

[13] Frank Jackson, "Consciousness," in *Handbook of Philosophy*, ed. Frank Jackson and Michael Smith (Oxford: Oxford University Press, 2006), 310–33; Frank Jackson, "Mind and Illusion," in *Minds and Persons*, ed. Anthony O'Hear (Cambridge: Cambridge University Press, 2003), 251–72.

successfully conceived of this alternate, qualia-free, world after all.[14] The creatures that appear to be zombies in this alternate world instead turn out to be duplicate beings to us, with real sensory experiences and qualia.[15]

Where does all this leave us with respect to the inhabitants of Westworld, like Dolores? Well, if the world *is* a purely physical world after all, and we humans inhabit it, then our thoughts and experiences will also be purely physical. After all, we are each of us aware of our own thoughts, our own experiences, and our own qualia. In such a world there is no barrier to having physical things with *minds*, since minds are physical things along with everything else in that world. This of course, leaves it an open possibility that Dolores herself, along with Bernard, and perhaps other advanced androids in Westworld, have a mental life including conscious experience. And if Dolores does have experience within her black and white room, then presumably she will soon have an experience of red when she exits her room, and so learn something new by experiencing a new quale: the quale *red*, which, along with other qualia, is itself a physical thing and so part of the physical world.[16] It also means that she will not be a zombie—not even a Hollywood zombie—because she experiences qualia through a variety of sensory modalities.

The ultimate question, of course, is: Would Dolores, as portrayed in *Westworld*, have a mind? That question, I think, is impossible to answer, because we aren't given the actual science required to understand how she is made, and how she interacts with the world, which would include details not only of her "sensory" systems, but also her learning history. Elsewhere I have argued that machines could have minds, but in the case of Dolores, I am a bit like a zombie: in the dark about her ultimate constitution.[17]

---

[14] There are many subtleties to these arguments, but ultimately, they boil down to the implausibility of dualistic ontologies, which require further, and as yet unstated, laws, objects and properties, and the perhaps even more implausible mental-to-physical laws that are required for interactions between the two realms; See Tye, *Ten Problems of Consciousness*, and Skokowski, "I, Zombie," for examples of problems with dualist ontologies.

[15] See Skokowski, "I, Zombie," and Tye, *Ten Problems of Consciousness*, for arguments regarding the conceivability of zombies.

[16] For the status of qualia in the physical world. see Dretske, *Naturalizing the Mind*; Tye, *Ten Problems of Consciousness*; Tye *Consciousness Revisited* (Cambridge, MA: MIT Press, 2009); Skokowski, "I, Zombie,"; Skokowski, "Is the Pain in Jane Felt Mainly in Her Brain?" *Harvard Review of Philosophy*, 15 (2007): 58–71; Skokowski "Temperature, Color, and the Brain," *Review of Philosophy and Psychology* 9, no. 2 (June 2018): 287–99.

[17] See Skokowski, "Structural Content" for examples of how machines could possibly have beliefs, and therefore minds; Paul Skokowski, "Structural Content: A Naturalistic Approach to Implicit Belief," *Philosophy of Science* 71, no. 3 (July 2004): 362–79.

# Bibliography

Chalmers, David. *The Conscious Mind*. Oxford: Oxford University Press, 1996.
Davidson, Donald. "Knowing One's Own Mind." *Proceedings and Addresses of the American Philosophical Association* 60, no. 3 (January 1987): 441–58.
Dretske, Fred. *Naturalizing the Mind*. Cambridge, MA: MIT Press, 1995.
Descartes, Rene. "The Passions of the Soul," 326–383, "Discourse on Method," 111–41, and "Mediations," 1–50. In *The Philosophical Works of Descartes*. Translated by Elizabeth, Haldane and G.R.T. Ross. Cambridge: Cambridge University Press, 1968.
Jackson, Frank. "Epiphenomenal Qualia." *Philosophical Quarterly* 32, no. 137 (April 1982): 127–36.
Jackson, Frank. "What Mary Didn't Know." *Journal of Philosophy* 83, no. 5 (May 1986): 291–5.
Jackson, Frank. "Mind and Illusion." In *Minds and Persons*, edited by Anthony O'Hear, 251–72. Cambridge: Cambridge University Press, 2003.
Jackson, Frank. "Consciousness." In *Handbook of Philosophy*, edited by Frank Jackson and Michael Smith, 310–33. Oxford: Oxford University Press, 2006.
Neander, Karin. "Swampman Meets Swampcow." *Mind and Language*, 11, no. 1 (March 1996): 118–29.
Papineau, David. "Doubtful Intuitions." *Mind and Language*, 11, no. 1 (March 1996): 130–32.
Skokowski, Paul. "I, Zombie." *Consciousness and Cognition*, 11, no. 1 (March 2002): 1–9.
Skokowski, Paul. "Structural Content: A Naturalistic Approach to Implicit Belief." *Philosophy of Science*, 71, no. 3 (July 2004): 362–79.
Skokowski, Paul. "Is the Pain in Jane Felt Mainly in Her Brain?" *Harvard Review of Philosophy*, 15 (2007): 58–71.
Skokowski, Paul. "Temperature, Color, and the Brain: An Externalist Reply to the Knowledge Argument." *Review of Philosophy and Psychology*, 9, no. 2 (June 2018): 287–99.
Tye, Michael. *Ten Problems of Consciousness*. Cambridge, MA: MIT Press, 1995.
Tye, Michael. *Consciousness Revisited*. Cambridge, MA: MIT Press, 2009.
Tye, Michael. "Phenomenal Externalism, Lolita, and the Planet Xenon." In *Qualia and Mental Causation in a Physical World: Themes from the Philosophy of Jaegwon Kim*. Edited by Terence Horgan, Marcelo Sabates, and David Sosa, 190–208. Cambridge: Cambridge University Press, 2015.

14

# New Visions of the Old West

## AI, Self, and Other in *Westworld*

Christopher Minz

HBO's *Westworld* (2016– ), a show indebted to cinematic Westerns, depicts a future in which a digitally designed and operated amusement park is populated by ultra-realistic AI robot hosts. The hosts become victimized by wealthy human guests for the visitors' pleasure. *Westworld's* narrative focuses on two android hosts, Dolores Abernathy (Evan Rachel Wood) and Maeve Millay (Thandie Newton), who, as they gain sentience, recall their traumas inflicted by the guests; through accrued self-knowledge, they override their maker's programs, which, through periodic erasure, were designed to prevent retaliation. Two men in search of identities, William (Jimmi Simpson) and The Man in Black (Ed Harris), repress Dolores's, Maeve's, and the hosts' uprisings. The activities of the AI specialists, capitalist bureaucrats and scientists who maintain the park's structures punctuate the action. The first season erupts in a cliffhanger—who or what will survive?

Central to the show is the notion that if we build AI without better knowledge of humanity, the outcomes look bleak. *Westworld* critiques our moment through playful destabilization and a repurposing of historical myths. The park guests feel compelled to embody the west's mythic superheroes and to act out their unsanctioned desires—rapacious, murderous, and others—normally blunted by society. These projected selves conflict with other states that can be understood through a Lacanian framework. The guests are already split subjects who've undergone the mirror phase (the only partially ambulant, spittle-clad child, who witnesses an imagined better self in a mirror and can never incorporate it). The hosts, too, are also stunted. Not given rights or history and bequeathed only the most schematic of identities, they are unable to act as reflections for the guests. Both AI and humans are rendered as disfigured and multiplied.

*Westworld's* vision of the West, drawn from the cinematic Western, also reveals a smudge on the mirror, a blemish that reminds viewers of who we are not, rather than who we imagine ourselves to be. These at first small disfigurations multiply and stain the *mise-en-scène*, and eventually, the park as a whole. The Lacanian Real, first rendered subtly—obdurate flies, disembodied voices, and disemboweled bodies—crescendos into an all-encompassing massacre.

In this chapter, I will reflect on *Westworld's* first season, focusing on the significance of this narrative outburst, but I'll also tie these disruptions to AI's dystopian potentials in late capitalism. According to Marx, even in the first stages of capitalism, people become more object-like, material rendered solely for use, and objects become more uncannily animate.[1] But we're now also in a period of late capitalism. As Fredric Jameson has observed, in this epoch, space and time collapse into a protracted present.[2] History and individual acts no longer matter. But *Westworld's* corporate leaders are fine with this. All they desire from the park is data, culled from the guests' and hosts' experiences, to be used for financial exploitation.

Not only *Westworld*, but also earlier Westerns like *D'où Viens-tu . . . Johnny?* (Nöel Howard, 1963) and *Once Upon a Time in the West* (Sergio Leone, 1968) portrayed a desire for an intimate engagement with the mythical West. Like these films' earlier characters, *Westworld's* seek fabricated, idealized selves. *Westworld* differs from these films, however, through the ways malevolence and repression fully saturate its texture. Nevertheless, all of these films attempt forms of mythmaking. Mythology manufactures (or in some cases re-manufactures) a potent set of aesthetic images and stories in which a contemporary society can retroactively found a social order. The Western functions as a type of cultural mirror stage for the American social self.

## Visiting *Westworld*—Those Small Remainders

At the end of the show's second episode, co-founder and mastermind Robert Ford (Anthony Hopkins) oversees an elaborate corporate unveiling of the park's next story line: "Odyssey on Red River." The sappy, overwrought advertisement,

---

[1] Karl Marx, Capital Vol.1 (London & New York: Penguin Classics, 1990), 150.
[2] Fredric Jameson, *Postmodernism or the cultural Logic of Late Capitalism* (Durham, NC: Duke University Press, 1991), 16–19.

promising "titillation" and "horror," repulses him, much to the chagrin of the park's corporate owners. In a sense, Ford kills the capitalists' narrative of the West. The scene alludes to an earlier Ford, who murdered outlaw Jesse James. *Westworld* ponders immersive engagements with Western mythologies: the Fords are analogs—myth-busters headed toward self-destruction.

In the scene, Ford stands among a line-up of deactivated hosts while the agents of capitalism and industry stand behind and above them, appearing as vacant and mechanical as the androids. A speaker warns the park guests that they will "find out who they are." But Ford corrects him, saying, rather, that guests "know themselves." They visit the park to find out who they *could* be. Here we have an allusion to the Lacanian mirror stage—a separation between an ideal and who we are.[3] At this moment, viewers assume they will witness some realization of an ego ideal, a most likely unattainable mode of being. Instead, we'll find, with the ensuing bloodbath, that the hosts' mechanical bodies provide more potential for an ideal life than the guests' human ones.

In *Westworld*, the mirroring process extends to production design, particularly through reflective surfaces. These help to map the show's aesthetics and social relations in a manner evocative of Jameson's "cognitive mapping."[4] The hosts' bodies are rebuilt time and again; they are programmed to be vulnerable, and as such suffer rape, torture, and murder. Built from material modeled on humanity's bone, blood, and viscera, android bodies remain the vanquishable other. As the Man in Black wistfully notes, earlier models were finessed from intricate machinery; the producers switched to human-like materials because "it costs less." The slippage here between humans' and hosts' bodies as commodities is obvious. *Westworld*'s corporate heads prefer to craft products out of inexpensive blood and viscera rather than shiny technologies. And the actual *Westworld* show's producers, too, perhaps to save money, stay with the flesh (*Westworld*'s intensity matches Sam Peckinpah's).

*Westworld* begins with a body. Over a black screen, a disembodied voice intones, "Bring her back online." Lights flicker, revealing a nude, seated woman in harsh shadow, head slumped, with one limp arm to her side and the other in her lap. Her room is comprised of glass, blueish-gray, reflective walls. The camera cuts quickly to a close-up of the android's face (Dolores'), cheek marred by

---

[3] Jacques Lacan, "The Mirror Stage as Formative of the I function (1949)," In *Ecrits*, trans. Bruce Fink (New York: Norton, 2006), 76.
[4] Jameson, *Postmodernism*, 52.

bruises—she appears human. As the camera fills the frame, a fly lands on her forehead and crawls across her face. As a voiceover queries her perceptions of time and space, the fly steps across her open, unflinching eyeball. The fly suggests something monstrous, inhuman, and refracted (reminiscent of Freud's descriptions in "The Uncanny," which drew from a German folk tale).[5] Dylan Trigg, discussing the phenomenology of the unhuman, claims that "the distinction of the unhuman is that it does not negate humanity, even though in experiential terms it may be felt as a force of opposition."[6] A similar example occurs early in *Westworld*'s second episode, when William—one of the guests—asks a woman if she's real. She responds, "If you can't tell, does it matter?" Whether or not William can tell, his unconscious mistrust may matter.[7] *Westworld* later attempts to clarify that the guests' violence against the hosts is not a defense against the uncanny. Guests indulge in their acts without feeling unsettled by a sense of alienation. Instead, only small, marked remainders indicate that something is amiss in the show's *mise-en-scène*—a speck, a blemish, a fly walking across an eyeball.

The aberrant eye recurs throughout *Westworld*, reminding viewers of a Jacob's ladder, a kind of staircase of dreams, built through reconstituted, mythic filmic images (for example, Luis Buñuel's razor slashing an eye against the eclipse of the moon in his seminal 1929 surrealist film *Un Chien Andalou*). These images of eyes are intimate and horrible, signs of humanity's vulnerability. In Stephen King's *Danse Macabre* (1981), he states: "we all understand that eyes are the most vulnerable of our sensory organs ... and they are (ick!) soft. Maybe that's the worst ..."[8] Carol Clover adds: "horror privileges eyes because, more crucially than in any other kind of cinema, it is about eyes. More particularly, it is about eyes watching horror."[9] While not primarily grounded in the horror genre (though there's a lot of going down corridors to encounter something monstrous behind doors), *Westworld* uses the eye as a picture window into the horror that unfolds before the viewer, and as a lens into the Western's history and its myths.

---

[5] In Freud's 1919 essay "The Uncanny" he discusses the folk-tale of The Sand-Man, who tears out children's eyes.
[6] Dylan Trigg, *The Thing: A Phenomenology of Horror* (Winchester & Washington: Zero Books, 2014), 6.
[7] In fact, as we suspect, the woman is indeed a host. And is seen, in the show's standard fashion, murdered later in the show in the park before being rebuilt again, likely for the "n"th time.
[8] Stephen King, *Danse Macabre* (New York: Everest House, 1981), 118.
[9] Carol J. Clover, "The Eye of Horror," in *Viewing Positions: Ways of Seeing Film*, ed. Linda Williams (New Brunswick: Rutgers University Press, 1997), 185.

One of *Westworld*'s most horrific moments occurs after Dolores's family is killed, and the Man in Black drags her away to assault her. Her programmed beau Teddy attempts to defend her, but he is shot and lies dying as the camera zooms in on his eye. This shot mirrors Dolores's first scene; both frame eyeballs in close-up. In Teddy's scene, we see, reflected in his pupil, the barn door's light as it cuts through the night's darkness, as the Man in Black drags her into the building. The camera's advance on Teddy's dying, open eye recalls Marion Crane's in Hitchcock's *Psycho* (1960). As Dolores struggles against the Man in Black, she is framed doubly: in Teddy's quivering pupil and in the camera lens' aperture. Then, the door slams shut, leaving everything in darkness. This moment re-creates John Ford's *The Searchers*'s (1956) final one. Here, John Wayne's Ethan Edwards forlornly walks into the distance, as the reunited family remains within the blacked-out cabin. *Westworld* subverts Ford: in *Westworld*, the homestead's barn provides the true horror.

While *Westworld*'s eyes provide windows to gaze upon the Wild West's horrors, the eyes' blemishes also comment on the digital age. While Dolores' unblinking eye signifies her inhumanity, the fly creates a buzzing remainder that continues as a thread. The hosts keep experiencing small glitches, confusing new program insertions called "reveries," which are intended to give them just enough character depth to engage the guests, but not enough to allow them to remember the atrocities they have endured. The hosts' minor returns of the repressed (in this case by AI programming) cause malfunctions and programmatic breaks that unnerve the guests. As with a psychoanalyst's witnessing of the return of the repressed, viewers observe blips or blemishes rather than a direct line from trauma to eruption. The fly works as a blot on the programmatic landscape of the hosts' behavior, standing in for the reveries' trace of underlying structural damage.

The fly is also significant for its intertextual references: Sergio Leone's classic *Once Upon a Time in the West* (1968) opens with a cacophony of diegetic sounds, including creaks and footfalls on the train station's floorboards, but a fly's buzz cuts through this noisy soundscape. In Hitchcock's *Psycho*, a fly lands on Norman Bates' presumable mother's hand, and she explains that she wouldn't hurt a fly. The mother, of course, is Norman in drag. The two comprise an uncanny duality, with external appearance and internal experience blurred. The more recent show *Breaking Bad* (AMC 2008–2013) also uses the fly to represent glitches, this time tied to methamphetamine production. Theorist Angelo Restivo urges us not to see the flies' appearances as mere pastiche, but rather to attend to "the formal equivalent of the fly itself... the *mise-en-scène* [provides] a surplus (or remainder) that *keeps returning* in other guises ... the repeated restaging of images from other films adds another, spectral dimension to the images."[10] To ensure the safety of the guests, *Westworld*'s hosts are programmed to not hurt any living things, including flies, which subsequently circle about. When hints of Dolores's sentience surface later, we most register them when we see her swat away a fly, as any cognizant human would. This sharp and vital moment signals a severing not only of her programming, but also of *Westworld*'s direct connection to its own cinematic past, giving this imagery a defiant spectral quality.

## Mechanical and Fleshly Bodies

From the particular of the blemished eye, let's zoom out and briefly examine bodies as fragments or wholes. *Westworld* posits a disconnect between flesh and voice, making the body's vulnerability evident. *Westworld* explores robot cognition through splitting the voice into two. In episode 2, when Maeve awakes during an operation on her supposedly offline (sleep-mode) self, we witness a disjuncture between her fleshly body and inhuman flesh. Maeve awakes, her abdomen sliced open. Two technicians, who had been trying to remove a bullet, appear above her, seemingly like horrific hallucinations, for Maeve's only been programmed to experience a nineteenth-century west. She drops from the table, all the while holding on to her belly's manufactured guts. As she stumbles through corridors, her nude body becomes mirrored in the abandoned android bodies behind glass walls, tossed about randomly. The scene culminates in a moment of bodily horror. Maeve stops before a large, plexiglass square room where android bodies, covered in blood, are hosed down. Maeve's nude body, alive and alert, disrupts the hosts' flesh. She is meant to be programmed as they are, but a short-circuit has caused her "memories" of past lives to give her body a new inflection, a disruptive meaning. As Angelo Restivo explains about Pasolini's films, "it is the presence of the body ... that introduces the disruption of the particular into the attempted totalization."[11] In *Westworld*'s sequence, Maeve's messy body disrupts the Western's idealized self, as well as the vision promised to Westworld's guests.

---

[10] Angelo Restivo, *Breaking Bad and Cinematic Television* (Durham, NC: Duke University Press, 2019), 19.
[11] Angelo Restivo, *The Cinema of Economic Miracles* (Durham, NC: Duke University Press, 2002), 79.

The hosts' programming resembles psychoanalysis's symbolic order: one must act and adhere, or be seen as glitching or psychotic. Yet the body remains the conduit through which the Real (or unconscious) flickers and erupts. This is most true in parapraxes and dreams. Maeve's body and memories, much like our parapraxes and dreams, surface as the resistant repressed.

We experience *Westworld*'s alien flesh in another surgical scene, this time from episode 3. As the technician works offline on a host, Ford looks on. The host's body is covered by a medical sheet. Ford then tears it away, exposing an elderly man's body. He scolds the technician: "it doesn't get cold; it doesn't feel shame!" He next picks up a scalpel and methodically slices the host's face from temple to cheek bone; "It doesn't feel a solitary thing that we haven't told it to." But the technician clearly remains disconcerted, because the flesh resembles his. Unconsciously, the staff member can't repress a reflex of horror. Yet Maeve's marred flesh and memories facilitate her rebellion. Maeve repeatedly kills herself to return to the surgeons, whom she hopes she can manipulate into helping her. She rejects the pursuit of the ideal-ego offered by bodily integrity. Unlike the guests, who covet passage to the mythological Wild West in search of an idealized identity, she seeks a route to autonomy through abnegation and otherness. Her different frontier hovers beyond the boundaries of her programming.

On the one hand, Maeve, even if her consciousness is unhuman, finds comfort in the emanations of her manufactured flesh. Trigg, echoing Merleau-Ponty, argues that flesh "designates an element anterior to experience yet at the same time implicated in experience."[12] In a sense, we never experience our own flesh as such, but only in relation to a sensation or affect. On the other hand, the Man in Black, a major stockholder in the Westworld park, cannot distinguish between his flesh and that of the hosts. According to him, that the hosts cannot severely injure or kill the guests represents a slanted game, a relation out of sorts. He wants the unhuman flesh to allow him to renew his own relation to his flesh. Doing so might enable a world wherein the androids can fight back and re-administer the world at large, and offer a wedge to his own power. The Man in Black senses that his capitalistic force only matters if he can exert power over something. In the first season's final moments, he watches as formerly deactivated nude hosts approach; then, suddenly, a gunshot rings out and a bloody wound appears on his arm. The camera draws in on his face as his surprise breaks into a smile.

---

[12] Trigg, *The Thing*, 125.

Although *Westworld*'s depictions of bodily conflict take center stage, the voice, especially in voiceover, conveys the hosts' emergent sentience and cognition. When the Man in Black searches for the maze's center—which he believes contains the park's key—the hosts, acting out of character, tell him, "It's not for you." (Westworld's androids have a special relation to the maze, because its imprint under their skulls leads them to greater cognition). The debates between Dolores and Bernard (a staff member who turns out to be a host), reveal a dialogue between Dolores and herself. Here, we hear two voices, one embodied, the other acousmatic. Slavoj Žižek claims that the voice, which interrupts the flow of the visual, "functions as a strange body which smears the innocent surface of the picture, a ghost-like apparition which can never be pinned to a visual object."[13]

In Dolores's poetic laments, her throat and lips rarely move. Her stillness contributes to a melancholic mood, evocative of Film Noir's voiceovers. It might be through both sound and image that *Westworld* takes on more genres than the Western, perhaps even much of cinematic history, signaled through momentary glitches, effects, and signs. Dolores's semi-disembodied voiceover reverberates against the hosts' conversations with unverifiable, offscreen characters. In truth, they are speaking to themselves, solidifying their notions of self-awareness. If we accept that the disembodied voice, whether it comes from without or within the hosts, stands for cognitive self-awareness, then we must assume it is subject to frailty. As Joan Copjec explains, "the disembodied voice, which conveys knowledge and power, and the embodied voice, which conveys the limitation of both, is underwritten by a simple opposition between the universal and the particular, the latter being conceived as that which ruins the possibility of the former."[14] Thus, the voice that signals cognition in the hosts also signifies another disruption. The hosts are bound by a body and a voice that must be kept in check. As Žižek states, "voice functions as a foreign body, as a kind of parasite introducing a radical split."[15]

Westworld's disembodied voices function as what Copjec calls "intemporal voices." She sees these voices as timeless, within yet beyond the scope of death. She argues that "this is not to deny that the voices are associated with death, but to note that this death brings no expiry; rather in them death persists. The voices

---

[13] Slavoj Žižek, *Enjoy your Symptom!* (New York & London: Routledge, 1992), 1.
[14] Joan Copjec, "The Phenomenal Nonphenomenal: Private Space in Film Noir," In *Shades of Noir*, ed. Joan Copjec (New York: Verso, 1993), 184.
[15] Žižek, *Enjoy Your Symptom!*, 2.

bear the burden of a living death, a kind of inexhaustible suffering."[16] We discover that the voice Dolores originally hears, before it is synthesized into her own, symbolizes cognition. The acousmatic voice belongs to Arnold Weber, Ford's former business partner. Prescient, Arnold predicted that park guests would torment hosts; to save them, he programmed them to murder each other and him (it's as if God desired that humans kill him). Even after death, Arnold's suicidal voice, now programmed into the hosts, resonates timelessly. The voice creates a cognitive map backward to an origin that is founded on death; it weaves back through the hosts' countless deaths and resets ad infinitum. *Westworld* depicts, via a mirroring of two conversing Doloreses, a reversal of the mirror of an ego and ego ideal. Both dwell in negativity and timelessness.

In a sense, there is another quilting point here that finally ties together the theoretical aspects I've been discussing as regards *Westworld*'s aesthetic implications, negativity, and timelessness: the diegetic music emanating from the player piano. The player piano provides a strange dissonance within the park's constructed world between the obviousness of having a mechanical piano in a setting comprised of artificial life, and the immediate gap within the diegesis of the piano player which would have been a staple of the cinematic Western saloon. This lack is part and parcel of the creeping capitalist tendency of the park itself, and lends an air of labor dissolution, as the player piano stands in for a general mechanization removing the need for a proletarian sector.[17] In a way we begin to see the layered consequences that the park itself hints at. In order to maintain the appearance of the cinematic mythological West (itself a commodity of the Hollywood industry), the particular sounds of the saloon upright are needed, while the same moment we are refused the visual of the human element of the performance of the music itself—commodity without the messy complexity of human labor.

The intricacies move beyond the gross capitalist implications and into the psychoanalytic and postmodern with the introduction of the specific music that is played by the player piano. The piano churns out contemporary (20th–21st century) pop music, mixed in with some more relatively arthouse indie songs performed in the Western saloon style. Most prevalent in the first season are indie rock icons Radiohead. The use of Radiohead's music provides a certain wink and nod to a particular group of iconoclasts that the band has always

---

[16] Copjec, "Phenomenal Nonphenomenal," 185.
[17] This recalls Kurt Vonnegut's first novel *Player Piano* (1952), a dystopian world of mechanization.

claimed as fans, and also lends credence to the program's thematic and aesthetic elements via the insinuation of Radiohead's common themes of alienation and plasticized aesthetic anxiety. However, what the use of these contemporary songs provides is a lynchpin of temporal and spatial incongruity and flattening, as described by Jameson's understanding of Postmodernism discussed above. We hear the recognizable sound of the saloon piano, and we are immediately aurally transposed into the Western, yet the songs are often recognizable as contemporary, so as to keep us rooted in the present. All this while watching a program that takes place well into the future. The temporal moment becomes, as Anne Friedberg describes it, "a ready panoply of other temporal moments, the *not-now* in the guise of the *now*."[18] We hear the past performing the present within the diegesis of the future. However, to tie this to the Lacanian and the mirror stage, we create an aural "will have been." For Lacan the mirror stage (as most aspects of psychoanalytic development) is given meaning retroactively, he argues the ideal-ego observed in the mirror "symbolizes the *I*'s mental permanence, at the same time as it prefigures its alienating destination."[19] Hence, the ideal-ego is only experienced fully after the fact and as a figure that in memory *will have been*. This if how the temporal and spatial flattening functions in *Westworld*'s diegetic music via the player piano. The West as constructed plays a music that is something that will have been once the diegetic world of the future occurs, and that these particular pop songs occur within a space in which they would have not existed as of yet, the viewer (or indeed the guests of the park) are captured within this knot of spatial-temporal compression.

## Cognitive Maps, Late Capitalism, and Freefloating Postmodernism

Thus far, this essay has focused on particulars—the hosts' bodies, affects, and voices. But these hosts also exist in a slowly unraveling system. Westworld attempts to construct a fully rendered, immersive environment under the auspices of late capitalism. All the bodies, human and android, exist aesthetically and materially within its diegesis. Westworld posits a question left relatively unexplored: what does it mean to situate visions of AI futures within a frontier

---

[18] Freidberg, "Cinema and the Postmodern Condition," 74. Italics in original.
[19] Lacan, "The Mirror Stage," 76.

mythology? Westworld's temporality isn't specified, but the world outside the park partakes of today's slick, anonymous corporate stylistics—cool whiteness, smooth walls, bright, wide screens. Yet perhaps whatever's exterior to Westworld is a façade. We never see the entrances, laboratories, or surveillance facilities necessary to maintain the park's order. Westworld's temporality shifts perpetually; the narrative manages multiple timelines withheld, except in part, from viewers, until they collide, when the past invades the present. The narrative of William, initially portrayed as a reluctant guest to the park, exists thirty years prior to Dolores and Maeve's awakening.

In one of the show's striking narrative reveals, the earlier, younger, good-natured William morphs into the cruel, obsessed Man in Black of the present. Jameson notes that one of late capitalism's key elements, its postmodernity, is a flattening of time that creates an information flow so insistent that humans are forced to exist in a perpetual present: "we now inhabit the synchronic rather than the diachronic ... our psychic experience, our cultural languages, are today dominated by categories of space rather than categories of time."[20] Atemporality is a phenomenon both of the Westworld theme park and the show itself. Guests continually step out of time to exist in a place of pure, simulated space. As Dolores poignantly laments in episode 7, she does not want to look forward or back, desiring to "just be in the moment I'm in." Ultimately, she reaches cognition by altering her spatial relation to the park, and to the temporality of the Western. This act, extending both her and our experiential confines, is radical.

Dolores wishes to fold her past traumas and her present experience into a temporal foundation from which she can assert linearity. Yet her desires work against the logic of late capitalism, in which commodity culture manipulates and re-molds the past in service of a sense of play and perpetual present. As Westworld's Wild West theme park creates an internal spectacle, the past loses all sense of use value. Jameson argues that postmodern society in late capitalism is "a society bereft of all historicity, one whose putative past is little more than a set of dusty spectacles ... the past as 'referent' finds itself gradually bracketed, and then effaced altogether, leaving us with nothing but texts."[21]

The frontier myth becomes a commodity itself, a playground for the obscenely wealthy. Within the park, capitalist structures reassert themselves. The hosts function as the proletariat, and the capital stays outside of the park. A daily ticket

[20] Jameson, *Postmodernism*, 16.
[21] Ibid., 18.

to visit Westworld costs $40,000 and privileges are capped at two weeks. While we viewers can watch the Westerns, the wealthy, to some extent, live the myth. The true West becomes long forgotten, subsumed by screen memories.

Inside Westworld, women serve primarily sexual functions, and men act as buddies or adversaries. Although capital appears to transcend gender (for example, a female guest flirts with a female prostitute), encounters remain instrumental. The hosts often resemble pictures on a tablet, a menu. Anne Friedberg suggests that the postmodern era has wrought a new type of Gaze, the "mobilized virtual gaze." Drawing from Walter Benjamin's concept of the *flâneur*, she notes our gaze "travels in an imaginary *flânerie* through an imaginary elsewhere and an imaginary 'elsewhen'"[22] The park's west becomes a set of menu items, an insidious, amplified version of Tinder. While the body has always been a commodity under capitalism, in *Westworld,* hosts' bodies, as potential proletarian working bodies, become devoid even of this hint of use value. In Jameson's words, this world is one where "exchange value has been generalized to the point at which the very memory of use value is effaced."[23]

In *Westworld*'s first season's final moments, Ford gives a melancholy, self-conscious speech. The stockholders have forced him to resign. Frustrated by his elaborate plans, they put an end to his seemingly old-fashioned methods for building and engaging with hosts. The stockholders want a cheaper spectacle. Ford knows he has but one last power play. As he finishes his speech, Dolores draws a pistol and shoots him in the back of his head before the horrified, wealthy crowd. The hosts begin to fire on guests, and the Man in Black, to his joy, is injured. A rebellion seems inevitable—the possibility of a new frontier.

While the frontier myth has remained the most monumental and static in American culture, *Westworld* contemplates its subversion: what would happen if historically dehumanized peoples (represented by the hosts) found a way to make that history more dynamic? The flies, eyes, disembodied voices, and bodies are all products of capital, subject to the same torments as the working class presumably outside the park. We can assume both are preyed upon by the wealthy, capitalist guests. Yet programmed into capital is an endgame, and when we allow the mythology, the West, to rear its head into our present as such, then we face the same bullets as Ford, the man who shot Jesse James, and Ford, the founder of Westworld.

---

[22] Anne Friedberg, "Cinema and the Postmodern Condition," in *Viewing Positions: Ways of Seeing Film*, ed. Linda Williams (New Brunswick: Rutgers University Press, 1997), 60.
[23] Jameson, *Postmodernism*, 18.

# Bibliography

Clover, Carol J. "The Eye of Horror." In *Viewing Positions: Ways of Seeing Film*. Edited by Linda Williams, 184–230. New Brunswick: Rutgers University Press, 1997.

Copjec, Joan. "The Phenomenal Nonphenomenal: Private Space in Film Noir." In *Shades of Noir*. Edited by Joan Copjec, 167–97. London & New York: Verso, 1993.

Davidson, Collette. "French rock star played on American persona." *The Christian Science Monitor*. December 11, 2017. https://www.csmonitor.com/World/Europe/2017/1211/French-rock-star-played-on-American-persona. (Accessed April 2, 2020).

Freud, Sigmund. "Screen Memories (1899)." In *The Standard Edition of the Complete Works of Sigmund Freud. Volume III: Early Psychoanalytic Publications*. Translated by James Strachey, 299–322. London: Vintage Books, 2001.

Freud, Sigmund. "The Uncanny." In *The Standard Edition of the Complete Psychological Works of Sigmund Freud, Volume XVII (1917–1919): An Infantile Neurosis and Other Works*. Translated by James Strachey, 217–56. London: The Hogarth Press, 1955.

Friedberg, Anne. "Cinema and the Postmodern Condition." In *Viewing Positions: Ways of Seeing Film*. Edited by Linda Williams, 59–83. New Brunswick: Rutgers University Press, 1997.

Jameson, Fredric. *Postmodernism: or the Cultural Logic of Late Capitalism*. Durham, NC: Duke University Press, 1991.

Lacan, Jacques. "Aggressiveness in Psychoanalysis (1948)." In *Ecrits*. Translated by Bruce Fink, 82–101. New York: Norton, 2006.

Lacan, Jacques. "The Mirror Stage as Formative of the I function (1949)." In *Ecrits*. Translated by Bruce Fink, 75–81. New York: Norton, 2006.

Marx, Karl. *Capital Vol.1, 150*. London & New York: Penguin Classics, 1990.

Restivo, Angelo. *Breaking Bad and Cinematic Television*. Durham, NC: Duke University Press, 2019.

Restivo, Angelo. *The Cinema of Economic Miracles: Visuality and Modernization in the Italian Art Film*. Durham, NC: Duke University Press, 2002.

Trigg, Dylan. *The Thing: A Phenomenology of Horror*. Winchester & Washington: Zero Books, 2014.

Zizek, Slavoj. *Enjoy Your Symtpom! Jacques Lacan in Hollywood and Out*. New York & London: Routledge, 2007.

15

# Scoring Music for *Westworld* Then and Now

## A Cognitive Perspective

Annabel J. Cohen

### Introduction: Two *Westworlds*, Two Scores

Since its 2016 debut, *Westworld*, the sci-fi–Western television/HBO series, has engaged audiences in a fictional world that questions the borders between past and future, freedom and determinism. This chapter explores *Westworld*'s musical dimension—the ways music draws a viewer into the show's world and helps explore its questions. The series takes its inspiration from Michael Crichton's 1973 film *Westworld*. Celebrated soundtrack composer and film educator, Fred Karlin, scored this early film, and his soundtrack is an excellent example of prototypical scoring techniques from the 1970s. Award-winning composer Ramin Djawadi scored the 2019 *Westworld* HBO series, as well as the popular *Game of Thrones* (2011–2019) and *Pacific Rim* (Guillermo del Toro 2013). In this chapter, I review scientific studies that demonstrate film music's ability to direct viewers' experiences in powerful and novel ways. I also unpack Karlin's work in the context of the early sci-fi–Western, as well as discuss Djawadi's style to create a more nuanced approach. Later, I turn to my own cognitive model to help explain the composer's power, the listener's susceptibility, as well as contemporary music's ability to convey *Westworld*'s AI. I close with a discussion of implications for other cybermedia.

### Harnessing the Brain's Representations of Musical Meaning and Emotion

One of music soundtracks' primary goals is to convey a director's vision. Unlike dialogue and sound effects, music resides typically outside the diegesis—it draws viewers into compelling fictional worlds.[1] Music also engages audiences emotionally in the film.[2] And by employing conventional codes, music establishes contexts of time and place. That this happens confirms the brain's interpretive skills with music across media.

Composers, music supervisors, film directors, and audiences bring tacit knowledge of music to films and media. Psychological research reveals that listeners prefer certain styles of music, can identify a piece's decade of popularity, and recall personal experiences associated with particular musical genres.[3] Participants' abilities to engage with these processes reveal complex knowledge structures established over years of musical exposure. Film composers exploit this familiarity with music in their scores.

Research shows that musical features convey meanings to listeners. For example, consider a melody's pitch direction. A melodic line that ascends in pitch will be judged by most individuals as conveying a happy meaning, and, conversely, a descending melodic line will be judged as conveying sadness.[4] Going a step further, functional magnetic resonance imaging (fMRI) of the brain during ascending and descending melodies has revealed three locations responsive to pitch direction.[5] One contributing factor to this brain activity, suggested by researchers of music, is the alignment of melodic contour with action. An ascending melody may be associated with one kind of motion, such as climbing upward, and a descending melodic line, with another, such as walking downward.[6]

---

[1] Claudia Gorbman, *Unheard Melodies: Narrative Film Music* (Bloomington, IN: Indiana University Press, 1987).

[2] Annabel J. Cohen, "Music as a Source of Emotion in Film," in *The Oxford Handbook of Music and Emotion,* eds. Patrick N. Juslin and John A. Sloboda (Oxford: Oxford University Press, 2010), 879–908.

[3] Carol Lynne Krumhansl, "Listening Niches Across a Century of Popular Music," *Frontiers in Psychology*, (March 8, 2017), at https://www.frontiersin.org/articles/10.3389/fpsyg.2017.00431/full (accessed February 21, 2021).

[4] William G. Collier and Timothy L Hubbard, "Musical Scales and Evaluations of Happiness and Awkwardness: Effects of Pitch, Direction, and Scale Mode," *American Journal of Psychology* (2001): 355–75.

[5] In an fMRI scanner, participants were tasked with click a button every time they heard a bidirectional, rather than unidirectional sequence. They then participated in a study in which they rated each sequence on a 7-point scale with *sad* and *happy* end points. Yune-Sang Lee et al., "Investigation of Melodic Contour Processing in the Brain Using Multivariate Pattern-Based FMRI," *NeuroImage, U.S. National Library of Medicine,* 57, no. 1 (July 2011): 293–300.

[6] A simple example of this association is provided in an early computer game which played an ascending 3-note pattern each time an avatar picked up an object and conversely a descending pattern when putting it down. Another game similarly paired ascending and descending 3-note patterns with going up and down stairs. See pp. 323–25 in Neil Lerner, "The Origins of Musical Style in Video Games 1977—1983," in *The Oxford Handbook of Film Music Studies*, ed. David Neumeyer (Oxford: Oxford University Press, 2014), 319–47.

Other neuroimaging research suggests that the same brain areas are activated whether one moves a body part (e.g., leg or tongue) or reads related verbal concepts (e.g., the words "leg," "face," respectively). The finding demonstrates that a listener's conceptual understanding is grounded in fundamental interactions with the environment, resulting in *embodied* meaning—be it verbal, musical, or motoric.[7]

Similar brain mechanisms for both perceived and experienced actions have been explained by studies of mirror neurons (in non-human primates' brains) which respond during an activity (such as picking up an object) or watching someone else carry out the activity (such as seeing someone else pick up the object).[8] An analogous human mirror-neuron system helps describe an audience's experience of performers' activities that unfold through media. In accordance with a mirror system, watching a film produces neural activation similar to that which would occur when carrying out the activity oneself, as if the viewer-listener were the actor on the screen. The concept of mirror system is not limited to a visual depiction of action—listening to music may engage the motor system as if the viewer were the musician performing.[9] Notions of musical embodiment offer further importance to the music presented in a film or TV program.

In addition to meanings conveyed by melodic direction, music conveys pleasantness and unpleasantness through consonance and dissonance. In one experiment, the same melody was harmonized in six different ways to represent six levels of dissonance.[10] Listeners heard these melodies while in a PET brain scanner that measured the amount of blood flow in various areas of the brain. The results showed increased regional cerebral blood flow in the paralimbic and neocortical areas for melodies with increasing or decreasing consonance. These areas are

---

[7] Olaf Hauk, Ingrid Johnsrude, and Friedemann Pulvermüller, "Somatotopic Representation of Action Words in Human Motor and Premotor Cortex," *Neuron* 41, no. 2 (2004): 301–7. Olaf Hauk, Ingrid Johnsrude, and Friedemann Pulvermüller, "Somatotopic Representation of Action Words in Human Motor and Premotor Cortex," *Neuron* 41, no. 2 (2004): 301–7. Olaf Hauk, "What Does it Mean? A Review of the Neuroscientific Evidence for Embodied Lexical Semantics," in *Neurobiology of Language*, ed. Gregory Hickok and Steven L. Small (London: Academic Press, 2016), 777–88; George Lakoff and Mark Johnson, *Metaphors We Live by* (Chicago, Ill: University of Chicago Press, 2003). See also Arnie Cox, *Music and Embodied Cognition: Listening, Moving, Feeling, and Thinking* (Bloomington: Indiana University Press, 2017).
[8] Giacomo Rizzolatti and Laila Craighero, "The Mirror-Neuron System," *Annual Review of Neuroscience* 27, no. 1 (2004): 169–92.
[9] Katie Overy and Istvan Molnar-Szakacs, "Being Together in Time: Musical Experience and the Mirror Neuron System," *Music Perception* 26, no. 5 (January 2009): 489–504.
[10] Anne J. Blood et al., "Emotional Responses to Pleasant and Unpleasant Music Correlate with Activity in Paralimbic Brain Regions," *Nature Neuroscience* 2, no. 4 (1999): 382–87.

typically engaged in emotional responses in general. In addition, some regions responded only to increasing dissonance, and others to greater consonance.

Many neuroscientific studies reveal that different properties of music activate different parts of the brain; however, few such studies have focused on the brain activity arising when listening to music while watching a film. One rare experiment using fMRI showed activation of the amygdala (a center of emotional processing) when brief emotionally neutral film were presented with emotional music. No amygdala activity, however, occurred when either the film or music were presented alone.[11] This study suggests that music which elicits emotions causes the brain to search for a source of its meaning.[12] Thus, composers can use emotionally laden music that encourages a viewer to discover more about a film character's motivation, action, or relation to plot. As one example from the *Westworld* film, an unsettling electronic sound with an irregular yet relentless beat accompanies the robot Gunslinger as it chases an innocent human. Although the music causes an unsettled feeling, audiences would tend to attribute the robot to the feeling's source.

## Film Composers, Resources, and Techniques

Film and TV composers draw on both external and internal resources, as well as multiple techniques for their craft. External resources, limited by budgets, include support from artisans and current technologies. Internal resources include composers' knowledge of styles and pieces of music, retrievable for matching an emotion or motivation for a filmic action. Internal resources also include imagination and creativity. Djawadi, for example, created the theme for *Westworld* having learned of its context before seeing any of the filming.[13] Composers, music editors, and directors know how to exploit an audience's musical knowledge. At the same time, composers are educators of audiences, teaching new musical meanings through exposure to novel, or sometimes older yet unfamiliar, music in a new context. Decision makers (directors, music editors,

---

[11] Eran Eldar, Ori Ganor, Roee Admon, Avraham Bleich, and Talma Hendler, "Feeling the Real World: Limbic Response to Music Depends on Related Content," *Cerebral Cortex*, 17, 12 (2007): 2828–40.
[12] Nicholas Cook, *Analysing Musical Multimedia* (Oxford: Clarendon Press, 1998).
[13] Mike Hilleary et al., "'Westworld' Composer Ramin Djawadi on Why Those Radiohead Covers Keep Coming," *Pitchfork*, November 18, 2016, https://pitchfork.com/thepitch/1370-westworld-composer-ramin-djawadi-on-why-those-radiohead-covers-keep-coming/ (accessed February 9, 2021).

and composers) rely on professional experience and training to intuit what music is needed and when. In a sense, they are mind readers, predicting how the audience will hear, interpret, and remember the film and its soundtrack.

Composers also draw on their proficiency with instruments as an internal resource. They may play instruments or be skilled at computer-based digital audio composition and production. Karlin was a professional trumpet performer, for instance. He could play several orchestral instruments and mastered the use of synthesizers and early electronic instruments. Djawadi's primary instrument is guitar, but he studied organ from the age of four to eleven years. His early experience with the organ acquainted him with a wide range of timbres and experiences of musical embodiment.

One technique often employed in creating the music soundtrack is a *temp track*—often composed of swatches of prerecorded music slotted into a rough edit of the film—which cues a director where to place scored music, and within which genre and style. This temp track can serve either the director and/or the composer at early postproduction screenings and through final editing.[14] Westworld's Jonathan Nolan (co-creator) and Christopher Kaller (supervising music editor) created a temp track for the series' opening and shared it with Djawadi.[15] Often, and in part for budgetary reasons, the supervising music editor and composer begin work only after completion of filming and editing.

The use of recurring *leitmotivs* (from the German for "leading motif") is an important scoring technique. In the simplest case, a leitmotiv links a musical theme or motif to a character or plot point. A film score may include numerous leitmotivs, their profiles chosen by the composer, though sometimes with the input of the film director and supervising music editor. Filmmakers assume that viewers retain information about the motif that first accompanies an incident or protagonist, and that each reoccurrence will bring to mind those first instances. Such hypothesized one-trial learning should be a mental challenge, but research by Marilyn Boltz provides some support for this.[16] Her study showed better memory for music that was cued by its accompanying visual scene as compared to a condition in which the music had no visual cue. The effect was specific,

---

[14] Fred Karlin and Rayburn Wright, *On the Track: A Guide to Contemporary Film Scoring*, 2nd ed. (New York, NY: Taylor & Francis, 2004), 28.
[15] Kingsley Marshall, "'Music as a Source of Narrative in HBO's *Westworld*," in *Reading Westworld*, ed. Alex Goody (Cham, Switzerland: Palgrave Macmillan, 2019), 97–118.
[16] Marilyn Boltz, "The Cognitive Processing of Film and Musical Soundtrack," *Memory & Cognition* 32, no. 7 (2004): 1194–205.

however, to situations in which the music and scene had similar meanings. Further research by Berthold Hoeckner and Howard Nusbaum showed that music cued visual scenes more effectively than visual scenes cued music, a phenomenon they refer to as the "*Casablanca* Effect."[17]

As part of the filmmaking team, the composer employs additional techniques to shape an audience's attention. Before synchronized film and sound, early film theorist and psychologist Hugo Münsterberg noted the ways music and film controlled an audience's ordering of concepts and events.[18] He claimed that film was more like music than other art forms, including theatre and photography. Structural and physical aspects of film and TV may, however, change over time. Cognitive psychologist James Cutting and colleagues measured several lower-level features of 160 English-language films released between 1935 and 2010 and corroborated earlier findings of a decline in average shot length.[19] They also found an increase in the amount of motion and a decrease in average luminance (amount of light). They offered a single explanation for these linear trends: "Filmmakers have incrementally tried to exercise more control over the attention of filmgoers."[20] Industry personnel for the *Westworld* series, with its themes of human control and free will, seem especially skilled at directing human attention.

Had the 1973 film *Westworld* been made in the twenty-first century, we might expect novel music which incorporated new styles and technologies. However, as Cutting has pointed out, changes in a film's lower-level features do not mean changes of style or genre at a higher level.[21] Over time, directors and their teams have simply gotten better at directing audience attention. Thus, we may expect similar functions served by music in the twentieth century film and the twenty-first century series *Westworld*, even though more external resources, as well as four additional decades of recorded music, were available to Djawadi.

---

[17] Berthold Hoeckner and Howard Nusbaum, "Music and Memory in Film and Other Multimedia: The Casablanca Effect," in *The Psychology of Music in Multimedia*, ed. Siu-Lan Tan et al. (Oxford: Oxford University Press, 2013), 235–63.
[18] Hugo Münsterberg, *The Photoplay: A Psychological Study* (New York: Dover Publications, 1970). Also published as *The Film: a Psychological Study; the Silent Photoplay in 1916* (New York: Dover Publications, 1970).
[19] James E Cutting et al., "Quicker, Faster, Darker: Changes in Hollywood Film over 75 Years," *i-Perception* 2, no. 6 (2011): 569–576.
[20] Cutting et al., "Quicker, Faster, Daker," 569.
[21] See also James E. Cutting, "Narrative Theory and the Dynamics of Popular Movies," *Psychonomic Bulletin & Review* 23, no. 6 (March 2016): 1713–43. This article reports the analysis of 175 popular movies in 12 studies that show normative aspects of parameters like shot duration and motion, and luminance. These reduced to five stylistic dimensions (music-conversation being one) and were distributed systematically across four acts.

I'll next discuss the music of the 1973 *Westworld* film as well as the music of the series. My cognitive model of how music works in multimedia will help explain *Westworld*'s two different musical approaches.

## Musical Challenges in *Westworld*, Then and Now

*Westworld* is an example of the sci-fi Western. Both genres, Western and science fiction, have different musical styles, and Karlin and Djawadi both had to deal with these two genres in one score. Musicologist Andrew Granade has traced the development of the two genres.[22] For the Western, he identifies three stages. The music of early Western films was typically "in a major key with duple meter; featured a bit of syncopation; and stressed simple, chordal accompaniments." Special effects included "triplet rhythms for galloping horses and tremolo strings for moments of mounting tension."[23] These conventions were influenced by Westerns of the silent era and music for staged melodramas of the late nineteenth and early twentieth century.[24] This period also included the singing cowboy movies of Roy Rogers and Gene Autry that used guitar, fiddle, banjo, and string bass, and attracted country and western music fans. The second stage began after 1939 with the John Ford films *Stagecoach* and *Drums Along the Mohawk*. The primary influences were American folk songs, hymns, and patriotic songs, scored with a style made popular in Aaron Copland's Western themed ballets *Billy the Kid* (1938) and *Rodeo* (1942), with perfect fourths rising in parallel motion (violating traditional harmony) and "thin orchestration."[25] The final stage is attributed to the influence of the late Ennio Morricone who added electric guitar, non-Western percussion, Mariachi-style trumpet, and pop stylized songs similar to groups like the Beatles. Morricone created a "totality of sound" that blended the sound effects of the West (e.g., horse hooves, gunshots) with the music track. Two further dichotomies added to the semantic cues. First, tonal music was associated with

---

[22] S. Andrew Granade, "'Some People Call Me the Space Cowboy': Sonic Markers of the Science Fiction Western," 2019, https://www.academia.edu/38629097/_Some_People_Call_Me_the_Space_Cowboy_Sonic_Markers_of_the_Science_Fiction_Western (accessed February 10, 2021). Note: An abbreviated version of this paper appears in *Re-Locating the Sounds of the Western*, ed. Mariana Whitmer and Kendra Preston Leonard (New York, NY: Routledge Press, 2019).
[23] Granade, "Sonic Markers," 7.
[24] Mariana Whitmer, "Melodramatic Music in the Western," *The Journal of Film Music* 5, no.1–2 (2012): 111–12.
[25] Granade, "Sonic Markers," 8.

*good*, and atonal music was associated with *bad*. Secondly, acoustic (orchestral) instruments were also associated with good, and electronic instruments with evil.[26] Granade notes the singability of music of Westerns and the use of syncopated ostinatos against a "rhythmically steady though leaping melody."[27]

Granade identifies only two periods for the development of music for science fiction. The first is characterized by electronic or atypical orchestral instruments, such as the theremin, as used by Bernard Herrmann in the 1950s (e.g., *The Day the Earth Stood Still*, Robert Wise 1951). The second is associated with the lush scores and full symphonic orchestra with "heavy brass" as introduced by John Williams in *Star Wars* (George Lucas 1977).[28]

## Karlin's *Westworld*, The Movie

For *Westworld* (1973), Karlin used the elements of the Western and science fiction traditions available to him, without conventions later introduced by Morricone and Williams. Table 15.1 lists all scenes with music. Gray shading identifies scenes with obvious electronic music; they typically accompany situations involving robots, particularly the Gunslinger. Early in the film, guests Peter and John, having arrived at the park, change into Western clothing, while non-diegetic country hoedown music featuring banjo, guitar, bass, and fiddles confirms the Western context. A trip to the town's saloon adds harmonica and honky-tonk piano, further confirming the Western diegesis. But the encounter with the robot Gunslinger (see table 15.1, 17:12) introduces electronic keyboard and synthesizer consistent with science fiction, which dominates as the film progresses (see table 15.1).

With a one-minute excerpt from *Westworld*'s score, I'll now show some ways Karlin exploits convention and originality to trace moment-by-moment actions within the image to enhance the narrative. Later, I'll analyze some of Djawadi's cues, such that one could claim that while Karlin's work is inventive, Djawadi, drawing on a longer history of film-scoring, older and newer technologies, a different musical upbringing, and the unique director team is more able to raise

---

[26] Quoting Lisa M. Schmidt, "The Popular Avant-Garde: The Paradoxical Tradition of Electronic and Atonal Sounds in Sci-Fi Music Scoring," in *Sounds of the Future: Essays on Music in Science Fiction Film*, ed. Mathew J. Barthowiak (Jefferson: McFarland & Company, Inc., 2010), 24.
[27] Granade, "Sonic Markers," 9.
[28] See also, Philip Hayward, "Sci-Fidelity: Music, Sound and Genre History," in *Off the Planet: Music, Sound and Science Fiction Cinema*, ed. Philip Hayward (Eastleigh, UK: John Libbey Publishing, 2004), 1–29.

Scoring Music for Westworld Then and Now    245

**Table 15.1** Music cues in *Westworld* film showing dichotomy of styles for Western and SciFi aspects. Gray shading highlights use of electronic music for robot (SciFi).

| Time | Scene | Description of music |
|---|---|---|
| 3:00 | Brief opening credits | One synthesized tone; then hovercraft noise and control-tower landing speech to/from pilot |
| 3:30–5:05 | End of hovercraft (1984) journey to Delos | Diegetic Muzak-jazz combo with piano and horns |
| 9:00–9:57 | *Having arrived in Westworld, Peter and John put on Western clothing, chose pistols | Upbeat western hoedown: banjo, guitar, bass, and fiddle |
| 10:15–11:12 | Stagecoach arrival carrying Peter and John to their hotel | Same as above |
| 13:28–17:03 | *On entering the town's saloon | Honky-tonk piano, first non-diegetic; at 14:20 camera pans to pianist. Ends when Peter confronts The Gunslinger, and everyone in the saloon clears out (including the piano player); Same instruments as above plus harmonica and honky-tonk piano |
| 17:12–17:24 | *Dual between Peter and robot, The Gunslinger | Sustained low then high electronic sound of Yamaha A5 keyboard, Arp 2600 synthesizer |
| 18:21–19:05 | Peter and John in their hotel room; Confirm guns will not shoot humans | Honky-tonk piano |
| 19:15–20:00 | Medieval world | Recorders, lute/mandolin (part of diegesis) |
| 20:29–26:09 | Brothel/saloon, ground floor, then upstairs | Honky-tonk piano waltz, banjo; solo piano for upstairs scene |
| 26:10–29:26 | *Street with corpses moved to truck; unload corpses on conveyer belt to laboratory; Reset and reconditioning of robots by technicians (Robot Repair) | "Electronic sounds and jagged dissonances" (p. 17); non–melodic, atonal; fast timbre changes (disrupt melodic grouping of consecutive tones; adds focus to timbre—ends with congruent mechanical high synthetic tone and onset of a light focusing on the eye of a prostrate female robot |
| 32:26 | Shot of Westworld | Banjo |
| 32:30 | Shot of Roman world | Flute |
| 32:33 | Shot of Medieval world | Trumpet |

| Time | Scene | Description of music |
|---|---|---|
| 38:11–40:30 | On street, John gives girl a covered "breakfast" tray for Peter in jail, Girl saunters across street, enters jail; sheriff inspects tray; girl passes tray to Peter; Peter reads note; puts on hat; girl leaves, crosses street; John mounts horse and leads second horse to the jail; jail wall explodes freeing Peter | Jaunty banjo music; sparse orchestration, percussion, xylophone, fiddle, harmonica |
| 40:48–41:06 | John and Peter race out of town on horses | Western music |
| 42:20–42:48 | Medieval world—Queen's dressing room | Baroque music with recorders |
| 43:40–44:04 | Queen's indiscretion | Modal, with period instruments, recorders, harpsicord, and viol da gamba |
| 47:01–47:30 | Electric cart picks up rattlesnake in desert | Electronic tones, same as robot repair |
| 48:57–49:37 | Medieval feast | Modal, with period instruments, recorders, harpsicord, and viol da gamba—visible lutist |
| 49:42–52:27 | Brothel barfight where John and Peter are playing cards | Western orchestration |
| 53:04–53:41 | Medieval seduction of Daphne | Harpsicord |
| 54:31 | Medieval scene—morning | Trumpet 2-note interval leitmotiv |
| 59:55–1:05 | Gunslinger chases Peter on horse (cuts to central control and Medieval world in chaos) | Synthesized tones with jazz rhythmic pulse Bass, Honky-tonk piano strings sound via strumming downward |
| 1:06–1:12 | On foot in Medieval world | Ostinato pattern, repeating gunshot motif |
| 1:13–1:14:35 | *The Gunslinger | Honky-tonk piano string strumming downward, electronic manipulation of sounds. 10/4 (3 3 4) meter violates duple or triple meter conventions |
| 1:18:19 | Doused in acid, Gunslinger reappears for final chase | |

| Time | Scene | Description of music |
|---|---|---|
| 1:19:20–1:19:57 | Peter arrives in Medieval world banquet hall; Queen and Man in Black are motionless in thrones; torches light hall; Gunslinger arrives | One sustained synthesized tone |
| 1:20:01–1:20:05 | Peter as seen on a fiery red grid by The Gunslinger | Piercing high pitched electronic noise |
| 1:21:58–1:22:41 | Peter reaches for torch and throws it on The Gunslinger | Electronic loud glissandi; Gunslinger 3 3 4 meter re-commences along with the desert ride motif |
| 1:24:59–1:25:08 | Burned Gunslinger appears again and falls face down | Repeating synthesized chord, low synthesized bass sustained note |
| 1:25:26–1:25:45 | The Gunslinger turns over, now faceless, revealing the electronics | Same electronic repeating noise and low piano background mixed with sounds of burning electronic parts of the robot |
| 1:25:56–1:28:38 | Peter looks on the "dead" burning Gunslinger | Low electronic sustained chord followed by several more low sustained "gongs" |

*4 Cues identified and described by Garade (2019, pp. 17–18)

questions concerning causes and effects as well as plumb concepts concerning free will and the gap between human consciousness and AI.

In a *Westworld* scene, Peter has just landed in jail, having "killed" the Gunslinger for the second time. John has arranged for a girl to bring Peter a "breakfast" tray. Figure 15.1 shows, beat by beat, four bars of music that underscore all of the actions, some subtle: Peter gazing downward, eyes open, turning head slightly, blinking, looking down; the tray with a note; covering the tray; and a close-up of the girl. Musically this continues a bluesy, Western, syncopated rhythm orchestrated with fiddle, banjo, harmonica, xylophone, and percussion. The scoring is economical: tremolo on a dissonant chord before it resolves (adding suspense) with a shot of Peter's open eyes; a pizzicato xylophone note accompanying the blink; a hollow percussion triplet underscoring the focus on the tray and note. As timbres, rhythms, and harmonies change, so too do contexts and meanings. In the next four bars, the girl looks up, while the sheriff sleeps on-duty in the background. Peter looks at the sheriff, takes the tray, and the girl then turns. During this sequence, the melodic motif gives way to a steady 4/4 beat of hollow percussion atop unsettled seventh chords, furthering the plot without

248 Cybermedia

**Figure 15.1** Fred Karlin's score for *Westworld* (Michael Chrichton, 1973: at approximately 38 minutes into the film) with actions and music instrumentation annotated for approximately one minute of action involving a jail setting in Westworld. From the Fred Karlin archives donated to the American Heritage Collection of the University of Wyoming.

words. The melody and rhythm say: there's a joke here, but on whom? Peter? The audience? The sheriff? Is Peter in trouble, or is this just the fun of Westworld? This multi-element yet spare musical score continues to add context and suspense until the denouement: an explosion in a jail cell, enabling Peter's jailbreak.

Karlin's skill can further be appreciated in an 8-second segment in which each of the three dormant worlds appear successively at the end of their overnight reset. While a different composer might think to score three short motifs uniquely representative of the West, Rome, and Medieval times, Karlin chose a single two-note motif: the opening notes of the traditional military bugle reveille (*sol-do* in the *do-re-mi-fa-**sol**-la-ti-**do*** scale, a musical interval of an ascending perfect fourth). He repeats each *sol-do* four times with decreasing loudness almost echo-like as the camera leaves one world for the next (see figure 15.2a, showing this decreasing loudness pattern three times over the 8 seconds). Karlin selected an appropriate timbre for each of the worlds: banjo/Western world, flute/Roman world, and trumpet/Medieval world. The subtle difference in the sounds is shown through a spectrograph (figure 15.2b). These musical choices can occupy both diegetic and non-diegetic domains; while the sounds reoccur in their respective worlds, they also signal that each of these runs on its own unique program.

Karlin also manipulated acoustical instruments electronically to reflect the partially-human characteristics of the robots. As he says: "I had a single violin—I played all this music incidentally to capture this horse galloping—it didn't sound like any violin, because I wanted that shrieking, primitive quality. So, it's a little electronic, but manipulated from an acoustic instrument."[29]

Karlin's music serves its traditional roles. The piano in the saloon contributes to the diegesis, as does music on period instruments in the Medieval banquet hall scenes. The Western-style country and banjo music provides context for the town of Sweetwater. Atonal music and dissonant, electronically controlled sounds code two spaces as sci-fi: the laboratory where robots are built and reconditioned as well as the master control room of Westworld's three domains. The regular duple (1 2–3 4) and triple (1 2–3) beat elements combine into an unfamiliar 10/4 time pattern (**1** 2 3 **1** 2 3 **1** 2 3 4) to create an irregularity (in Western cultures) due to an unexpected "extra" last beat of the 4/4 (i.e., 1–2 3–4) when only the repeated 3/4 (1 2–3) is expected. This irregular halting rhythm is well suited to the chase by the evil Gunslinger robot.

[29] From a 1994 interview published in *Soundtrack Magazine* 13, No. 52, republished at https://westworld.fandom.com/wiki/Westworld_(1973_film_soundtrack) (accessed February 10, 2021).

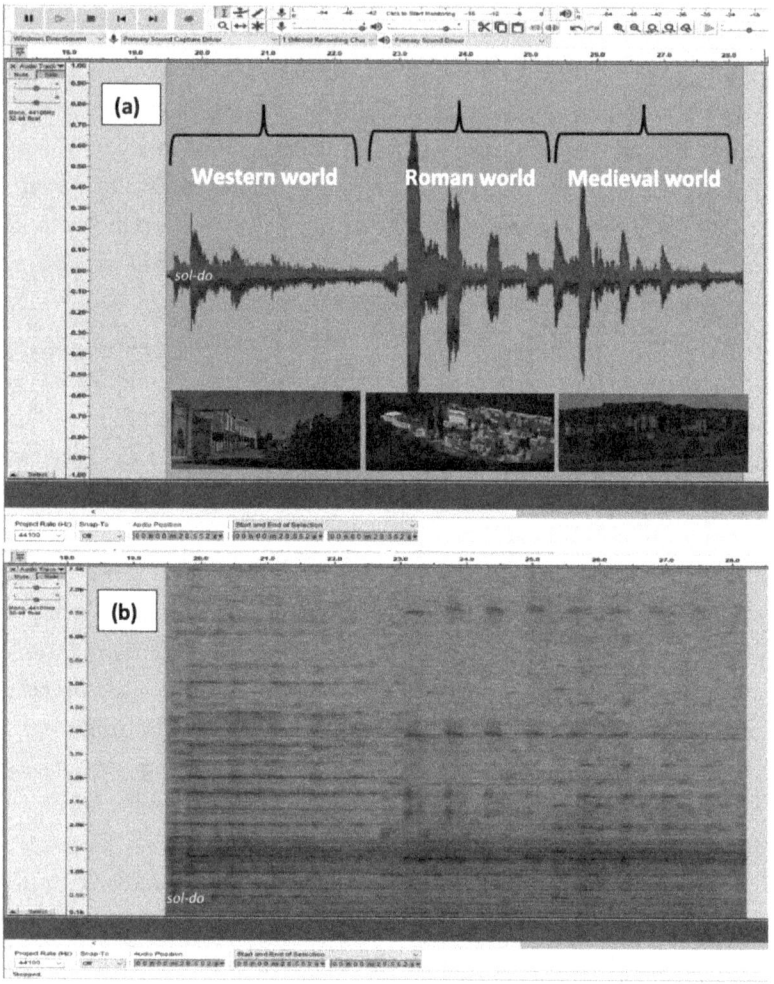

**Figure 15.2** Fred Karlin's music "Dormant worlds" from *Westworld* (Michael Chrichton, 1973) played during shots of the three worlds of Delos presented in succession over 8 seconds at about 30 minutes into the film. The same brief melody, sol-do (B to E) repeats 4 times during each shot represented under each): (a) wave form showing the intensity (loudness) pattern for each instrument (b) spectrogram, showing the frequencies from 100 to 7500 Hz presented by the banjo for Western World, a flute sound for Roman World, and a cornet for Medieval World. The two-note musical motif is the same that begins the traditional bugle call in the military to wake up the soldiers at the start of their day. The sol-do on the figure corresponds to the first of the 12 repetitions depicted in (a) and (b).

## Westworld, The HBO Series

In interviews with director Nolan, supervising music editor Kaller, and producer Stephen Semel, film theorist, composer and director Kingsley Marshall reveals Djawadi had enormous resources for the TV series' score—a stark contrast to Karlin's one-man band.[30] Kaller was hired full-time for 18 months during the creation of the series, and composer Djawadi was also employed from the start. The producers' and practitioners' commitment to the series' score is atypical.

I'll focus on Season 1's episodes, which are most closely aligned to the *Westworld* film. One difference between Karlin's and Djawadi's approaches arises from the treatment of the piano. In Karlin's *Westworld*, the piano is a bar-room instrument requiring a human pianist. In Djawadi's *Westworld*, the piano is now a *player piano* that makes music either by a human pianist or by internal mechanics, including its piano roll. While the instrument is chronologically accurate, the player piano's repertoire of songs draws anachronistically from the late twentieth century. The film's directors decided on this use of instrument and source music and relied on Djawadi for their successful execution. Granade identifies the music's three purposes: "first, it activates the audience's personal memories of the song, increasing engagement; second, it builds metadiegetic thematic connections when the music recurs as orchestrated underscore; and third, it works diegetically by controlling the hosts."[31]

From a cognitive psychological perspective, a viewer's associations to music are often unique and based on experience, though these associations are also shaped by demographics and culture. Individual knowledge of music can be personal: it may be hard for a younger or older person to appreciate that a song, such as Amy Winehouse's "Back to Black" (2006), is so revered by a middle-aged demographic. In the *Westworld* TV series, sources as diverse as early twentieth-century classical composer Claude Debussy's *Rêverie* (1890) or *Clair de Lune* (1913), Italian opera, and Radiohead's "Fake Plastic Trees" (1995), will most likely not be familiar to all audiences, and prior familiarity won't much hinder an emotionally rich filmic experience.[32] Familiarity with music, however, can occasionally detract from its effectiveness, as older associations complicate the film's new context. Films that receive awards for best song seldom win awards for

---

[30] Marshall, *Music as Source of Narrative*.
[31] Granade, "Sonic Markers," 26–7.
[32] Ann-Kristin Herget, "Well-Known and Unknown Music as an Emotionalizing Carrier of Meaning in Film," *Media Psychology* (2020): 1–28.

best film or best director. It seems a song that stands out on its own may disengage the audience from the story.[33] A positive relation is found, however, between awards for film *score* and film direction or best film, suggesting that in the *Westworld* series, pop songs like "Back to Black" that lack words and are reworked for player piano, with or without full orchestration, have different resonance. Here, prior association to known songs may add to emotional impact, via an unconscious level. So believes director Nolan, who is credited with choosing the hits Djawadi rendered for player piano.

Granade also states that the player piano establishes context and supports the narrative through repetition. When the plot moves in unexpected directions, the music provides a familiar foundation. This music may also provide clues about Westworld inhabitants and its puzzles. For example, Marshall notes that the unheard lyrics of "Back to Black" are "relevant to Maeve's awakening."[34] The song reflects that the hosts have "'died a hundred times' over the course of their narrative, but lack access to these experiences." Marshall notes the *Clair de Lune* cue brings a further extra-textual connection to the Paul Verlaine poem that inspired the piece, which refers to Italian masked theatre.[35] The piece makes its first appearance in episode 5, Season 1, in a guitar version when Dolores and William are walking through Pariah, and then performed by Dr. Ford in the bar visited by William and Teddy. "Back to Black's" unvoiced lyrics are probably more available to most viewers than Verlaine's poem and may be more effective, but research on the roles of unheard text and lyrics would be needed to determine these relations.

Granade's third use is unique to the *Westworld* TV series: music's control over the hosts. Music's control of android behavior might strike the audience as science-fictional, but in truth, music controls human behavior in the quotidian. Music in relation to humans' daily life is paradoxically one of both control and freedom. On the freedom side, music enables social activities that may otherwise be difficult. Love songs enable the expression of affection for someone without having to use words. Dance music enables people to come in closer proximity than may otherwise accord with social norms. Music may activate a part of the brain engaged in emotional rather than logical interpretations of the world.

---

[33] Dean Keith Simonton, "Film Music: Are Award-Winning Scores and Songs Heard in Successful Motion Pictures?" *Psychology of Aesthetics, Creativity, and the Arts* 1, no. 2 (2007): 53–60.
[34] Marshall, *Music as Source of Narrative*, 110.
[35] With (translated) lines "Peopled with maskers delicate and dim, that play on lutes and dance and have an air, of being sad in their fantastic trim," Marshall, *Music as Source of Narrative*, 111.

Music plays a role in emotional regulation[36] and coping in adolescence.[37] Music can provide energy and change mood. Music's centrality can be discerned in patients with Parkinson's disease: in the presence of rhythmic music, many regain a more normal walking stride.[38] Music is used by athletes to reduce perceived exertion and increase output.[39]

Given music's control over human behavior, the *Westworld* soundtrack's control of the hosts by music might be regarded as a caricature of the power of music on humans. Viewing an everyday dance party from above would show that activity starts and stops in accordance with the starts and stops of the music. Yet, when music controls the androids in Westworld, it seems "out of this world." *Westworld*'s piano music additionally controls the robots and viewers on two levels of activity—the robots in the diegesis and the audience's attention.

Whereas Djawadi never disappoints with his original, catchy melodies (e.g., the infectious "Sweetwater" theme) and periodic lush orchestration, it is the piano that attracts our attention, beginning with the opening credits and extending through most of Season 1. When the hosts become skilled at manipulating the music, they gain agency over their world. Interestingly, this dynamic extends to film directors and the popular music industry at large: power is expressed and retained through musical control.

As an example of cybermedia, *Westworld* uniquely provides a musical opportunity to explore free will, consciousness, and reality. The guests at Westworld must distinguish humans from android hosts in a world where the consequences of human actions have neither moral nor permanent physical implications. As the hosts acquire consciousness and free will, audiences may consider their own relation to free will, consciousness, and reality.

The music helps to convey these concepts and relationships by transporting the audience member on a non-diegetic vehicle into the diegesis, but then suddenly halting, changing direction, or metaphorically crashing that vehicle.

---

[36] Margarida Baltazar and Suvi Saarikallio, "Toward a Better Understanding and Conceptualization of Affect Self-Regulation through Music: A Critical, Integrative Literature Review," *Psychology of Music* 44, no. 6 (2016): 1500–21.

[37] Dave Miranda, "A Review of Research on Music and Coping in Adolescence," *Psychomusicology: Music, Mind, and Brain* 29, no. 1 (2019): 1–9.

[38] Miek J de Dreu, Gert Kwakkel, and Erwin E. H. van Wegen, "Rhythmic Auditory Stimulation (RAS) in Gait Rehabilitation for Patients with Parkinson's Disease: a Research Perspective," in *The Oxford Handbook of Neurologic Music Therapy*, ed. Michael H. Thaut and Volker Hoemberg (Oxford: Oxford University Press, 2014), 69–93.

[39] Costas I. Karageorghis and David-Lee Priest, "Music in the Exercise Domain: A Review and Synthesis (Part I)," *International Review of Sport and Exercise Psychology* 5, no. 1 (2012): 44–66.

Marshall notes that "slowing or distorting familiar cues is a device used within the series to signal an imminent disruption of the park's narrative."[40] The music generally is either diegetic (a pianist playing in a saloon) or non-diegetic, like when, after a day of bank holdups and brawls in Sweetwater, electronic, atonal music provides a background to the nightly pick-up of "dead and wounded" hosts. Why and how non-diegetic music is understood by audiences is a topic of much interest to film-music scholars, including Claudia Gorbman,[41] Jeff Smith,[42] and Robynn Stilwell,[43] as well as scientists such as Siu-Lan Tan and colleagues, who have conducted related studies.[44] It's notable that a film director would take pains to make a scene hyperrealistic, and then add music within which it didn't belong.[45]

Djawadi explores the diegetic/non-diegetic divide, disrupting the audience's absorption in Westworld through music that intimates, "Something is not right." Characters and audiences also aren't hearing the same music, after all. I've described Karlin's two-note series as straddling diegetic and non-diegetic worlds. Djawadi's player piano has a similar double role. In the saloon, the pianist is part of the diegetic action; the bar room's characters are aware of the piano music only to the extent that anyone would be aware of hired musicians. Yes, as a non-diegetic element, the visual image of the player piano roll also signals scene changes—or, as Marshall describes, a restarting of the programmed behavior loop.[46] But the player piano signals something deeper, as a mechanical system capable of executing one of the most human of tasks, the performance of complex piano music. Music, like language, distinguishes humans from other living creatures. And yet a mechanical device like the player piano can reproduce and sometimes even exceed human musical virtuosity. Moreover, the piano, and the classical

---

[40] Marshall, *Music as a Source of Narrative*, 209.
[41] Gorbman, *Unheard Melodies*.
[42] Jeff Smith, "Once More into the Breach: Interrogating Ben Winters's Nondiegetic Fallacy," in *Voicing the Cinema: Film Music and the Integrated Soundtrack*, ed. James Buhler and Hannah Lewis (Urbana, Illinois: University of Illinois Press, 2020), 260–77.
[43] Robynn Stilwell, "The Fantastical Gap Between Diegetic and Non-Diegetic," in *Beyond the Soundtrack*, ed. Daniel Goldmark, Lawrence Kramer, and Richard Leppert (Berkeley and Los Angeles: University of California Press, 2007), 184–202.
[44] Siu-Lan Tan, Matthew P. Spackman, and Elizabeth M. Wakefield, "The Effects of Diegetic and Nondiegetic Music on Viewers' Interpretations of a Film Scene," *Music Perception* 34, no. 5 (January 2017): 605–23.
[45] *Blazing Saddle's* appearance of the Count Basie orchestra on the desert has become a classic example. For further discussion see Jeff Smith, "Bridging the Gap: Reconsidering the Boundary between Diegetic and Non-Diegetic Music," *Music and the Moving Image* 2, no. 1 (2009): 1–25; and Annabel J. Cohen, "Resolving the Paradox of Film Music Through a Cognitive Narrative Approach to Film Comprehension," in *The Social Sciences of Cinema*, ed. James C. Kaufman and Dean Keith Simonton (Oxford, UK: Oxford University Press, 2014), 57—83.
[46] Marshall, "Music as a Source of Narrative," 105, paraphrasing music editor Kaller.

music it affords, has always been a symbol of high society and sophistication. This is represented in the TV series by the music of Debussy. Yet the piano also has served as "cultural go-between, as a medium through which social spheres that stood in opposition to each other could nonetheless nourish each other."[47] Pianos in the early twentieth century brought together people of many walks of life, for example, piano teachers going into homes of aristocrats. In the *Westworld* series the piano provides the foundation for traversing between diegetic and non-diegetic worlds, and between the experience of attention and inattention, between consciousness and unconsciousness, and free will and determinism.

The musical elements of the opening credits (see figure 15.3) present surprising shifts between non-diegetic and diegetic roles of the piano, startling the audience, and over the course of the episodes allowing for contemplation of the film's themes such as that of freedom and control. The computer video production company, Elastic, worked with the film's directors and composers to create "something that you'll enjoy watching the first time, and the more you watch the show, the more you'll see how these images reflect the story and its significance."[48] Lasting one minute and 45 seconds, it first shows a giant automated device (Bar 4) producing a fiber that becomes a piano string. Then, more fiber creates the skeleton and muscles of a horse, and finally an android. The cello theme begins an ascending motif—$A_3$ $C_4$ $D_4$ $E_4$ (Bars 3-4). The piano (Bars 5-6) echoes the motif in reverse (descending order—$A_3$ $E_4$ $D_4$ $C_4$). The cello "call" is echoed by the piano "response," teaching the piano, in a sense. The piano, however, reverses the tones, getting it wrong. Further, piano-key hammer action does not allow continuous control over the string. Once a piano note is hit, there is no turning back—the program must run its course. The bowing motion afforded by the violin family, by contrast, provides continuous control over the tone.

The ascending cello line repeats (Bar 7). The piano produces a jarring note ($B \cong_3$, Bar 9) outside the key, creating a dissonant tritone (6 semitones) with the previous final note ($E_4$) of the cello motif. The piano then offers a distorted reversed

---

[47] James Parakilas, *Piano Roles: Three Hundred Years of Life with the Piano* (New Haven, CT: Yale University Press, 1999), 4.
[48] Quoting Patrick Clair the director the *Westworld* titles for all three season in an interview with Jennifer Vineyard, "The Secrets Behind *Westworld*'s Opening Title Sequence," *Vulture*, October 5, 2016, https://www.vulture.com/2016/10/westworld-title-sequence-secrets.html (accessed February 9, 2021); see also Staff "Patrick Clair on 'Westworld' Title Sequences, Creativity & Robotics," *ARTpublika Magazine*, June 1, 2020, https://www.artpublikamag.com/post/the-beautiful-the-grotesque-emmy-winning-motion-graphics-designer-patrick-clair-talks-westworld-titl (accessed February 9, 2021).

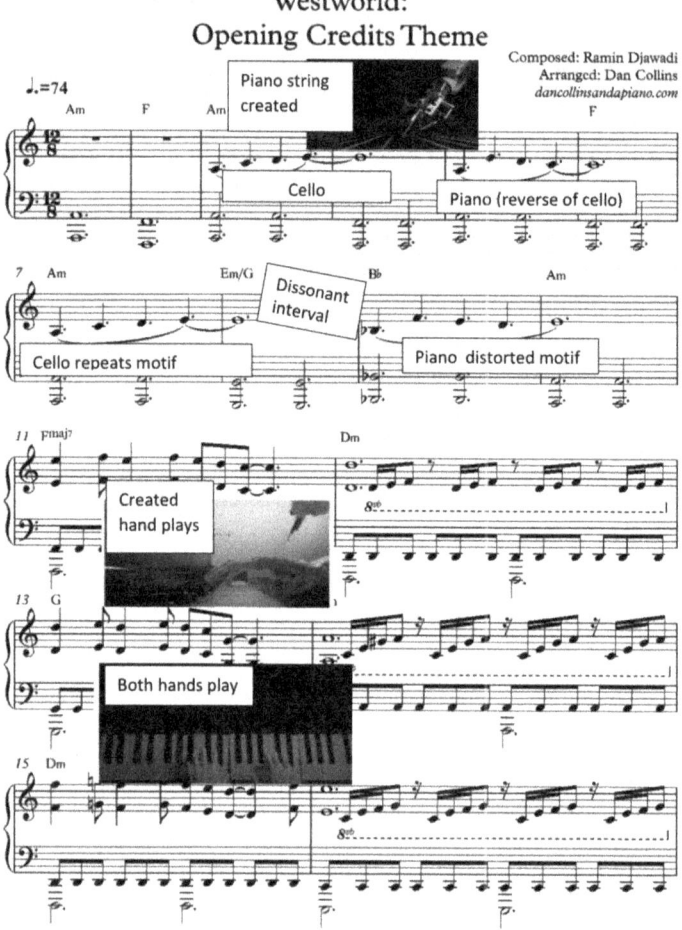

**Figure 15.3** Music composed by Ramin Djawadi for *Westworld* (HBO, 2016). The musical score, notated for piano by Dan Collins with images from the opening credits of *Westworld* depicting themes of the series. Here, the music emphasises the application of artificial intelligence in the creation of Westworld's Hosts by representing the human-like quality of the pianist that can be readily replaced by a piano role and appropriately mechanized piano. Score transcribed for piano by and used with kind permission of Dan Collins (dancolllinsandapiano.com).

variation of the cello theme (Bar 9–10). With the introduction of a new surprising jaunty phrase (Bar 11), astonishingly, a skeletal right hand at a piano keyboard appears performing the music, transporting the piano from the non-diegetic world into the diegesis. A shot with *both* skeletal hands follows (Bar 13) repeating the jaunty theme a scale-step lower. Following an interlude based on the first

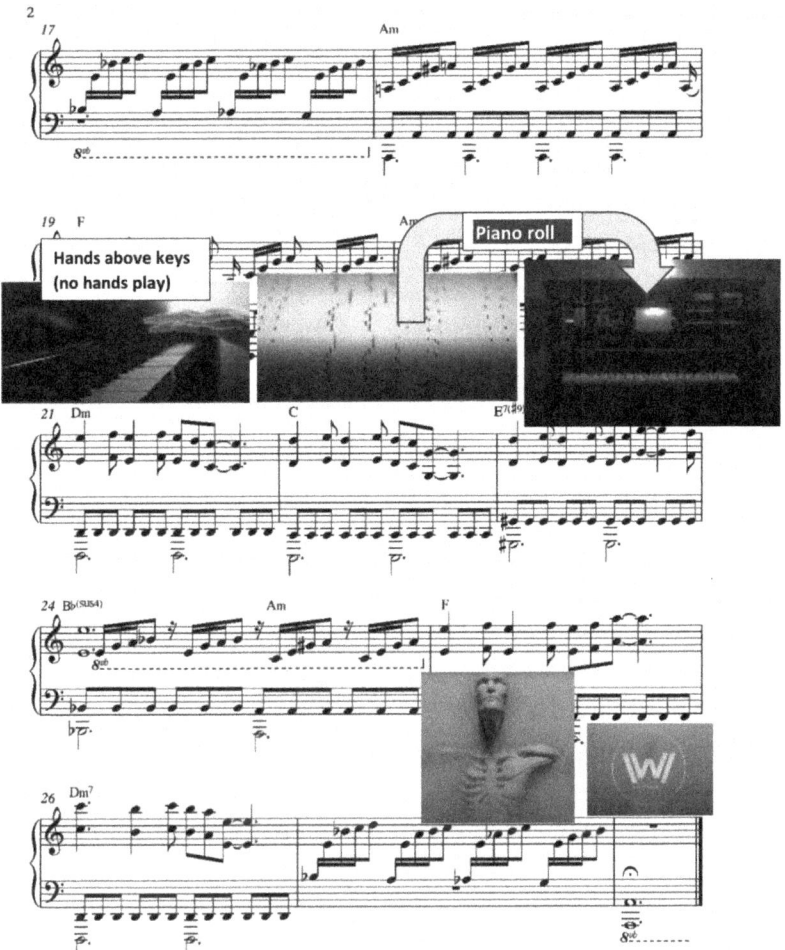

**Figure 15.3** continued

theme but in double time, the jaunty theme returns with more surprises—the two hands raise (or are raised) from the keyboard (Bar 21), but the keys continue to move consistent with continuing piano music. It has been stunning to see a skeletal hand created and then play a piano, but now, even more astounding, the piano plays by itself! The dual roles of the piano in the diegesis (automatic moving keys) and non-diegesis (providing emotional associations with eerie, jaunty, melodic music) are clarified with a further surprise, a shot of a player piano roll (Bar 22) with its perforated paper controlling the specific piano strings, whose location coincides with the aligned moving black holes in the paper. A shot from a longer vantage point includes the entire player piano (Bar 23).

The player piano reminds us that twenty-first century electronics are not needed to control a complex musical instrument that produces a beautiful musical theme. Player pianos became commercially available at the turn of the twentieth century.[49] The piano rolls resemble the early punched paper tape cards that operated the first digital computers. A parallel can thus be drawn between the player piano and artificial intelligence (AI), where AI refers to the carrying out of human tasks by non-humans.[50]

In an interview prior to Season 3's release, Jonathan Nolan and Lisa Joy identified fashion as the most challenging aspect of representing what the forthcoming series' world would be like in 50 years. Although music wasn't what came to his mind, it is at least equally challenging to think how music, particularly popular music, would change over the next decades. The half-hour interview included background music from *Westworld* and also showed Joy seated at an electric piano performing Debussy's technically and conceptually challenging *Clair de Lune*. The use of the player piano in the *Westworld* TV series draws attention to the question of human creativity and free will. The 1973 *Westworld* film missed this opportunity, and not because the player piano had not yet been invented. The idea was cleverly picked up by the creators of the twenty-first century series and added to the depth of cognitive processing in which audiences may engage, and consequently to the meaningfulness of the experience of this modern sci-fi Western.

## A Cognitive Model: CAM-WN

I propose the Congruence Associationist Model with Working Narrative (CAM-WN, see figure 15.4) to account for several aspects of how music contributes to a viewer's experience of film and video.[51] It offers a perspective on the diegetic/non-diegetic distinction and emphasizes commonalities in the mental processing of music and other types of media (i.e., visual, text, speech, sound effects).

---

[49] Timothy Dean Taylor, *Music and Capitalism: A History of the Present* (Chicago: The University of Chicago Press, 2016).
[50] Kurt Vonnegut's book *Player Piano* (1952) is a story about workers displaced by technological advance. In one chapter, two persons who have lost their jobs due to automation converse in a bar, that has a player piano, and they disparage the fact that this device has taken a job away from a human pianist. Jonathan Nolan claims that the use of the player piano in *Westworld* pays homage to Vonnegut and this idea.
[51] Annabel J. Cohen, "Film Music and the Unfolding Narrative," in *Language, Music, and the Brain: a Mysterious Relationship*, ed. Michael A. Arbib (Cambridge, MA: The MIT Press, 2013), 173–201.

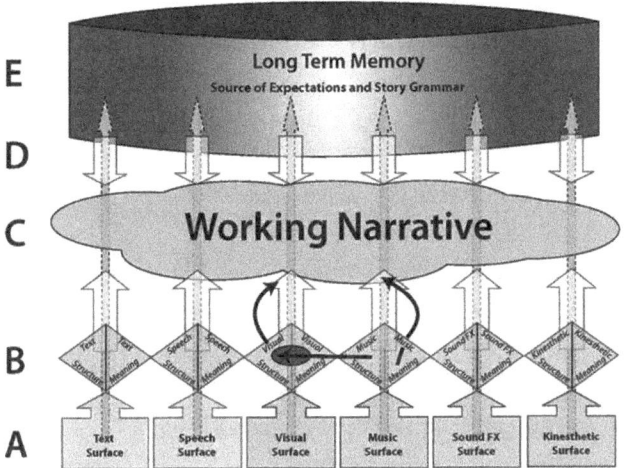

**Figure 15.4** Congruence-Associationist Model with Working Narrative (CAM-WN) for understanding film-music communication. Illustration by Annabel Cohen. See text for explanation. [from figure 7.8, p. 195, Cohen, Annabel J. "Film Music and the Unfolding Narrative," in *Language, Music, and the Brain: a Mysterious Relationship*, edited by Michael A. Arbib, 173–201. Vol. 10, *Strüngmann Forum Reports*, series ed. Julia R. Lupp. Cambridge, MA: The MIT Press, 2013. [Used with permission]

The model first assumes that a film audience member wishes to become absorbed in a story, and in a way similar to reading a novel or newspaper. Here, the viewer relies on two sources of information. One is the moment-by-moment unfolding of audiovisual content—sights and sounds that impinge on the human sensorium. At Level A, in the model (see figure 15.4), these packages of physical energy are represented as "surfaces." Sound is divided into three types of acoustical information (speech, music, and sound effects) and light is divided into scenes and text. These categories are in accordance with five channels identified by film theorist Christian Metz.[52] Kinesthetic information is included as an additional surface that accommodates stimulation arising from recently automated theater seats (e.g., D-Box, MX4D), which immerse viewers through movements forward, backward, and sideways. Kinesthetics also account for more subtle embodied aspects of perception, such as sensations associated with motion and pressure that might correlate with the viewer's perception of filmic action.[53]

---

[52] Christian Metz, *Film Language: A Semiotics of the Cinema* (Chicago, Ill: Univ. of Chicago Press, 2007). Originally translated from the French by Michael Taylor in 1974.
[53] David A. Rosenbaum, *Human Motor Control*, 2nd ed. (Amsterdam, NE: Academic Press is an imprint of Elsevier, 2016), 129–30.

At the top of the diagram, Level E describes the viewer's higher-level understanding of sensory input. This level can detect when notes are in or out of tune, what sentences mean, and events, like what falling downstairs might sound like. Supporting this is knowledge of the story's, music's, and language's grammar,[54] as well as its sequential action.[55] The model (which necessarily leaves out content) describes only story grammar but can be taken for representing grammars in general.

It is assumed that a viewer's coding of Level A's information is carried out early on to Level B for all sensory surfaces. This coding at Level B is of two types. The first is structural (e.g., patterning in time, or configural such as smooth of jagged visual edges) and the second is semantic or meaning (e.g., positive/negative valence, or item recognition). The model divides Level B's "Structure" and "Meaning" through diamond-shapes representing these two analyses of Level A's sensory surfaces. It's assumed that text, speech, visual scenes and actions, music, sound effects, and kinesthetic information all have both meaning and configural patterns.

As recently revealed by Gao et al., a brain network determines the congruence of meaning coming from different sources of information.[56] There may also be such a network for determining the degree of multimodal configural congruence. As Bregman has identified in his notion of Auditory Scene Analysis,[57] similar motion patterns from two independent streams suggest a single source of the two streams (e.g., increasing pitch and decreasing loudness in the auditory realm, deepening color and increasing size in the visual realm, moving lips and changing speech sounds as a multimodal example). CAM-WN assumes that in the face of limited attention, the brain focuses on congruencies within and across different surface origins as the source of meaningful actions. At Level B, an arrow points from Music structure to Visual structure, reflecting heightened significance attached to cross-modal (here music- and visual structural) congruencies. For example, a galloping rhythm will direct attention to a galloping horse rather than a horse that is walking or trotting.

[54] Siu-Lan Tan, "Scene and Heard: The Role of Music in Shaping Interpretation of Film," in *The Routledge Companion to Music Cognition*, ed. Richard Ashley and Renee Timmers (London: Routledge, Taylor & Francis Group, 2019), 364–76.
[55] Jeffrey M. Zacks, "Event Perception and Memory," *Annual Review of Psychology* 71, no. 1 (April 2020): 165–91.
[56] Chuanji Gao et al., "An FMRI Study of Affective Congruence across Visual and Auditory Modalities," *Journal of Cognitive Neuroscience* 32, no. 7 (2020): 1251–62.
[57] Albert S. Bregman, *Auditory Scene Analysis: The Perceptual Organization of Sound* (Cambridge, MA: MIT Press, 2006).

Thus, at Level B, decoding of the multisensory surfaces from Level A takes place along two lines: configural structure and semantic or associationist meaning. Configural structure refers to such things as temporal patterning (e.g., the motion pattern of a galloping horse, the intensity or timing patterns characterizing musical rhythms; or on a longer time span, the boundary between the beginning of one event and the start of another). It also can identify spatial relations (e.g., a crash heard from the left-hand side of the room). Analysis at Level B produces a pool of semantic and structural information. An assumption of the CAM-WN models is that the brain does not care where the information is coming from, at least for some aspects, and instead is interested in redundancies. If auditory and visual temporal patterns align, then this visual information will have priority over visual information that is not aligned with the auditory pattern.

A further feature of CAM-WN is that throughout the system of levels, some coding is quick and coarse-grained, and other coding is slower and refined. The fast and coarse/ slow and refined concept is consistent with various theories of perception, memory, and attention.[58] In CAM-WN, the output of quick coding ascends to the higher brain centers (dotted vertical arrows beginning at A), to cue the cognitive processes (at E) in their task of deciphering a story underlying the multisensory barrage of cues from A and B. Here the brain can be understood as an inference machine, sometimes referred to as a Helmholtz machine (Clark, 2013)[59] in recognition of Helmholtz's concept of "unconscious inference" that is fundamental to everyday perception. The idea is that we see what we expect to, based on experience.[60]

Two types of processes now co-occur. One is the continuing bottom-up analysis of the multisensory barrage (Levels A to B) and the other is the top-down inferencing (Level D) based on weak cues that have leaked quickly through from bottom-up fast encoding. This inferencing is based on experience and story, music, language, and sequential-behavior grammars. To achieve the creation of

---

[58] Christoph Herrmann, Mattias H. J. Munk, and Andreas K. Engle, "Cognitive Functions of Gamma-Band Activity: Memory Match and Utilization," *Trends in Cognitive Sciences* 8, no. 8 (2004): 347–55.

[59] Andy Clark, "Whatever next? Predictive Brains, Situated Agents, and the Future of Cognitive Science: Behavioral and Brain Sciences" 36, no. 3 (June 2013): 181—204.

[60] The role of experience in the interpretation of media is also highlighted by Noah Fram, "Expecting the Twist: How Media Navigate the Intersections Among Multiple Sources of Prior Knowledge," in *Cybermedia: Explorations in Science, Sound, and Vision*, ed. Carol Vernallis, Holly Rogers, Selmin Kara, and Jonathan Leal (London: Bloomsbury Press, 2021), 329–346.

the best story, the brain makes the best match between the bottom-up fine-grained analysis of the multisensory information and the top-down inferences. The matching process takes place at Level C. Because of the dynamic nature of this ongoing process, from the opening to the final credits of the film, this best match is referred to as the *working narrative*. In other words, the experience of *Westworld* is the audience member's working narrative.

How can this model be helpful? As an example, consider a scene with the player piano projecting Winehouse's famous 'Back to Black'. If the audience member has never heard the song before, through processing at Level E, the piece may sound like slightly modernized traditional honkytonk music that might have been played in an early twentieth century Western setting. If the audience member is a fan of the song, they may feel connected to the scene, and open to overlooking the mismatch of a modern piece played in a twentieth century saloon. They may even remember unheard lyrics that reinforce *Westworld*'s themes. These individual differences will lead to slightly different working narratives or experiences of the episode (Level C). One's accuracy of sensory encoding and prior knowledge will limit or enhance one's working narrative.

As another example of the use of CAM-WN, consider how diegetic awareness is explained by the top-down matching process. The best match between sensory analysis (outcomes of Levels A, B) and top-down inference regarding the story (Levels E, D) wins at Level C. Once the story diegesis is established, some information will enter consciousness (Level C) and other information will not. In the opening credits, attention is directed to the creations of the piano, horse, and android. The non-diegetic music offers acoustical, music-grammatical, and emotional information. The music begins with a relatively low pitch, establishing an A minor tonality; the B♭ at bar 9 violates the tonality. Its underlying ostinato tempo is a relentless four sets of triplets per bar, with a melody at bar 12 that doubles the meter to 24–16th beats per bar. The unsettling emotion arising from this combination of features finds its explanation (Eldar et al., 2007) in the visual objects depicted. Initially, the acoustical information (Level B) finds no match from Level D. There is no reason to be visualizing a cello or a piano. The acoustical information therefore does not form part of the initial diegesis. It is outside the working narrative ... that is, until we see the skeletal hand at the piano. Now the top-down inferences (Level D), based on the image of a hand playing the keyboard, can match the acoustical information arising from Level A to B and the match between the image of the piano and the sound of the piano leads to consciousness of the piano sound, at Level C, the working narrative. At level B, temporal congruency of the visual hand motion and fingers moving on the keys and the auditory pattern of the melody direct attention to the hand, making it prominent in consciousness. The meaning of the music (relentless, unsettling, ominous) picked up from the semantic analysis at B and matched by meanings inferred from the higher level at D also match, bringing into consciousness an emotionally charged engagement for this *Westworld* scene. The sudden switching from the non-diegesis to diegesis in the opening credits is but one example of how the music is used throughout the film, and this use, assisted by the exploitation of twenty-first century computer-based editing studios, coupled with the imaginative creative team, offer more sophisticated and nuanced play with the audience's mind. The audience experiences a moving in and out of the diegesis that's more striking than what's offered from the *Westworld* film, produced over four decades earlier.

The CAM-WN model reminds us that the brain can engage with an analysis that counters our intuitions. For example, it might seem obvious that music has an acoustic element, but the brain might separately "perceive" an acoustic element and emotional element and encode music as such, using only what it needs to make sense of the story. The brain can discard the acoustic element if there is no reason for a musical instrument to be in the scene, while at the same time preserving the emotion carried by the music, which belongs there. This seems consistent with Smith's (2020) discussion of the diegetic/non-diegetic distinction from the perspective of the "spectator's narrative comprehension. The crux of the issue is not the kinds of departures from everyday experience of which film-worlds are capable, but rather how viewers could ever process any story information without some tacit application of real-world knowledge."[61] Here I suggest that "story information" is Level A of the CAM-WN model and "real-world knowledge" is Level E. Smith continues, noting film perception's reliance on normal cognitive processing, including heuristics and then proposes that the film score should be considered alongside film's other formal systems, where functions carried out by music can be carried out by other means (and vice versa). His idea and the CAM-WN model both concur about the common coding principle (for meaning and structure) of all filmic categories and

---

[61] Smith, "Once More into the Breach", 274.

prediction processes. Finally, Smith recognizes that norms associated with sound design enable viewers to make inferences about the sound's relation to the narrative and, as well, that departures from these norms can disrupt "normal conceptual schemas."[62] These features are compatible with the CAM-WN model, as exemplified by the admission of the piano sound into the *Westworld* working narrative.

Scientists are currently delineating the neural networks underlying various sensory, perceptual, and cognitive processes. Consistent with my approach, research on event perception identifies the human sensitivity to common event structures. One line of research examines human segmentation of events, often using video representations, as reviewed by Zacks (2020)[63] although boundary formation in melody has also been explored in segmentation tasks.[64] Further studies with eye-tracking, neuroimaging, and computational modeling may soon tell us more. Interdisciplinary research teams should also help us determine what brain regions are involved with processing complex multimedia.[65] The imaginative as well as the traditional uses of music in *Westworld* offer fertile ground for empirical research.

## Summary and Conclusion

Drawing from cognitive psychology and neuroscience, this chapter explores two major film composers' scoring of similar cybermedia themes. Common functions of music within science fiction and Western genres were evident, despite passages of time and different formats. However, psychological research suggested that in the twenty-first century, teams of directors and audio-based practitioners may be getting smarter about holding audience attention. Newer music technologies, such as better computer music systems, and clever new uses

---

[62] Ibid., 276.
[63] Zacks, *Event Perception*.
[64] Bradley Frankland and Annabel J. Cohen, "Parsing of Melody: Quantification and Testing of the Local Grouping Rules of Lerdahl and Jackendoff's 'A Generative Theory of Tonal Music'," *Music Perception* 21, no.4 (2004): 499–543.
[65] Siu-Lan Tan et al., "Future Directions for Music and Sound in Multimedia," in *The Psychology of Music in Multimedia*, ed. Siu-Lan Tan et al. (Oxford: Oxford University Press, 2013), 391–405. See also, Annabel J. Cohen, "Congruence-Association Model and Experiments in Film Music: Toward Interdisciplinary Collaboration," *Music and the Moving Image* 8, no. 2 (2015): 5–24.

of traditional instruments (such as *Westworld*'s player piano) have expanded narrative depth and links across non-linear plots. The Congruence-Association Model with Working Narrative (CAM-WN) was discussed as an aid to understanding the brain's multiple tasks of integrating audio and visual tracks of a film, with the conclusion that the experience of a film or TV drama might be understood as a working narrative created from the match between bottom-up multisensory analytical processes and the top-down inferences based on remembered experience and the grammars of story, music, and language. The bottom-up/top-down matching process explains why the audience typically does not as much 'hear' but nevertheless feels the music of film soundtracks, as this acoustical information, unlike the emotion of the music, is not matched by the top-down inferences.

The twenty-first century demands more from film makers than ever before, given the many kinds of media that compete for audiences. In winning audience attention to contemporary cybermedia, Jonathan Nolan and Lisa Joy, with the help of their creative team, have exploited the player piano as a vehicle to transport the audience in and out of the narrative, simultaneously heightening their consciousness of their own consciousness. Both the *Westworld* movie and series use music in traditional ways to transport viewers into a fictional world and establish its context; however, the *Westworld* TV series uniquely exploits music as a means of exploring the concept of control of the androids, by the androids, and of the audience. The chapter lays a foundation for further studies of these musical phenomena in cybermedia.

## Acknowledgment

Co-editors, Selmin Kara, Holly Rogers, Jonathan James Leal, and Carol Vernallis's close readings, constructive critiques, and editorial work are appreciated. Special gratitude is owed to Carol Vernallis for her careful and conscientious labors to both increase the accessibility of and maintain the integrity of the CAM-WN model in the face of word count constraints. Feedback from Noah Fram is also acknowledged. Thank you to University of Wyoming's American Heritage Center for, in the pandemic's midst, granting access to Karlin's score, and Dan Collins, too, for permission to reproduce his score of *Westworld*'s opening credits. Gratitude is also expressed for previous support from the Social Sciences and Humanities Research Council of Canada.

# Bibliography

Baltazar, Margarida, and Suvi Saarikallio. "Toward a Better Understanding and Conceptualization of Affect Self-Regulation through Music: A Critical, Integrative Literature Review." *Psychology of Music* 44, no. 6 (2016): 1500–21.

Blood Anne J. et al., "Emotional Responses to Pleasant and Unpleasant Music Correlate with Activity in Paralimbic Brain Regions." *Nature Neuroscience* 2, no. 4 (1999): 382–87.

Boltz, Marilyn G. "The Cognitive Processing of Film and Musical Soundtracks." *Memory & Cognition* 32, no. 7 (2004): 1194–205.

Bregman, Albert S. *Auditory Scene Analysis: The Perceptual Organization of Sound.* Cambridge, MA: MIT Press, 2006.

Clark, Andy. "Whatever Next? Predictive Brains, Situated Agents, and the Future of Cognitive Science: Behavioral and Brain Sciences." *Behavioural and Brain Sciences* 36, no. 3 (June 2013): 181–204.

Cohen, Annabel J. "Music as a Source of Emotion in Film." In *Handbook of Music and Emotion: Theory, Research, and Applications*, edited by Patrik N. Juslin and John A. Sloboda, 879–908. Oxford: Oxford University Press, 2010.

Cohen, Annabel J. "Film Music and the Unfolding Narrative." In *Language, Music, and the Brain: A Mysterious Relationship*, edited by Michael A. Arbib, 173–201. Cambridge, MA: The MIT Press, 2013.

Cohen, Annabel J. "Resolving the Paradox of Film Music Through a Cognitive Narrative Approach to Film Comprehension." In *The Social Sciences of Cinema*, edited by James C. Kaufman and Dean Keith Simonton, 57–83. Oxford, UK: Oxford University Press, 2014.

Cohen. Annabel J. "Congruence-Association Model and Experiments in Film Music: Toward Interdisciplinary Collaboration." *Music and the Moving Image* 8, no. 2 (2015): 5–24.

Collier, William G., and Timothy L Hubbard. "Musical Scales and Evaluations of Happiness and Awkwardness: Effects of Pitch, Direction, and Scale Mode." *American Journal of Psychology* 114, no. 3 (2001): 355–75.

Cook, Nicholas. *Analysing Musical Multimedia*. Oxford, UK: Clarendon Press, 1998.

Cox, Arnie. *Music and Embodied Cognition: Listening, Moving, Feeling, and Thinking.* Bloomington: Indiana University Press, 2017.

Cutting, James E. "Narrative Theory and the Dynamics of Popular Movies." *Psychonomic Bulletin & Review* 23, no. 6 (2016): 1713–43.

Cutting, James E, Kaitlin L Brunick, Jordan E Delong, Catalina Iricinschi, and Ayse Candan. "Quicker, Faster, Darker: Changes in Hollywood Film over 75 Years." *i-Perception* 2, no. 6 (2011): 569–76.

de Dreu, Miek J, Gert Kwakkel, and Erwin E. H. van Wegen. "Rhythmic Auditory Stimulation (RAS) in Gait Rehabilitation for Patients with Parkinson's Disease: a

Research Perspective." In *Handbook of Neurologic Music Therapy*, edited by Michael H. Thaut and Volker Hoemberg, 69–93. Oxford, UK: Oxford University Press, 2014.

Eldar, Eran, Ori Ganor, Roee Admon, Avraham Bleich, and Talma Hendler. "Feeling the Real World: Limbic Response to Music Depends on Related Content." *Cerebral Cortex* 17, no. 12 (2007): 2828–40.

Fram, Noah. "Expecting the Twist: How media navigate the intersections among multiple sources of prior knowledge." In *Cybermedia: Explorations in Science, Sound, and Vision*, edited by Selmin Kara, Jonathan Leal, Holly Rogers, and Carol Vernallis, 329–346. London, UK: Bloomsbury, 2021.

Frankland, Bradley and Cohen, Annabel J. "Parsing of Melody: Quantification and Testing of the Local Grouping Rules of Lerdahl and Jackendoff's 'A Generative Theory of Tonal Music.'" *Music Perception*, 21, no. 4 (2004): 499–543.

Gao, Chuanji, Christine E. Weber, Douglas H. Wedell, and Svetlana V. Shinkareva. "An FMRI Study of Affective Congruence across Visual and Auditory Modalities." *Journal of Cognitive Neuroscience* 32, no. 7 (2020): 1251–62.

Goldstein, E. Bruce. *Cognitive Psychology: Connecting Mind, Research, and Everyday Experience 5th Ed.* Boston, MA: Cengage, 2019.

Gorbman, Claudia. *Unheard melodies: Narrative Film Music*. Bloomington, IN: Indiana University Press, 1987.

Granade, Andrew. "Some People call me the Space Cowboy": Sonic markers of the science fiction western." Expanded version of a paper appearing in *Re-locating the Sounds of the Western*, edited by M. Whitmer & K. P. Leonard. New York, NY: Routledge, 2019.

Hauk, Olag. "What Does it Mean? A review of the Neuroscientific Evidence for Embodied Lexical Semantics." In *Neurobiology of Language*, edited by Gregory Hickok and Steven L. Small, 777–88. London, UK: Academic Press, 2016.

Hauk, Olaf, Ingrid Johnsrude, and Friedemann Pulvermüller. "Somatotopic Representation of Action Words in Human Motor and Premotor Cortex." *Neuron* 41, no. 2 (2004): 301–7.

Hayward, Philip. "Sci-Fidelity: Music, Sound and Genre History." In *Off the Planet: Music, Sound and Science Fiction Cinema*, edited by Philip Hayward, 1–29. Eastleigh, UK: John Libbey Publishing, 2004.

Herget, Ann-Kristin. "Well-Known and Unknown Music as an Emotionalizing Carrier of Meaning in Film." *Media Psychology* (2020): 1–28.

Herrmann, Christoph, Christoph Munk, Mattias H. J., and Andreas K. Engle. "Cognitive Functions of Gamma-Band Activity: Memory Match and Utilization." *Trends in Cognitive Sciences* 8, no. 8 (2004): 347–55.

Hilleary, Mike, Sam Sodomsky, Noah Yoo, Matthew Strauss and Alphonse Pierre, Jazz Monroe, Allison Hussey, Matthew Ismael Ruiz, and Matthew Strauss. "'Westworld' Composer Ramin Djawadi on Why Those Radiohead Covers Keep Coming." *Pitchfork*, November 18, 2016. https://pitchfork.com/

thepitch/1370-westworld-composer-ramin-djawadi-on-why-those-radiohead-covers-keep-coming/. (accessed February 5, 2021).

Hoeckner, Berthold, and Howard Nusbaum. "Music and Memory in Film and Other Multimedia: The *Casablanca* Effect." In *The Psychology of Music in Multimedia*, edited by Siu-Lan Tan, Annabel J. Cohen, Scott Lipscomb, and Roger Kendall, 235–63. Oxford, UK: Oxford University Press, 2013.

Karageorghis, Costas I., and David-Lee Priest. "Music in the Exercise Domain: A Review and Synthesis (Part I)." *International Review of Sport and Exercise Psychology* 5, no. 1 (2012): 44–66.

Karlin, Fred, and Rayburn Wright. *On the Track: A Guide to Contemporary Film Scoring*. 2nd ed. New York, NY: Taylor & Francis, 2004.

Karlin. "Westworld | Soundtrack Suite." YouTube, January 31, 2020. https://www.youtube.com/watch?v=-vaGveA_gxQ. (accessed February 5, 2021).

Krumhansl, Carol Lynne. "Listening Niches across a Century of Popular Music." *Frontiers in Psychology*, March 8, 2017. https://www.frontiersin.org/articles/10.3389/fpsyg.2017.00431/full. (accessed February 5, 2021).

Lakoff, George, and Mark Johnson. *Metaphors We Live By*. Chicago, Ill: University of Chicago Press, 2003.

Lee, Yune-Sang, Petr Janata, Carlton Frost, Michael Hanke, and Richard Granger. "Investigation of Melodic Contour Processing in the Brain Using Multivariate Pattern-Based FMRI." *NeuroImage, U.S. National Library of Medicine* 57, no. 1 (July 2011): 293–300.

Leonard, Kendra Preston, and Whitmer, Mariana. *Re-locating The Sounds of the Western*. New York, NY: Routledge. 2019.

Lerner, Neil. "The Origins of Musical Style in Video Games 1977–1983." In *The Oxford Handbook of Film Music Studies*, edited by David Neumeyer, 319–47. Oxford, UK: Oxford University Press, 2014.

Marshall, Kingsley, and Antonia Mackay. "'Music as a Source of Narrative in HBO's Westworld'." In *Reading Westworld*, edited by Alex Goody, 97–118. Cham, Switzerland: Palgrave Macmillan, 2019.

Martin, C., Polaire, M., Wray, C., Semel, S., O'Toole, N., Witz, D., & Goldberg, J. (Producers), & Nolan, J., & Joy, L. *Westworld*. HBO Entertainment, Kilter Films, Bad Robot Productions, Jerry Weintraub Productions, Warner Bros. Television, 2016.

Metz, Christian *Film Language: A Semiotics of the Cinema* [originally translated from the French in 1974 by M. Taylor]. Chicago, Ill: Univ. of Chicago Press, 2007.

Miranda, Dave. "A Review of Research on Music and Coping in Adolescence." *Psychomusicology: Music, Mind, and Brain* 29, no. 1 (2019): 1–9.

Münsterberg, Hugo. *The Film: A Psychological Study; the Silent Photoplay in 1916*. New York: Dover Publications, 1970 [1916].

Overy, Katie, and Istvan Molnar-Szakacs. "Being Together in Time: Musical Experience and the Mirror Neuron System." *Music Perception* 26, no. 5 (2009): 489–504.
Parakilas, James. *Piano Roles: Three Hundred Years of Life with the Piano*. New Haven, CT: Yale University Press, 1999.
Rizzolatti, Giacomo, and Laila Craighero. "The Mirror-Neuron System." *Annual Review of Neuroscience* 27, no. 1 (2004): 169–92.
Rosenbaum, David A. *Human Motor Control*. 2nd ed. Amsterdam, NE: Academic Press is an imprint of Elsevier, 2010.
Salimpoor, Valorie N, Mitchel Benovoy, Kevin Larcher, Alain Dagher, and Robert J Zatorre. "Anatomically Distinct Dopamine Release During Anticipation and Experience of Peak Emotion to Music." *Nature Neuroscience* 14, no. 2 (2011): 257–62.
Schmidt, Lisa M. "A Popular Avant-Garde: The Paradoxical Tradition of Electronic and Atonal Sounds in Sci-Fi Music Scoring." In *Sounds of the Future: Essays on Music in Science Fiction Film*, edited by Mathew J. Barthowiak, 22–43. Jefferson, NC: McFarland & Company, Inc., 2010).
Simonton, Dean Keith. "Film Music: Are Award-Winning Scores and Songs Heard in Successful Motion Pictures?" *Psychology of Aesthetics, Creativity, and the Arts* 1, no. 2 (2007): 53–60.
Smith, Jeff. "Bridging the Gap: Reconsidering the Boundary Between Diegetic and Non-Diegetic Music." *Music and the Moving Image* 2, no. 1 (2009): 1–25.
Smith, Jeff. "Once More into the Breach: Interrogating Ben Winters's Nondiegetic Fallacy." In *Voicing the Cinema: Film Music and the Integrated Soundtrack*, edited by James Buhler and Hannah Lewis, 260–77. Urbana, Illinois: University of Illinois Press, 2020.
Staff. "Patrick Clair on 'Westworld' Title Sequences, Creativity & Robotics." *ARTpublika Magazine*, October 15, 2019. https://www.artpublikamag.com/post/the-beautiful-the-grotesque-emmy-winning-motion-graphics-designer-patrick-clair-talks-westworld-titl. (accessed February 5, 2021).
Stilwell, Robynn. "The Fantastical Gap Between Diegetic and Non-Diegetic." In *Beyond the Soundtrack*, edited by Daniel Goldmark, Lawrence Kramer, and Richard Leppert, 184–202. Berkeley: University of California Press, 2007.
Tan, Siu-Lan. "Scene and Heard: The Role of Music in Shaping Interpretation of Film." In *The Routledge Companion to Music Cognition*, edited by Richard Ashley and Renee Timmers, 364–76. London: Routledge, Taylor & Francis Group, 2019.
Tan, Siu-Lan, Annabel Cohen, Scott David Lipscomb, and Roger Allen Kendall. "Future Directions for Music and Sound in Multimedia." In *The Psychology of Music in Multimedia*, edited by Siu-Lan Tan, Annabel J Cohen, Scott D Lipscomb, and Roger A Kendall, 391–405. Oxford: Oxford University Press, 2013.
Tan, Siu-Lan, Matthew P. Spackman, and Elizabeth M. Wakefield. "The Effects of Diegetic and Nondiegetic Music on Viewers' Interpretations of a Film Scene." *Music Perception* 34, no. 5 (2017): 605–23.

Taylor, Timothy Dean. *Music and Capitalism: A History of the Present*. Chicago: The University of Chicago Press, 2016.

Vineyard, Jennifer. "The Secrets Behind Westworld's Opening Title Sequence." *Vulture*, October 5, 2016. https://www.vulture.com/2016/10/westworld-title-sequence-secrets.html. (accessed February 5, 2021).

Vonnegut, Kurt. "*Player piano*." New York, NY: Dial press trade paperbacks (Penguin Random House), 2006 [1952].

Whitmer, Mariana. "Melodramatic Music in the Western." *Journal of Film Music* 5, no.1–2 (2012): 111–12.

Zacks, Jeffrey M. "Event Perception and Memory." *Annual Review of Psychology* 71, no. 1 (2020): 165–91.

# Part Five

# Interface, Desire, Collectivity

# 16

# Director Terence Nance Discusses *Random Acts of Flyness*

Carol Vernallis, Jonathan Leal, Holly Rogers, Elizabeth Reich
and the contributors of *Cybermedia*

**Jonathan Leal**  Terence, would you like to share a bit about where you're at?

**Terence Nance**  Yes, thank you for inviting me, especially with the pandemic, when everyone feels isolated. I've been trying to maintain my energetic, spiritual, and physical sovereignty. Physical isolation, current events, and personal life matters seem to call for an elevated level of discipline. For me, physical therapy is one part of that. Everyone seems to be trying to stay functional at the end of the day.

**Jonathan**  Yes, there's the discipline of self-care, and, if you can separate them, the artistic practice.

**Terence**  Yes, they're inseparable. I like to think my and everyone's body are portals. I neglected mine in my youth. Today, I was daydreaming, trying to reimagine the last time my mom or dad told me to go to bed. I wondered about having kids, who might be 13 or 14, and how I might hint to them, "It's time to go to bed, man." My choreographer-friend Justina talks about how startling it is to see her child's behavior modeled from hers. I landed on that I should just ensure I go to bed when I'm actually tired, or when my portal (body) says so. This is part of my meditation.

**Carol Vernallis**  Sleep comes up a lot in *Random Acts of Flyness* (2020).

**Terence**  We were thinking about it, because no one was sleeping while we were making the show. It was paranormal shit when you're trying to make something in a capitalist environment, but the irony is that it made us hyperaware of the necessity of taking rest and what it is to feel entitled to it. The codename we used, because we weren't allowed to disclose the title on forms, was "Black Rest."

**Elizabeth Reich**  I'm writing about how you use "Redbone" during *RAoF*'s lullaby sleep sequence with the mom holding the baby. It reminded me of *Beloved*—I thought about the Black wake—the horror of stay woke, but also you best stay woke.

**Terence**  Oh, that's "Master Teacher," which samples "Redbone." The money to license "Redbone" was not in our budget. Erykah Badu and Georgia Anne Muldrow coined the usage of the word "woke" in that song. In their album, *New Amerykah Part One* (2008), they repeat "I stay woke, I stay woke, I stay woke." People say Black people were using that language before the song, but it became a pop cultural thing after.

**Carol**  At *RAoF*'s end, Solange hums two oscillating pitches of "stay woke" as a refrain. I feel viewers at this moment might sense they're holding on to the past, the present, and the future. Threading sound across one of your episodes would be nice to hear about.

**Terence**  Yeah. I think that one came out of a conversation about how mantras and mudras can be deployed for ourselves. Solange was then pioneering that in her music—she had a bunch of mantras in songs off her album, *When I Get Home*, and I became drawn to finding everyday mantras. Our use of mantras and songs hopefully mirrored our conversation, how everyone in the room or people in our family used them.

"I stay woke" for us, is a big thing. The balance between being vigilant and the entitlement to put that down is important, especially when you're trying to be in a family environment or hug your lady. But out in the world, you've got to stay super vigilant. Mantras can be key for the transition. My dad would come home and put on an album. A lot of people around me grew up with that relationship to music. *RAoF*'s sixth episode with mantras and repetition probably carries this over.

**Holly Rogers**  In all six episodes, I was struck by their use of repetition (I like the word mantra here); yet you never hear the same thing twice. Someone repeats something over and over again, and it becomes more profound, disturbing, or upsetting as it goes on.

**Terence**  It wasn't at first in my conscious mind, but by episode 6, we were like, yeah, we're doing mantras. I've noticed it especially in relationships—I'm hypersensitive to having to repeat things and cyclical arguments: why are we revisiting this, either in small, banal ways or in larger ones? That suffering, that I'm supposed to move to the other side of is part of me. For building resistance and moving past difficulty, repetition is important. Even for basic shit, like in order for your back to improve, you need to do your physical therapy every day. It literally says "reps" on your device. Oversimplification has that element too, you're going to forget the lesson, you got to keep repeating it, and that's just the way the world works.

**Jonathan**  And there's a kind of accrual. The more you repeat, the more it acquires, the more sediment attaches itself to it. Could you talk about your own musical practice, and how it relates to your writing and filmmaking? I love the EP you put out, *Things I Never Had* (2020).

**Terence**  Yeah, my process of making music is about engaging in a flow state and trying to find the moment of being a channel without judging whatever's coming through. I think everybody knows what that looks like, as if there's someone up there doing it, yes?

Unlike music, filmmaking privileges preparation, planning and predetermining, because of the resource conversation: you don't want to get there and waste money or time. I think it was important for my ancestors to make sure I was a musician first, so that when I come into the music thing, I didn't get colonized by the culture of moviemaking, which privileges conscious mind, masculinist, A to-B linear modes of making things. If we make space for what it should be, it'll come through.

**Jonathan**  That's beautiful. It's similar to what Quincy Jones says about making the beats for "Thriller": he tries to leave enough room, space, for the spirit to walk through.

**Elizabeth**  DuBois states that Black people have the challenge of carrying the burden of staying vigilant, but they also have the gift of double-consciousness, and he expresses these concepts through sound. How does double-consciousness show up in your work, and how does this reflect the communities of care you assemble with your collaborations?

**Terence**  Well, I came into making art with a head start. I was male, from a lower-middle class family of artists, but didn't have to skip a meal. While I had that head start on others, I was also behind, because I didn't know when you need some money to make a thing, you're not going to be talking to a Black person. Even when talking to a Black person, your concept of their work is colonized. I use colonial project as a descriptor of white supremacy I've been dealing with, because it often feels extractive. You're just a little mine, you're a little chocolate factory, you're an ivory coast. They come to you when they need the chocolate, and they pay you a penny and sell the chocolate for a dime, whatever it is.

I feel like I came into that understanding late, because I was hypnotized by the story of musicians. Musicians are able to catch on because of who they are, their work, and their style. I assumed people noticed you as an artist because you're interesting or have things to say, and they'll keep asking you to do more of it. But, in truth, the process is more collaborative, especially for the generation of Black musicians who preceded me. All of the R&B people when I was growing up were groups, like SWV, Dru Hill, and Boyz II Men, but by the time I got to high school, they were all individuals. The groups didn't go away; rather, they left the marketing conversations. I had to de-colonize myself about how the great-men-of-history stories narrativize cultural production, and I think DuBois's veil speaks to this. To make anything of consequence or with any level of emotional clarity, I needed to decolonize myself.

Colonization has deeply infiltrated the art world. When I went to art school, you were expected to sit in your studio and paint the painting. Whatever I made, I made alone. The evolution away from that began when Shawn Peters came up to me one day and said, "I got a camera, you want to shoot anything, we should do it." If a work is truly decolonized, it shouldn't be associated with Terence Nance or any individual, but rather, the choir made it. The same is true along the lines of gender, and the ways domination or submission becomes institutionalized. We're working through this. People get hurt on things I make and sustain damage. That's why I'm into pedagogy; we're intending to make actionable all the constant repairing that has to happen, and we acknowledge we're only at the story's beginning.

**Bevil Conway**   I'd like to ask you about your film editing.

**Terence**   I'd say it's ancestral, like all these things. My dad was a TV news cameraman. When I was much younger, in the late '70s, he went to Afghanistan, when the Mujahideen were fighting with or against the Soviets, as well as to West Africa a few times. As we got a little older, there's four of us, he started to work for local American news affiliates. In the '80s and '90s, all the shooters had to edit their own packages. So, I probably understood what editing was ancestrally from that, and I think he got it from his dad, who was an avid watcher of news.

And then I went to church every day growing up. I was in the audiovisual ministry, which was all run by kids, but it was also a major operation, because our Community Methodist Church was 3,000 members. A guy there, Arthur Porter, taught me, probably 11–12 years old at the time, how to run the linear editor. We were making these tapes for those who couldn't physically make it to church. That was my first practical experience in editing.

**Bevil**   *Random Acts of Flyness* is all digital, right?

**Terence**   No, there's film in it, and also everything else you could think of. We have a team of editors, and I do the final cuts. Some I only spend 2 weeks on, others 6–8 weeks. Josh Bohoskey and I work closely on the color grading to achieve a high level of granularity for each section, each world that gets created.

**Laura Wagner**   In your interview with Liz [Reich], you talk about space clearing and space claiming gestures. You say that, when it comes to Afrofuturism, we're thinking about where Blackness can be in relation to the past and present, and about how to expansively imagine Africanness. Then you add that it is not really your kind of space clearing move. What sort of spatial interventions are you engaged with now?

**Terence**   Yeah, I'm not interested in anything that feels claimy now. I'm most interested in sovereignty and exploration of the portal: friends, family, community, our bodies, and in a more inward-directed, granular level. Considering literal space, I've been daydreaming about what outer space signifies, and how distant I am from it. I wonder if I'm antagonistic to the concept of outer space. A friend told me, "We're in outer space right now. This is also space, you know?" But terra-centrality, the earth as center, overly animates space exploration. I find myself perturbed by the whole gesture of exploration as progress or escape. But paradoxically, I also believe our destiny is in the stars. I believe interdimensional, interplanetary entities are alive and 100 percent in contact, and our plane of existence is much larger than Earth. Maybe future travel won't feel like exploration.

**Sara Ferrando Colomer**   I'd like to ask you about *RAoF*'s Nuncaland scene, which for me represents freedom and healing. How did you use dance to represent this?

**Terence**   Nuncaland is one of my favorite parts, and it was made in collaboration with the whole room. Naima Ramos-Chapman, who grew up trained as a dancer, was my principal collaborator. I prompted her to make a musical, assuming dance would be integral, but she doesn't write songs. I don't remember if it was my idea or someone else's, but at that time, I was struck that Peter Pan symbolized men's arrested development, their fear of maturing into destroyers. I think Peter Pan, through subtext, articulates destructive masculinity. My prompt for her was to use a musical form, and have fun with it, and initially, at least for me consciously, the scene had no direction. That number was one of the hardest ones to make: at first, there wasn't any music, and even the words were on the fly. Shout out to Damani Pompey who's the choreographer. At the time, I was listening to this taped interview of my heroes Malidoma and Sobonfu Somé. Malidoma said that with dance and Dagara culture, when you're dancing and moving, you're not thinking, "I'm going to pick my right foot up and put it down there." He claimed dance conveys release, and I believe whatever shame and guilt I had internalized growing up can be exorcised through dance.

**Noah Fram**   I'm interested in the ways you use music, especially throughout *RAoF*, to shape how we feel about images. The tension between the musical and visual references are evocative. How do you fold music in with image?

**Terence**   We're walking in spirit. We're not thinking about it. It's ancestral, from the spirit realm. I think that's my gift—juxtaposing songs, melodies, rhythms in certain moments of time, related to images. Others call it editing, but editing's a weird word. Editing, understood within literature and journalism, reduces something. Juxtaposing, appearing, building, or sculpting is more appropriate. I'm drawn to collaborators who share a particular relationship to listening.

I think there are types of people whose nervous system are activated by certain rhythms, songs, and melodies. Others are attuned to taste or math. My hyperawareness-es are light and music. If I walk into a room and it's lit a certain way, I can either feel really good, or I have to leave immediately. I have a somatic response to different light situations, as well as with sound. That's also what drives the arranging process's meticulousness, but it's not like a meaning-making exercise in the moment. It's a channeling exercise, and then we figure it out a bit later or as we refine a thesis.

**Eric Lyon**   I stumbled on *RAoF* on my own, randomly, and it just completely blew my mind, especially "Bitch Better Have My Money." Part of what struck me was how it seemed like a piece of music. It's exactly 3 minutes, the length of a pop song, and yet it packs so much into it. How did that work of art get made?

**Terence**   I don't even remember. [laughs] That's a good question. Shit. I think the joke of its name and Rihanna's song. But some part of the genesis is from Ta-Nehisi Coates's transformative "The Case for Reparations." Part of the concept of Black people or my self-concept of our liberation is about revealing information. And then there's another layer; the tragedy is that everyone knows the information and shit's still fucked up. Coates's work is thorough, long, and deeply irrefutable. It's a prizefighter; it has all your little rebuttals and it's airtight, sewn up. People say, "Well, my grandparents didn't own slaves, I was a poor white person, I was Irish, whatever, I was an indentured servant," and he buttons that up. The article is not just about slavery, but also housing discrimination today in Chicago or in your parent's generation, and the wealth gap this produces. What are you going to say? You didn't grow up in Chicago, Baltimore, or any of these cities? Part of that vibe is hip hop to me, like the moment Nas released "Ether." I don't think Jay-Z put out a similar song after that. He was like, "You got it, my dude." I think even Rihanna's song, which has multiple writers, was derived from the energy of Coates's article.

And then there's a little joke about Silicon Valley and what technology should be purposed for. "Oh, you've made a more convenient rideshare," but the technology exists explicitly just to make money from the app, so that transactions are more financially convenient. A "Bitch-Better-Have-My-Money" app would only take Google two weeks—they'd buy Ancestry or 23andMe tomorrow or both, as well as the data, and then deploy the app. The sheen of hip hop and tech are both relevant, so the scene feels part of pop culture.

**Simon Levy**  Because I have no technical background in art of any kind, I can ask a very subjective question based on my own impressions. At the beginning of episode 6, there was to me something I can't get out of my head, a powerful scene with a lullaby, where you can only see people's eyes and mouths, and a woman singing to what seems to be a full-grown man or woman. It felt simultaneously terrifying, beautiful, and tragic. Someone online commented it might have been a representation of a slave ship's hold, because everyone's so tightly packed together. First, I was struck by the beauty and scariness of seeing people's eyes and mouths as they speak and sing, and one of the people who's just smiling and looking up. The other thing that occurred to me are these horrible racist tropes. I know Spike Lee has shown some of this—you can't see a Black person in the dark until they smile—to comic effect in *Bamboozled*. So many things I can't get out of my head from that scene.

**Terence**  You're definitely circling it. That's my favorite part. That's when the whole shit comes together. That expression, which is I think very much from all of us who made the show, is beyond words. Whatever you felt, hopefully I can frame it for you accurately. I hadn't thought about the slave ship thing, but that's cool.

What's most present for me right now to say about that scene is I practice Ifá. In Ifá, there's Olodumare, who are the state of being of all being, from which everything comes. You, also as a person, contain every sentient piece of existence. In Ifá, no human can see them, but they are sentient. And then your soul creates your Ori (your higher self, possessing three components). One remains up there with Olodumare and doesn't let you do anything without permission, but they are you, your portal. There's also Ori in your head; the word literally means head. So, your Ori goes through these primordial waters where you forget your destiny. When you're "alive," you have to go through all this shit, you have to remember it and stay on it, and you constantly are consulting with your head. This destiny thing, as part of a conversation of Olodumare, is probably the most relevant for that scene. I think there's truth to everything, even if it feels evil, crazy, or inconvenient. I think the whole blackface thing, there's truth to it. There's something really—getting blacker is really attractive and beautiful and funny and scary in all the ways that I want to be, like visually blacker. And blue-black. It feels like the place I actually am. If I could say what my energy is, if I could make a portrait of it, it's that. That is me. I think part of the gesture is being a little more flagrant about it, you know?

**Carol** Tomorrow, I'm interviewing Eric Weidt, David Fincher's colorist. I feel your voice is so open, possible, and expansive, and David's is telescoped—every single pixel is surveyed. What might the two of you teach each other?

**Terence** As I just said, no matter how much as I disagree with the thing, I'm starting to learn there's truth to it. Fincher looms large over my life in a weird way—I closely watched his films, *Seven* (1995), *The Game* (1997), and *Fight Club* (1999), in high school. *Fight Club* at the end destroys the financial system. His repurposing of the Shiva thing is fantastic. I draw on his work about the white male body in pain. Everything you read about Fincher suggests he's dialed in to shit that seems like it doesn't matter, but I learned from him about the reverence and respect for the moment, how valuable it is.

When we came to HBO, I said I'm not going to develop anything with anybody, this is what it is. I borrowed that approach from Fincher.

**Elizabeth** What's next Terence?

**Terence** We have *Random Acts* season 2 this year, and I'm also making an album. I'm finishing it up now and putting it out over the next few months.

**Jonathan** Fantastic. And what an amazing session today! Thank you for speaking with us, Terence.

**Terence** No problem. Thank you so much for your questions. I appreciate it.

17

# The Gift of Black Sonics

## Interface and Ontology in *Sorry to Bother You* and *Random Acts of Flyness*

Elizabeth Reich

In the long six years since Eric Garner's and Michael Brown's murders and the start of the Ferguson uprising, the U.S. has seen unremitting and increasingly militarized police suppression of Black protest.[1] Daily demonstrations against pervasive Black death continue today, often under the banner of the non-hierarchical activist network Black Lives Matter, which rose to public notice in 2013 when its three founders, Alicia Garza, Patrisse Cullors, and Opal Tometi, began circulating videos of anti-Black police brutality with the hashtag #BlackLivesMatter (#BLM). The work of #BLM, both in the streets and online, has created new networks that extend earlier ones: the Underground Railroad; the Black presses and basement circulations of early Black film; SNCC (Student Non-Violent Coordinating Committee) and its telephone-tactics;[2] community-based bail funds; and continued voter registration drives. Nonetheless, when #BLM delivered viral videos of police executions of Black civilians, the internet-driven communities that formed around them struggled to effect needed political change. Like twenty years earlier, when the failure of video-recorded eyewitness testimony to secure convictions for Rodney King's beatings resulted in uprisings, Ferguson–era artists and activists have met resistance in their efforts to make Black moving images sufficiently meaningful to non-Black publics. Filmmakers in particular have had to contend with the limits of Black

---

[1] Keeanga-Yamahtta Taylor, *From #BlackLivesMatter to Black Liberation* (Chicago: Haymarket Books, 2016).
[2] I've learned some history of these counterstrategies used by SNCC from conversations with activist and filmmaker Judy Richardson.

representation, and the innate problems in using moving images to create awareness of ongoing challenges to Black freedom.[3]

This essay shares a premise with contemporary Black studies scholars from Fred Moten and Alexander Weheliye to Kodwo Eshun: one answer to the shortcomings in the visual register *must* be Black sound. Whereas historically Black activists and artists have sought to demonstrate Black "humanness" through writing, photography, and film, here I argue that via sonic strategies, Black people have been able to move beyond the faulty category of the human for more expansive world-making projects. This sonics, Weheliye points out, is markedly in the domain of the technical that Blackness has historically been defined against (by virtue of being "behind" or pre-historical [according to Hegel] and unsophisticated); in the same vein, Moten and Eshun have pointed out that epistemologically and ontologically Black people survive as the "commodity that speaks" and the non-"human species."[4] In what follows, I argue that it is by technological, sonic means, like Ralph Ellison's 1952 *Invisible Man*'s "lower frequencies," that Black media makers in the post-Ferguson era have articulated Black meaning and power; and particularly, Black peoples' technologies of times, spaces, and communications that subvert and exceed the limits of mainstream, hegemonic paradigms for earth-bound and temporally-progressive "human" life.

In this chapter, I explore a critical cluster of recent moving-image productions sharing common speculative aesthetics, political concerns, and even writers and casts—from Steve McQueen's *Twelve Years A Slave* and Jordan Peele's *Get Out* (2017) and *Us* (2018), to Terence Nance's fantastical HBO series, *Random Acts of Flyness* (2018), with a special emphasis on activist Boots Riley's dystopian *Sorry to Bother You* (2018).[5] These productions, I argue, have specifically deployed sound to critique the un-representability of Black life in a world in which Black living has been so thoroughly endangered. In doing so, they have produced something new via sonic technics: an alternative ontology of Blackness extending

---

[3] Courtney Baker, *Humane Insight: Looking at Images of African American Suffering and Death* (Champaign: University of Illinois Press, 2015); Michael Boyce Gillespie, *Film Blackness: American Cinema and the Idea of Black Film* (Durham: Duke University Press, 2016); Jacqueline Goldsby, *A Spectacular Secret: Lynching in American Life and Literature* (Chicago: University of Chicago Press, 2006).
[4] Fred Moten, *In the Break: The Aesthetics of the Black Radical Tradition* (Minneapolis: University of Minnesota Press, 2003); Alexander G. Weheliye, *Habeas Viscus: Racializing Assemblages, Biopolitics, and Black Feminist Theories of the Human* (Durham: Duke University Press, 2014), 3; Kodwo Eshun, *More Brilliant than the Sun* (London: Quartet Books, 1998).
[5] Special thanks to Stephen Yeager and Mary Zaborskis for helping me identify this critical cluster.

beyond the "afterlife of slavery,"[6] bounded neither by the painful rhythms of what Christina Sharpe calls "wake"[7] nor even by the "break,"[8] within which Moten imagines an ecstatic irruption in hegemonic time and sound. Rather, I show that the new expansive worlds of Blackness can only be apprehended from our perspectives *as* "irruptions," momentary or anomalous events, though they are much more. This is in part because we most often find ourselves stuck in Cartesian perspective and Western historical accounts, what Walter Benjamin has described as "empty, homogenous time" (the sovereign's "history") from which we cannot perceive Black life fully.[9] In fact, mediated Black sonics compel and instantiate life and movement *across* and *beyond* such unlivable times and spaces, insisting on Black lifeworlds of uncontainable dimensionality. In what follows, I offer readings of *Sorry to Bother You* and *Random Acts of Flyness* to show how sound acts as an interface between characters, audiences, and Black time-space. I show that this interface is at once technical, temporal, and embodied—similar to philosopher Bernard Stiegler's epiphylogenetic man, who is defined by the tools he uses and the records he keeps; and that it is also a specifically Black and epigenetically shaped body-technology like the now-famous "veil" W. E. B. Du Bois describes in the "Forethought" to his 1903 text, *The Souls of Black Folk*.[10] Via and in the interface, Black life resounds forcefully and reproduces itself in art, politics, and culture.

## Black Sonic Technics and the Interface

In *Souls*, alongside stories and analyses of Black life in the United States, W. E. B. Du Bois posits the essential workings of a "veil" he himself lives within and manipulates to make meaning. He states that "the worlds within and without the Veil of Color are changing, and changing rapidly, but not at the same rate, not in the same way"; their disjunction creates the Black Americans' "double life, with

---

[6] Saidiya Hartman, *Lose Your Mother: A Journey along the Atlantic Slave Route* (New York: Farrar, Straus and Giroux, 2008) 6.
[7] Christina Sharpe, *In the Wake: On Blackness and Being* (Durham: Duke University Press, 2016).
[8] Moten, *In the Break*.
[9] Walter Benjamin, "Theses on the Philosophy of History," in *Illuminations*, ed. Hannah Arendt, Trans. Harry Zohn (New York: Schocken Books, 2007), 253–64. This is "state time" and the time that prevents and precedes the reading against the grain of history.
[10] W. E. B. Du Bois, *The Souls of Black Folk: Essays and Sketches* (Chicago, A. G. McClurg, 1903. New York: Johnson Reprint Corp., 1968).

double thoughts, double duties, and double social classes, [that] must give rise to double words and double ideals."[11] In this formulation, the veil stands out as a sign and signifier of the psychic and sociopolitical dissonance between the hegemonic world in which African Americans must live as second-class citizens (at best) and the extra-temporal world of Black life. The "veil," which is the product of a "double-consciousness" that is molded by the pain of holding such dissonance or paradox, is also Black people's most effective technology for resistance—a place to hide as well as a tool or technics that can be manipulated to survive. Frantz Fanon describes his version of Du Bois's double-consciousness as a "triple" consciousness resistance,[12] because as a technology of resistance the veil in fact presupposes and creates a third component of itself. Inherent hardship in African American identity "tempt[s] the mind to pretense or revolt, to hypocrisy, or radicalism,"[13] turning the veil into an interface through which such revolt, hypocrisy, and radicalism meet and find expression. Consequently, the veil becomes a synecdoche for a Black people shaped by inherited trauma as well as the consequent challenges and gifts of being Black in America.

Du Bois begins by promising to "lift" the "veil" that occludes (white) apprehension of Blackness, so that readers might "faintly see the deeper recesses" of "the spiritual world in which ten thousand thousand Americans live and strive."[14] In Du Bois' formulation, the veil is a result of Black suffering, a part-flesh interface that manipulates a relationship with whiteness, and a mechanism for producing (self) knowledge that inheres in "souls" and "bone and flesh"[15] of Black people. The veil thus performs a tripartite function: ideological, technical, and epistemological. Bernard Stiegler theorizes in *Technics and Time, 1* that the human becomes human only as a prosthetic being along with his use of tools. Tools, like the "veil," mark and reveal distinctly human time (part of what Stiegler calls techné). Further, they reflect the metaphysics of a human that co-evolves with his tools: as an always-technological being. Bringing Stiegler's theorization of technics to Du Bois's similar but racially specific articulation of the veil from nearly a hundred years earlier, I understand the veil as a Black technics that Black peoples may manipulate but do not fully control; a technology that cloaks a brutally vulnerable people and also the dynamic, a-linear, extra-temporal space in which Black people live.

[11] Du Bois', The Souls of Black Folk: Essays and Sketches, 128-9.
[12] Frantz Fanon, Black Skin, White Masks (New York: Grove Press, 1967), 96.
[13] Du Bois', The Souls of Black Folk: Essays and Sketches, 129.
[14] Du Bois', The Souls of Black Folk: Essays and Sketches, 8.
[15] Ibid., 8.

Here, I draw on Stiegler's broader contemplation of recorded memory and "epiphylogenesis"[16]—through which human memory records become externalized and understood as almost always prosthetic—in order to show how we can productively understand Du Bois's veil anew: as a quasi-genetic Black technology that has epigenetically produced the Black body itself.[17] The biological term epigenetics describes a process by which heritable changes, such as an individual's or group's life experiences, affect gene expression, and so Stiegler describes epigenetic transformation as a stage of producing records of human living that precedes epiphylogenetic experience. For Stiegler, humans remember *in time* by externalizing their memory via records, tools, writing, even computers, *and then experiencing*, for the first time, such memory and ancestral knowledge through that record. This is the process of epiphylogenesis, the ultimate expression of technical-humanity, and it only occurs with particular media forms that have belatedness built into their technology. In the case of sound, Stiegler argues it be most apprehended neither as it is sung nor as it is recorded, but as it is listened to from a recording. The recording technology itself creates its own genetic and historical embeddedness that returns the melody to it.

In this chapter, I argue that *Souls* offers a more theoretically consequential concept specific to Black bodies—*already doubly technical* by virtue of their use as labor by white people—by making non-linear Stiegler's epiphylogenetic process, and identifying Black life as extra-temporal because of its epigentic flesh. Du Bois describes the third feature of the veil *as flesh* that, like a caul which has somehow remained long after birth, occludes seeing and also "gift[s] with second-sight."[18] As an inherited effect of Black trauma, the flesh-veil produces Black "see[ing that is both] ... through the revelation of the other world" *and beyond sight, a radar more than vision.*[19] The *flesh* thus is the experiencing of the record *and* the recording both.

Hence, Du Bois uses sound to describe the veil, opening each chapter of his book with bars of melody that cannot be translated but "echo" and "haunt" the book.[20] "Sorrow Songs" and other Black sounds Du Bois depicts are the

---

[16] Steigler, Bernard. Technics and Time, 1: The Fault of Epimetheus. Translated by Richard Beardsworth and George Collins (Stanford: Stanford University Press, 1998), 135.
[17] Today, for instance, we might speak of how epigenetics encodes genocide, ensuring changed bodily outcomes for generations within a demographic or community of people. Janell Ross and National Journal, "Epigenetics: The controversial science behind racial and ethnic health disparities," *The Atlantic* March 20, 2014. (see https://www.theatlantic.com/politics/archive/2014/03/epigenetics-the-controversial-science-behind-racial-and-ethnic-health-disparities/430749/ for reference.
[18] Du Bois, The Souls of Black Folk: Essays and Sketches, 13.
[19] Ibid.
[20] Ibid., 8.

unrecordable transgenerational tissue of Black life and its doubled-technicity. Without its genetic quality and "gift," Du Bois' veil would perform *only* the adverse a priori conversion enacted by Hortense Spillers' slaver's whip: transforming Black flesh into a "captive," marked by untranslatable signs of becoming never/no-longer-human but slave-gendered, a "hieroglyphics" on the body that lives on in a "zero" time.[21] But because of "the gift," the veil can do more. Where it is sonic, I argue, this Black technology moves people and consciousness beyond the times and spaces of anti-Blackness, signaling a world of existence in a dimensionality not fully audible and certainly not clearly visible from the homogenous here and now.[22] So Black sonic strategy survives Du Bois by more than a hundred years to haunt Black cinema as well, undoing Stiegler's teleology of epiphylogenetics: replacing dominant technological production with the Black body itself as the authoritative record of Black life.[23]

Though neither via consideration of the veil specifically nor Stiegler's theory, Moten underscores how Black expression that might be apprehended as exteriorized (like a record) is in fact always also what we could call a Black somatics. In his later work, Moten focuses, too, on Black hapticality, writing that Black diasporic peoples—with their collective and collectivist origins in the packed "hold" of the slave ship—are philosophers of "the feel."[24] For him, as for Stiegler, human (sonic) expression is part of human ontology. For Moten, the extra-timely nature of Black sound occurs in "the break," a technical and also epistemological time-space media scholar James Snead defines as the cultural (and culturally-compelled) difference of Black living from hegemonic Western (Enlightenment) practices of signification and telling time.[25] But it is also in this break—a time-place *for* Blackness like that which Afrofuturist Black Audio Collective scholar Eshun describes as the "counterfuture" and, before them both, extra-planetary musician Sun Ra called the "alterdestiny" of the Black planet

---

[21] Hortense J. Spillers, "Mama's Baby, Papa's Maybe: An American Grammar Book," *Diacritics* 17, no. 2 (1987): 67.
[22] For a powerful example of sound working to create new worlds and destroy anti-Blackness, see Sun Ra's (Black Afrofuturist musician) 1972 film *Space Is the Place* (New York: Plexifilm, 2003).
[23] Arthur Bradley, "Originary Technicity: Technology & Anthropology," in *Technicity*, eds. Arthur Bradley and Louis Armand (Prague: Litteraria Pragnesia, 2006). I'm indebted to John Landerville's suggestion that I use the specific term "epiphylogenesis," rather than just "technics."
[24] Fred Moten, "Black Optimism/Black Operation," October 19, 2007, 7. https://lucian.uchicago.edu/blogs/politicalfeeling/files/2007/12/moten-black-optimism.doc (accessed February 8, 2021).
[25] The "break" is the key term for Moten's seminal text, where it serves as a time and space for Black improvisation. Moten's "break" extends what James Snead has called "the cut" (a sonic event and reiterative temporality in Black music that is intrinsic to Afrodiasporic culture). Moten, *In the Break*, 6; James A. Snead, "On Repetition in Black Culture," *African American Review* 50, no. 4 (2017): 648–56.

"after the end of the world"—it is here that Blackness performs its potent, instantiating, and politically-transformative "operations."[26]

Because my focus is, like Moten's, on Black embodiment and its most sensory and haptic practices, I also follow Moten's commitment to "move by way of a ... resistance to ... anti-essentialism."[27] This resistance I understand to be a recuperation of Blackness as a physicality that interacts with its ontology. Moten describes his movement by way of resistance as toward "trying to own ... the underprivilege of being-sentenced to this gift of constantly escaping and to standing in for the fugitivity ... that is an irreducible property of life, persisting in and against every disciplinary technique while constituting and instantiating not just the thought but the actuality of the outside that is what/where blackness is—as space or spacing of the imagination, as condition of possibility."[28] Moten links "trying to own ... the underprivilege" with a carceral "gift" of a Blackness that is "fugitive" both because of and despite the destructive effects of encompassing, ongoing "disciplinary techniques" (emphasizing in the word play the disciplining of bodies via biopolitical structures like prison, military, census taking, medicalization; and also the brutal divisions enacted by [quasi-]academic discourse). Moten's "trying" is paradoxical because it reveals the commodification and imprisonment of Blackness, which must escape and therefore should *not* be owned—and so it also describes the goal as the *"underprivilege,"* the resistance, and the condition of possibility, rather than the stable property of "self." For him, as for Du Bois', it is the dissonance between white hegemony and Black experiencing retroactively that creates "essential" Blackness: the "actuality of the outside that is/ what Blackness is." Blackness: a simultaneity folding back on itself to instantiate itself, combining the above conditions *while also "constituting" itself as* "not just the thought but the actuality" of the space wherein it exists, an "outside," "space," and "condition of possibility." Thus, I understand the "gift" Moten describes to be like that of Du Bois' veil: body and strategy both, instantiating and operating within the "essential timelessness of the always already existing," the "alterdestiny" that is immanent now as living, full Black life "what/where blackness is."[29] But whereas Moten recuperates essentialism that has been rejected by critical race studies, my

---

[26] Eshun, Kodwo. "Further Considerations on Afrofuturism," *The New Centennial Review* 3, no. 2 (2003): 288; Sun Ra and his Intergalactic Solar Arkestra, *Space Is the Place* (New York: Plexifilm, 2003); Moten, "Black Optimism/Black Operation."
[27] Moten, "Black Optimism/Black Operation," 2.
[28] Moten, "Black Optimism/Black Operation," 4.
[29] Ibid.

theorization of the genetic technics of Blackness operates within the fault-line between genetic and cultural arguments about race. I am not claiming that we should reify race as real or biologically determined, but rather that Black people who have so long been the objects of medical experimentation have already theorized and actualized a quasi-genetic, coded, and technological mechanics to deploy against anti-Blackness in freedom struggle.[30] In what follows, I understand the "gift" of Du Bois's veil and Moten's fugitivity as just such a mechanics: an interface determined by its self-same flesh and code. I read sound in *Sorry to Bother You* and *Random Acts of Flyness* as constituting this interface and as describing and instituting Blackness as a collaborative and communal category of being. Growing with the Black body and preserving and reproducing Black experience and political resistance *in time*, Black sonic collectivity and its interface create an ontology for Blackness alternative to that we have traditionally observed in cinema: participation in apocalyptic and ecstatic lifeworlds at the lower frequencies, in cosmic waves, and in freedom we at best can only begin to imagine.

## *Sorry to Bother You*: White Voice and Black Silence

At the opening of *Sorry to Bother You*, the audience observes a world that deals in ironically transparent deception mediated by sound, and in which Black sonic manipulation negotiates between a potentially liberatory practice and a repressive technology of Black false consciousness. At telemarketing company RegalView, employees, particularly if Black, make sales; if ambitious, they manipulate their voices to gain advantage by sounding "white." Protagonist Cash gets his job at RegalView after fabricating false awards and a fake resume revealed by a reference check. The reference, Cash's friend Sal, is absent except sonically, where his voice and diction on the answering machine fails to pose as Cash's former boss. Cash nonetheless gets the job for his efforts at falsification, and he begins the work of selling undesirable products to undesiring consumers over the phone via scripts provided by the company (figure 17.1). Cash—named for a measure of exchange—is Moten's "commodity that speaks," but as such he fails. All of his efforts to manipulate the sonic are unsuccessful—and we soon realize Cash fails because he believes that he has the agency to manipulate *the terms of*

---

[30] For more on race, medical experimentation, and in particular Henrietta Lacks, see Keith Wailoo, *How Cancer Crossed the Color Line* (New York: Oxford University Press, 2010).

The Gift of Black Sonics 291

**Figure 17.1** Cash taking calls at RegalView in *Sorry to Bother You* [dir. Boots Riley, 2018].

*his life*. He does not, both because he is Black and because he is a worker—such are the intersectional politics of director Boots Riley, well-known Black anticapitalist activist and musician.

*Sorry to Bother You* insists that the audience, like Cash, suffers from the false belief that because we are agents of our actions and choices, we can perceive the world accurately. Watching, we laugh at Cash's mediocrity and enjoy the realistic acting within a familiar but nonetheless speculative diegetic world. In one absurdist scene, Cash's comfortable studio apartment, where he is making-out with his girlfriend, loses its third wall (normally the audience constitutes and thereby suspends disbelief about the missing *fourth* wall) as it retracts into the ceiling. Despite knowing a wall is missing by virtue of our presence, we viewers have already invested in the solidity of the false representation of an apartment. But Black life does not have access to such solidity. Cash's domicile is an aboveground, attached garage on a trafficked street in the middle of an unaffordable, fast-gentrifying Oakland. With semi-naked Black bodies now visible to folks walking to work, the domestic has become a showroom for the sexual commodity after all—and the audience has paid its ticket to watch. Cash suffers from false consciousness—rather than the double consciousness gifted by the sound-veil—because he believes his situation to be one of marginal empowerment, when he and his predicament are so clearly on display.

Cash struggles at his new job until an older co-worker (played by Donald Glover, Sr.) advises him to use his "white voice" with customers, inaugurating the thematic and praxis that will define Black life as double, performative, and epiphylogenetically technical for most of the film. Ironically, Glover's ability to be Black *for* white directors and audiences in the past has been a survival strategy analogous to the one his character advises Cash adopt. Cash accepts that white sounds and meanings will dominate and shape his life if he doesn't at least acknowledge their hegemony and learn how to Blackly deploy them. Cash and his co-worker's "white voices" manifest as eerie voiceovers: their mouths say the words, but the voices are utterly someone else's. Glover's uncannily synched white voice is that of Patton Oswalt, and Cash's, which is ultimately even more successful in the world of the film, is spoken by David Cross.[31] Phenomenally extra-diegetic, these voices somehow still supplant a profoundly Black silence (figure 17.2). Black minds and mouths move but cannot sound out agency in the ontology of *Sorry to Bother You*. As the film goes on, Cash is forced to use his white voice more and more, until, at times, even his Black sounds are no more than those anticipated by whiteness (as when mid-film he is required to "rap" and can only yell "nigger shit, nigger shit, nigger shit" to the beats of '90s rap. He is successful because "nigger shit" is all that whiteness expects).[32]

The scene and its jarring effects enact the falsification of race and racial representation, as well as the non-indexicality of the Black film image. In a TV news interview, Boots Riley describes Langston's mentoring of Cash as "explain[ing] ... [t]here is no real white voice. Everything we're doing is a performance and ... the white voice ... is this sort of mythical thing that says, 'Everything's okay.' And that's the performance of whiteness."[33] The Black actor and his movements and sounds refer to things from the diegesis and the world of the audience both, but they can't be matched to their referents because the image is a false one, constructed by script, director, film stock, actor, camera, production and reproduction, projection, and the space and community of the theater. The product was also created *to deceive*, to sell a Blackness that is *not* irruptive or powerful, and which cannot vocalize the dissonance within the ontology of race and racialization.

---

[31] Both actors' voices are uniquely recognizable as both began their careers in stand-up comedy before appearing in bigger film and television projects like *Ratatouille* and *Arrested Development*. More to the point, these specific actors freight a broad nerd culture cache stemming from their ubiquity across niche comedy shows, video games, and comics. Cross's voice is featured in the video games *Grand Theft Auto: San Andreas* and *Halo 2* (for which he won the G-Phoria award for Best Voice Male Performance).

[32] John Landreville's insights about the voices synched to Glover's and Stanfield's characters, and the white operations they perform, helped me to understand the potency of the Black silence here.

[33] CBS, "Director Boots Riley on the 'mythical white voice' in *Sorry to Bother You*," *CBS This Morning*, July 16, 2018. https://www.cbsnews.com/news/director-boots-riley-on-the-mythical-white-voice-in-sorry-to-bother-you/ (accessed February 8, 2021).

**Figure 17.2** Cash improvising Blackness-in-white-terms for partygoers in *Sorry to Bother You* [dir. Boots Riley, 2018].

As Cash's "white voice" dominates Cash, it becomes *the* voice of the film, speaking against a unionization movement as well as selling slave labor and the institution of slavery itself, formalized in brick-and-mortar structures and a global corporate conglomerate. However, by the film's conclusion, Cash's voice calls for a class-based resistance. This voice and its movements through race, time, space, and the fundamental violence that structures their formation under what Cedric Robinson calls "racial capitalism," perform the work of the veil and Black technics, serving quite literally as the cloaking, revealing, and spatializing apparatus of interface.[34] Toward the end of the film, we learn that Blackness is a genetics that itself can be modified, and therefore that Cash's sounds were potent mechanisms of soon-to-be silenced critique as he undergoes genetic transformation. I also read these Black sounds as exemplary of my corrective to Stiegler's "epiphylogenetic memory" construction: while in hegemonic human-technics "epiphylogenetic memory" is produced external to organisms, that is, exosomatic[ally],"[35] Black sound recording leaves its clearest record in Black embodiment.

Because Blackness and its sounds are, according to *Sorry to Bother You* and the moment in which its aesthetics and logic are both embedded, always also *in* the body, Black epiphylogenetics and the Black cyberveil is *endosomatic*, a true "neurotechnology."[36] Thus, moving itself and the audience across the world of Blackness, sound becomes a deconstructive tool compromising racial capitalism.

[34] On "racial capitalism" see, Cedric J. Robinson, *Black Marxism: The Making of the Black Radical Tradition* (Chapel Hill: University of North Carolina Press, 2000).
[35] Bernard Stiegler, *The Neganthropocene*, Trans. Daniel Ross (London: Open Humanities Press, 2018), 216.
[36] Ibid.

Through its Black operations, sound collects us with the fugitive flesh of other laborers, no longer scarred by hieroglyphics, and now able to transmit messages of solidarity and struggle via phones, whistles, and potentially non-diegetic communication. Though the white voice has been the sound of anti-Black violence in the film, Black sounds and new technologies become a hopeful resistance by the end. Perhaps no surprise—since Boots's band, The Coup, first conceived and produced the story in a 2012 concept album *as* sound.

At the end of *Sorry to Bother You*, Cash has become a strange half-horse worker in the midst of revolt, perhaps for pay or perhaps as an ultimate consequence of participating in anti-Black economics. Sonic action (including throwing old cell phones into a bucket at a naked performance) has instantiated diverse false and/or expansive modes and frequencies for Blackness. iPhone and genetic technologies have variously delivered messages of Black and laboring ontology, and they have provided a shared interface by which an intersectional politics of rebellion, masking, and subterfuge has begun to mobilize in the last scenes. But we still don't know what form of being Cash is or to what end he may manipulate his cyberveil: to carry himself elsewhere; to recapitulate a form of status quo with a radical embrace of his ontology; or to lead the revolution.

## Cash for Brains: Movement, Genetics, and the Black Interface

Elsewhere, in Jordan Peele's *Get Out*, a transmedia version of Cash as Andre King, a Black man with only a piece of his parasympathetic nervous system left where his brain used to be and a white brain and consciousness living in its place, has already lost that which he in fact never owned (according to Afropessimist theory).[37] Essentially, he has been made a passenger in his own head. Protagonist Chris Washington encounters this hollowed-out Andre as the only other Black person at a white gathering, which foreshadows the way whiteness will treat his own Black body later in the film. Across *Get Out*, psychotherapeutic hypnosis enacted through the sounds of a teaspoon moving circularly, rhythmically on porcelain, is the sign of how white power freezes Black bodies, both in the film and extradiagetically. In one scene, both the

---

[37] Afropessimism is a theoretical paradigm associated with Orlando Patterson's concept of Black social death described in Orlando Patterson, *Slavery and Social Death: A Comparative Study* (Cambridge: Harvard University Press, 1982).

Figure 17.3 Chris receding into the sunken place in *Get Out* [dir. Jordan Peele, 2017].

discourse and sign of white imperialism (psychology, with its origins in a highly racialized and anti-Black teleological Hegelianism, as well as the traded Asiatic tea-cup) operate sonically to paralyze and sink Chris into the visually rendered "sunken place" that is an inversion of the agency and pleasure provided by the contrast between dark and light in the film theater (figure 17.3). It is also a version of the paralytic terror Fanon describes while waiting for a film's debased image of Blackness to appear—itself the representation of a paralyzed and amnesiac Black soldier from *Home of the Brave* (a 1949 American film about racial integration mediated by a psychiatrist who handily converts racial trauma to shell shock). Fanon describes his state in what he and, after him, Kara Keeling, have called "the interval"—the time of an ontologically death-bound Blackness— in which his body becomes "epidermaliz(ed),"[38] sutured to his skin and its Black meaning in the social.[39] In *Get Out*, Chris escapes Fanon's momentary fate by realizing that he must refuse white sounds. Watching anti-Blackness while physically restrained, Chris retains his Black consciousness by blocking his ears with cotton, the sign and product of slavery. One of the dominant Black sounds in this film in which most Black sonic communication is blocked is the song

---

[38] The term is Franz Fanon's from *Black Skin, White Masks*. Frantz Fanon, *Black Skin, White Masks* (New York: Grove Press, 1967).
[39] Kara Keeling, "'In the Interval': Frantz Fanon and the 'Problems' of Visual Representation," *Qui Parle* 13, no. 2 (2003): 91–117.

"Redbone" by Childish Gambino (Donald Glover, Jr.), intoning "Stay Woke." And because Chris does resist white hypnosis a second time, he survives.

In the conclusions to both *Get Out* and *Sorry to Bother You*, Black genetic essentialism is marked as the praxis and sign of Black survival. Chris has retained his whole brain; the other Black folks in *Get Out* have a small but essential and irruptive component of theirs—that can be awakened into consciousness and sensory-motor agency by a flash—while Cash's has been fundamentally altered by a gene therapy. This alteration may release him from the silence and empower him, albeit as a non-human worker rather than a specifically Black man, to incite revolution; or it may remove him forever from the possibility of transforming into the kind of freer, more operational and optimistic Black embodiment that his sometimes-girlfriend, Detroit, names.

## *Random Acts*, Flyness, and the Sounds of Black Collectivity

Terence Nance's first film, *Oversimplification of Her Beauty*, was a complex visual and sonic desynching of actor, character, plot, and aesthetics—and failed to propel Nance to the success of the other directors in his cohort listed above. Yet in 2018, with HBO series *Random Acts of Flyness*, Nance chose to produce a complex collection of interwoven Black sounds and images, developed collaboratively and through sampling (a practice common to Black music[40]) and layering of Black artists' audio-visual works.[41] *Random Acts* distinguishes itself from Nance's earlier productions with a more direct political address, more extensive collaboration, and the stated goal of transforming Black being. The tagline is #shiftconsciousness. Opening with references to police brutality, and including sections on queer embodiment as well as musical numbers and hyper-virtual and speculative aesthetics, *Random Acts* instantiates its own shifting and collective consciousness that invites a multiplicity and simultaneity of sonic, visual, and identarian perspectives. As a sonic cyberveil, *Random Acts* and *its* #shiftconsciousness allow for Black improvisation, transformation, and, most

---

[40] Tricia Rose, *Black Noise: Rap Music and Black Culture in Contemporary America* (Hanover: University Press of New England, 1994); Kodwo Eshun, *More Brilliant than the Sun* (London: Quartet Books, 1998).

[41] For more on Nance's work as a collaborator, particularly one who works with sound and image separately, see my discussion of Nance's documentary, *The Triptych*, which he also distributed via Black communal circuits through Afropunk in 2013. Elizabeth Reich, "Documentary Strain, Black Artists, and the Afrofuture: Terence Nance's The Triptych," *World Records* 3, article 5 (2018): 46–60.

importantly, a form of Black sociality supported also by what Nance has called Black "sovereignty."[42]

Despite *Random Acts* incredible journey through a transporting tapestry of Black voices and visions, the show's final episode begins disconcertingly with a scene of Black vigilance during a dark time-place of some kind of middle passage.[43] Any "joyful tarrying" of Black-sonic-sociality, which Ashon Crawley has described as an enlivening and freeing sonics of Blackness collectivity, seems here curtailed by disturbing visual and sonic rejoinders. The first is the image of a living pattern, perhaps the barely perceptible motion of fish or krill in a great, dark ocean. Elemental, slightly-twitching beings are everywhere, filling the liquid blackness with movement and tiny crescents of light. As the camera moves in slowly, the crescents grow apparent as an unending multitude of terribly white smiles, glowing teeth forced into visibility by the rigid, restrained, and fixed lips of Black faces stuck in masks of terrified, falsified smiles in an inky, destroyed darkness. This initially visual journey into the veil (by way of the veil) presents as a brutal emotional confrontation with the horrors of slavery or post-slavery or middle passage, all, the time-space in which Toni Morrison's Sethe slits her daughter Beloved's throat to save her from the enduring unliving.[44] But the full journey inside is sonic, guided by the sounds of the mother singing. The melody carries the image and weaves the scene and its ongoingly slavery/death-driven temporality across the episode, haunting the other would-be histories and futures of Black life with this terror. It travels with us as we watch the show and works to transform not only our viewing subjectivity but also our possible places and our beings within the episode as well.

Rocking her large, frightened daughter there, perhaps below the waves,[45] the woman and her voice become an exosomatic interface we as audience cannot fully penetrate but which nonetheless penetrates us with its Black time (figure 17.4).

---

[42] With "Black sociality" I'm drawing on concepts like Black study and fugitivity from Fred Moten and Stefano Harney's *The Undercommons*, as well as the embodied, aural, and perhaps more joyful idea of "choreosonics" in Crawley. Stefano Harney and Fred Moten, *The Undercommons: Fugitive Planning and Black Study* (New York: Minor Compositions, 2013); Ashon T. Crawley, *Black Pentecostal Breath: The Aesthetics of Possibility* (New York: Fordham University Press, 2017). In Cybermedia's group interview with Terence Nance on January 14, 2021, Nance described "sovereignty" as the privilege to rest and put down vigilance. It's my sense that this would be connected with the sociality I'm theorizing here, though not a prerequisite for it.

[43] Such an evocation here seems Afropessimistic, depicting Black life "in the wake," in an "afterlife of slavery," and dominated by "social death." Frank B. Wilderson, *Afropessimism* (New York: Liverlight Publishing Co. 2020); Sharpe, *In the Wake*; Hartman, *Scenes of Subjection*.

[44] Toni Morrison, *Beloved* (New York: Alfred A. Knopf, 1987).

[45] Sharpe, *In the Wake*.

**Figure 17.4** A mother sings a lullaby in an alternative sunken place: in the hold, in the wake, or somewhere below the waves in *Random Acts of Flyness* [dir. Terence Nance and produced by MVMT A24, 2018].

While the two women are trapped by whatever substance constitutes the darkness and also the tight space between the unfathomable others in the same place, the singing traverses the times and diegeses of the show and takes us simultaneously outside of and deeper within the episode through historical cultural references. The mother-knowledge in the song is itself two songs in one voice, much like Childish Gambino's "Redbone," though here it is Erykah Badu's earlier 2008 "Master Teacher" transforming the familiar lullaby into "Hush little baby, don't stay woke." Shifting between sounding hope for "a beautiful world" and imagining the end of anti-Blackness in lyrics "What if there was no niggas/Only master teachers," and the vigilance a Black woman must practice in "staying woke" here and now, Badu interweaves Black sonic movement toward freedom.[46] In the traditional lullaby, the child is supposed to sleep and be lulled by belief in the maternal presence through calming promises,[47] but in Badu's song, there is a high stakes

---

[46] Erykah Badu, "Master Teacher," Track 8, *New Amerykah, Part 1* (New York: Universal Motown, 2008).
[47] The category of maternity is one interrogated in Hortense Spillers "Mama's Baby, Papa's Maybe" and what she calls the "grammar" of America. If the slave's "flesh" is gendered by slavery, then its gendering binds it to reproduce slaves: Papa may be unknown, but "Mama" is a false consciousness. By definition a woman cannot produce commodities; and a commodity cannot be a Mama. The mother in this scene desperately attempts to perform her maternal function, even through sound—a register in which she might hold more agency and promise. But, because the category of motherhood is drowned in the passage, her song too is haunted by a Black-death-ontology.

tension: how "woke" must Blackness be? In *Sorry to Bother You*, sounds of false consciousness reproduce slavery; in *Get Out*, soporific sonics and their paralytic sleep bring about the de-braining and death of Black being. But no one can survive without rest, and a Black mother has to listen, look, and pick her moment. In *Random Acts*, she urges her child to relinquish knowledge, the comfort of the lullaby, and even the sure knowledge of parentage, because who knows who Mama's baby is really, amid a sea of wakes and the drowned of middle passage, and the many, many times and spaces of the veil. Here the technology of sound travels, and it warns against death; but its message and technics are impossible to fully discern because they are manifold, drawing Blackness through time but also into danger, out of danger but also into a sound-time that is always imperiled.

The mother's lullaby forms not only a sonically and semantically legible intertext for *Random Acts*, but also the ways in which Black sound (and image too) circulate, repeat, cut, and recirculate, like the repetition that Snead describes as fundamental to Black culture and its sonic rhythms. And both in the mother-daughter dyadic body and in Badu's *New Amerykah*, the song serves epiphylogenetically to record and instantiate the Black "gift" of the veil-interface: here Black communal knowledge of needed vigilance on the way to "a beautiful world I'm trying to find."[48]

Despite showing scenes of awful pain, *Random Acts of Flyness* resists the kinds of sonic violence we observe in *Sorry to Bother You* and *Get Out*, giving its characters simultaneous existences across multiple dimensions of the diegesis and the world that has produced them. It aims to set Blackness free and shift viewers' consciousness toward freedom—in part, however paradoxically, through the kind of intertextuality performed by the lullaby. For Shaka McGlotten, another Black studies theorist, such layering is a functional strategy that confounds Black legibility to those tracking Black movement and resistance, allowing Black people to travel. Here, the layering, though itself an instantiation of the burden of the veil, is also liberatory and sounds out the "choreosonics" that Crawley describes as producing a Black lifeworld elsewhere via sound-movement. Here, there is collective "untranslatable" sounds and "joyful tarrying" enabled by the veil's sonic transport—sonic freedom rather than just fugitivity.[49]

---

[48] Badu, "Master Teacher."
[49] Shaka McGlotten, "Black Data," in *No Tea, No Shade: New Writings in Black Queer Studies*, ed. E. Patrick Johnson (Durham: Duke University Press, 2016), 262–87; Crawley, *Black Pentecostal Breath*.

Sonic *unintelligibility* to whiteness—the opposite of *Sorry to Bother You*'s overlaid "white voices"—is *also* the province and strategy of Blackness. And this is not fundamentally new: As Henry Louis Gates, Jr., described in his theory of the "trickster," sounds and meanings that shape Black communication, Black "signifyin'" practices, from word games like "the dozens," freedom songs like "Wade in the Water," and sonic poems like Hughes' *Go Ask Your Mama*, are fundamental to veiled Black communication technologies: sonic play and layering providing safe passage for Black survival, even in view and hearing of perpetrators of anti-Black violence.[50]

Episode 2 of *Random Acts* presents a more utopian interface: a fully empowering "choreosonics" and "elsewhere." The episode opens in a tight frame that nonetheless opens expansively, exceeding the possible world. In the room, the camera ceaselessly circles, like the Black sound both Moten and Snead describe, producing a more-than-360-degree and more-than-three-dimensional interior. There Nance's character and his on-screen girlfriend Nadja talk, as she plays a video game on her phone in which she is briefly inscribed as a character. Nadja rejects Nance's male character's heteropatriarchal terms and leaves both the game and the apartment for the nighttime Brooklyn street. There, walking, she moves toward the camera and suddenly beyond the diegeses into a meta-layer of the show, directly engaging the audience through voice, eyeline match, and camera work as though we and she were onlookers to the scene that had played out before. Nadja is brought back into the narrative level of the show by a stranger's voice, but, after dealing handily with the guy cat-calling her, she continues through the night to Coney Island, where she enters an arcade with both analog and futurist digital games. Here, the sound layering sound defines the heterotopic layers of diegesis. Soon, it carries the audience and Nadja through the diegetic space and technologies of transit and into a collaborative journey about gender, engendering, and how to sonically instantiate and articulate a female body with full agency.

Nadja's technicity joined with the operations of the video game deliver a new dimension of Nadja who, by the end of the episode, performs a Black feminist cartology of virtuality that we, the audience, can only arrive at via sound. But first, the time and space travels of Najda and onlookers are interrupted by two structuring stories, which themselves irrupt through sound, and then inaugurate new sonic-spaces wherein stories about manipulation of gender,

[50] Henry Louis Gates, Jr., *The Signifying Monkey: A Theory of African American Literary Criticism* (New York: Oxford University Press, 2014).

perspective, and temporality unfold. Their sonics lift a veil that occluded the primary narrative by presenting new interfaces that themselves create "shiftconsciousness." At the same time, these interfaces technologically produce alternative interiors exterior to other diegeses, visual spaces, sounds, and technics of the episode. Their "shiftconsciousness" transforms characters and audience members alike.

The first of these "stories" disrupts the sonics and optics of intersectional race and gender via a series of what Nance would call "documentary" renderings—he labels their actors "documentary subjects" in the credits. In these, women describe their relationship to historicist formations of self, subjectivity, and, in particular, femaleness in a hall of mirrors and before an iPhone or camera of their choice (figure 17.5). Their critiques are necessarily, even if not directly, also critiques of any "true" or documentable racial ontology—*à la* Spillers and with reference to the earlier part of the episode and later sonics of Episode Six. These renderings are presented suddenly synchronous with the temporality of the audience by virtue of their "truth value" and interruption of a markedly "fictional" film story in-the-telling. The women's conversations, which they are, along with the *Random Act*'s crew, themselves documenting (including in previously unseen and heterotopic spaces [though within the same time], in first-person and via various forms of hand-held recording technologies), are conveyed by

**Figure 17.5** Documenting femaleness in *Random Acts of Flyness* [dir. Terence Nance and produced by MVMT A24, 2018].

way of an alternate camera that captures their efforts and some of the images produced by their interactions with the mirrors and screens they use. Exosomatically expressive, the spaces in which the people appear nonetheless seem expansive because of the reflections with differences.

In each clip, voices tell a subjective truth impossible to portray by camera because it is a shifting consciousness of self and self-in-relation, and relies upon an extrapolation of a partial vision of selfhood seen through the eyes of an other—a literal double consciousness rendered through a visual technics of the veil, then recorded, reinserted into a more dominant temporal and thematic structure, and relayed after the fact as a form of internal interface structuring the outer, sonic one. A woman speaks about being happy with her skin, which is dark brown and now stretched to incorporate newly-developed breasts. Another talks, often in a voiceover, about performances of her gender, none feeling quite right, and others' responses to them feeling *all* wrong. One Black woman struggles with her self-image, particularly around her figure and feelings of fitting in. The women's truths focus on how their identities and gendered selves—and relationships to their bodies—are in flux and can only be *mis*apprehended when fixed to images. At best, the camera can show them in-process, with bodies bound to visual referents and stuck within veils they manipulate but cannot fully control. The sounds and words reshape our perception of these bodies and their meanings, turning the technology of our superficial sight into a kind of insight disrupted by Black sound, time, and womanhood.

The women's scenes portray a double consciousness expanded and multiplied into a gender- and time-traversing consciousness by the technics of an immanent sonic interface, which the women use to let others in, retain self-definition, and transform from within the time-space-images that surround them in the episode. This veil is the powerful technology that might bring freedom, if it can "tarry."

At the end of the episode, a far more playful, musical, and ironically literal presentation of double consciousness, the veil, and Black sound technology splits Blackness from itself, portraying its total control of the transformation enacted by (being) the veil. The scene is a long and shifting one, with a young Black Latinx person chasing their shadow across the neighborhood into an apartment building where they insist they must enter to find themself. To enter they speak the right words and watch their sounds matched by a body seemingly unaware of their own—but they are successful! They arrive at a party, where their shadow runs wild and out of the apartment into a building basement reminiscent of the

**Figure 17.6** The boys of Nuncaland wooing a potential queen in *Random Acts of Flyness* [dir. Terence Nance and produced by MVMT A24, 2018].

one in which the Invisible Man loses himself and consciousness in white paint. The young shadow-chaser seems to think their shadow is another being, and, at times, the shadow does exceed its creator. All around are sounds of Spanish Harlem and the party.

The free shadow is also linked with the sonics of the story, narrating with its movement an entirely separate musical motif that distinguishes it from the visual and diegetic environment and delivers us to multiple possible elsewheres. For Nance, the split in consciousness and embodiment depicted via the shadow and its slow sonic translation into a character in a re-scripted *Peter Pan* also reflects the shadow violence of gendered roles and masculinity.[51] In the heterotopic, multi-ethnic party-space, the shadow's choreosonics convey us to "Nuncaland." Nuncaland is curiously neither an Afropessimist end, as its name would suggest, nor the Latinx version of a utopian Peter Pan Neverland. This is because rather than a foreclosed time-space or fantasy, Nuncaland manifests as a *shared* technical collectivity: where the construction of Blackness under white hegemony can stop, and a new, multiracial, gender-fluid sonic collaboration produces new communities and beings with overlapping temporalities and interfaces. After transporting musical numbers, the boys and Pan of Nuncaland sing their

[51] Cybermedia group interview with Terence Nance, January 14, 2021.

masculinity into a non-binary un-gendering that reforms them into flesh rather than captive (in Spillers' schema)—and in their midst a more mature Black woman plays the consenting but elusive role of Wendy (figure 17.6). The scene presents a thematic and formal answer to the many clips of the women speaking about their genders and gender transformations, in part because its Wendy sheds the many forms of mediation and reflection described by earlier women. Formally exiting the sonics and aesthetics of fantasy by refusing to dance or sing, and rather commanding the camera in a direct address, Wendy clearly answers Pan's request to stay forever with "No." She leaves the musical, Nuncaland, Pan, and the visual frame and traverses into the digital time of video-gaming and the transhistorical time of fugitivity. Here, Nadja and Black women manipulate the veil to exercise sonics against heteropatriarchy and racialized, gendered immobility and choose whether to lull their daughters into survival, wokeness, or nunca.

In the last scenes of *Random Acts*, Nadja returns to her room with the Nance character, but this time she operates primarily inside of the game that she controls, a virtual reality in which life looks like a virtual schematic of living. But within the technics of this game, Nadja holds power to organize space and time and manipulate tools of protection and free expression. Interfacing with the various forms of Blackness that constitute the human and non-human and elsewise, the virtual Black woman appears with most agency in this extradiegetic space. She was always-already virtual anyway, moving across domains, enacting her existence. She veils and unveils collectivities in the wake and the wordless bars of sounds that enabled Du Bois's readers to *travel*, evidencing the show's efforts to shift consciousness in the unchartable time and to explore the unstable performance of "otherwise possibilities."

"Otherwise possibility" is, for Crawley, "a break with the known" expressed and instantiated by "the joyful noise of tarrying ... that highlights ... the intentional refusal to produce coherence, the intentional standing outside the circle of language consciousness."[52] Crawley's noisy and often untranslatable "choreosonics" move Black bodies and Black hopes via Black technologies of sound, but less across historical time than beyond the time-spaces of limited and foreclosed existence. In *Sorry to Bother You*, the invocation and performance of the sonic veil provides transit and access, but also serves as the sign of white supremacy at work; in Nance's evocative and transformative multidimensional

---

[52] Crawley, *Black Pentecostal Breath*, 4–5, 167.

worlds of *Flyness*, given by characters and artists inside and beyond the diegesis, sonic and layering produce a resignifying process that opens, along with it, a sound space of new Black knowledge(s).

## Conclusion: Simultaneity and a Black Essentialism

While I've argued above that the Black sound-veil has a genetic component, *Get Out* imagines a more radical rejection of race as a socially-constructed phenomenon, proposing that an essential Blackness is locked in the brain stem and the nervous system and cannot be excised or overwritten by whiteness. Rendering Black bodies as tools for production has also produced an indomitable genetic core "Blackness" that can communicate something essential to other Black people: to be free and to sound the warning to kin.

A year later, the shifted consciousness Nance hopes for and instantiates with sound across *Random Acts* is represented as both liberation and the ultimate ontological horror in Jordan Peele's *Us*, where "kin" is the danger. In *Us*, Lupita Nyong'o, who played death-bound slave Patsy in *Twelve Years a Slave* (who there could only reproduce her daily chance of life as a weight in cotton), is both a free Black woman and her own ontologically-, genetically-, and temporally-tethered body-double (implied to be a product of genetic experimentation through parallel shots of rabbits in cages in labs). In both films, her characters move in the slowed, dragging time *of* slavery—not yet in the afterlife. Shot duration in *12 Years* and *Us* emphasizes sonic and physiological overlap and entrapment, and forms of movement that cannot progress. And the doubling of Nyong'o's character in *Us* is further doubled by her lagging history—wherein, at the beginning of the film, in a room of mirrors and inaccurately labeled no-way-out "exits," antagonist Red takes the place of protagonist Addy, compelling Addy to become Red, and, by virtue of leading a silent, shadow life, forgets her own self and self-story in that silence. Above-ground, Red seems to have escaped her tethered ontology until shadow-Addy returns, still unknowing, but prepared to untether herself. The tethered are voiceless and removed bodily and mentally from cause and effect, agency, or logic in the plot, as well as from the diegetic objects and events to which they must respond. Nonetheless, shadow-Addy experiences a momentary bodily integration that transforms her consciousness when her above-ground-self dances ballet. The classical music from the dance is slowed and interwoven with the film's theme—1995 hit "I've Got 5 On It"—during the later untethering struggle between the two adult women. This mixed-sound and time accompanies shadow-Addy explaining after the diegetic fact that she "knew" during the ballet that the untethering was beginning. This originary scene—in broken flashback—acts as the interface between the women, between film meaning and audience understanding, and among Black times, places, and possibilities. It is sonic; its cuts and interpenetrations of time and characters are structured by music that conveys the swells, dips, and soars of free movement and a fully haptic, choreosonics of dance during the girl's first recital. This sound-body-experience inaugurates for shadow-Addy another time we enter only for brief sequences, but which we come to learn has co-evolved as a technics since the recital: the time of Black operations and training for the untethering. The untethering appears as a belated enactment, in sound-flashes of organized, progressive militarization disaggregated temporally, logically and spatially.

At the close of *Us* the earlier car trip set to the first presentation of "I Got 5 On It" with Red and her son snapping offbeat to its sounds continues. Now untethered and driving, the Wilson family stays offbeat, having made no translatable progress: they remain in Sneed's "cut" and Moten's "break" with their entire story a fiction in the lifeworld of the music to which they travel. Rather, they have on-screen and via de-synched movement and times, enacted a horrifying encounter with Blackness and an escape from those burdens of history, epigenetics, and Afropessmist time embodied by their now-destroyed, tethered twins. Still, the blackened sound has served not only as a reconstructive interface, *but also as the sign of its already-occurrence*. With the integration of Black consciousness accompanying the music and dance of the ballet, the tethered Addy came to know her dis-integration and her paradoxically bound ontology. And she resisted. The sonics of the film deliver its own lifeworld of already-having-happened and ongoing Blackness which records itself in the afterwards. Before the untethering, it is "already after the end of the world, don't you know that yet?"—to quote Sun Ra. And Blackness is already living out its alterdestiny, leaving behind only its myth on the white planet for those of us who remain and spend cash money to watch.

# Bibliography

Badu, Erykah. "Master Teacher." Track 8. *New Amerykah, Part 1.* New York: Universal Motown, 2007.

Baker, Courtney. *Humane Insight: Looking at Images of African American Suffering and Death.* Champaign: University of Illinois Press, 2015.

Benjamin, Walter. "Theses on the Philosophy of History." In *Illuminations,* edited by Hannah Arendt. Translated by Harry Zohn, 253–264. New York: Schocken Books, 2007.

Bluemink, Matt. "Stiegler's Memory, 1: The Problem with Husserl." *Blue Labyrinths,* June 1, 2015. https://bluelabyrinths.com/2015/06/01/stieglers-memory-1-the-problem-with-husserl/. (Accessed February 8, 2021.)

Bradley, Arthur. "Originary Technicity: Technology & Anthropology." In *Technicity,* edited by Arthur Bradley and Louis Armand. Prague: Litteraria Pragnesia, 2006.

CBS. "Director Boots Riley on the 'mythical white voice' in 'Sorry to Bother You.'" *CBS This Morning,* July 16, 2018. https://www.cbsnews.com/news/director-boots-riley-on-the-mythical-white-voice-in-sorry-to-bother-you/. (Accessed February 8, 2021.)

Chun, Wendy Hui Kyong. *Control and Freedom: Power and Paranoia in the Age of Fiber Optics.* Cambridge: MIT Press, 2006.

Childish Gambino. "Redbone." Track 2. *Awaken, My Love!* Glassnote. 2016.

Crawley, Ashon T. *Black Pentecostal Breath: The Aesthetics of Possibility.* New York: Fordham University Press, 2017.

*Cybermedia* Group Interview with Terence Nance. Unpublished. January 14, 2021.

Du Bois, W. E. B. (William Edward Burghardt). *The Souls of Black Folk: Essays and Sketches.* Chicago, A. G. McClurg, 1903. New York: Johnson Reprint Corp., 1968.

Eshun, Kodwo. *More Brilliant than the Sun.* London: Quartet Books, 1998.

Eshun, Kodwo. "Further Considerations on Afrofuturism." *The New Centennial Review* 3, no. 2 (2003): 287–302.

Gates Jr, Henry Louis. *The Signifying Monkey: A Theory of African American Literary Criticism.* New York: Oxford University Press, 2014.

Gillespie, Michael Boyce. *Film Blackness: American Cinema and the Idea of Black Film.* Durham: Duke University Press, 2016.

Goldsby, Jacqueline. *A Spectacular Secret: Lynching in American Life and Literature.* Chicago: University of Chicago Press, 2006.

Harney, Stefano and Fred Moten. *The Undercommons: Fugitive Planning and Black Study.* New York: Minor Compositions, 2013.

Hartman, Saidiya. *Lose Your Mother: A Journey Along the Atlantic Slave Route.* New York: Farrar, Straus and Giroux, 2008.

Hartman, Saidiya. *Scenes of Subjection: Terror, Slavery, and Self-Making in Nineteenth-Century America.* New York: Oxford University Press, 1997.

Lim, Bliss Chua. *Translating Time: Cinema, the Fantastic, and Temporal Critique.* Durham: Duke University Press, 2009.

Keeling, Kara. "'In the Interval': Frantz Fanon and the 'Problems' of Visual Representation." *Qui Parle* 13, no. 2 (2003): 91–117.

Luniz, "I Got 5 On It." Track 3, *Operation Stackola*. Noo Tryve, 1995.
McGlotten, Shaka. "Black Data." In *No Tea, No Shade: New Writings in Black Queer Studies*, edited by E. Patrick Johnson, 262–287. Durham: Duke University Press, 2016.
Molloy, Tim. "The Strange Story Behind 'I Got 5 on It,' the Secret Weapon of Jordan Peele's 'Us.'" *The Wrap*, June 25, 2019. https://www.thewrap.com/i-got-5-on-it-luniz-jordan-peele-us/. (Accessed February 8, 2021.)
Morrison, Toni. *Beloved*. New York: Alfred A. Knopf, 1987.
Moten, Fred. *In the Break: The Aesthetics of the Black Radical Tradition*. Minneapolis: University of Minnesota Press, 2003.
Moten, Fred. "Black Optimism/Black Operation." October 19, 2007. https://lucian.uchicago.edu/blogs/politicalfeeling/files/2007/12/moten-black-optimism.doc. (Accessed February 8, 2021.)
Nance, Terence. Dir. *Random Acts of Flyness*. United States: MVMT A24, 2018.
Patterson, Orlando. *Slavery and Social Death: A Comparative Study*. Cambridge: Harvard University Press, 1982.
Peele, Jordan. Director. *Get Out*. United States: Monkey Paw Productions, 2017.
Peele, Jordan. Director. *Us*. United States: Monkey Paw Productions, 2019.
Reich, Elizabeth. "Documentary Strain, Black Artists, and the Afrofuture: Terence Nance's *The Triptych*." *World Records* 3, article 5 (2018): 46–60.
Riley, Boots, director. *Sorry to Bother You*. United States: Twentieth Century Fox Home Entertainment, 2018.
Robinson, Cedric J. *Black Marxism: The Making of the Black Radical Tradition*. Chapel Hill: University of North Carolina Press, 2000.
Rose, Tricia. *Black Noise: Rap Music and Black Culture in Contemporary America*. Hanover: University Press of New England, 1994.
Ross, Janell. "Epigenetics: The controversial science behind racial and ethnic health disparities." *National Journal*, *The Atlantic* March 20, 2014. https://www.theatlantic.com/politics/archive/2014/03/epigenetics-the-controversial-science-behind-racial-and-ethnic-health-disparities/430749/. (Accessed February 8, 2021.)
Sharpe, Christina. *In the Wake: On Blackness and Being*. Durham: Duke University Press, 2016.
Snead, James A. "On Repetition in Black Culture." *African American Review* 50, no.4 (2017): 648–656.
Spillers, Hortense J. "Mama's Baby, Papa's Maybe: An American Grammar Book." *Diacritics*. 17 no. 2 (1987): 64–81.
Stiegler, Bernard. *The Neganthropocene*. Translated by Daniel Ross. London: Open Humanities Press, 2018.
Stiegler, Bernard. *Technics and Time, 1: The Fault of Epimetheus*. Translated by Richard Beardsworth and George Collins. Stanford: Stanford University Press, 1998.
Sun Ra and his Intergalactic Solar Arkestra. *Space Is the Place*. New York: Plexifilm, 2003.

Taylor, Keeanga-Yamahtta. *From #BlackLiveMatter to Black Liberation*. Chicago: Haymarket Books, 2016.
Wailoo, Keith. *How Cancer Crossed the Color Line*. New York: Oxford University Press, 2010.
Weheliye, Alexander G. *Habeas Viscus: Racializing Assemblages, Biopolitics, and Black Feminist Theories of the Human*. Durham: Duke University Press, 2014.
Wilderson, Frank B. *Afropessimism*. New York: Liverlight Publishing Co., 2020.
Wynter, Sylvia. "Unsettling the Coloniality of Being/Power/Truth/Freedom." *The New Centennial Review* 3, no. 3 (2003): 257–337.

18

# Technology, Chaos, and the Nimble Subversion of *Random Acts of Flyness*

Eric Lyon

Shift Consciousness

A young, enthusiastic Black man bounds out onto the stage. He's wearing jeans, a black turtleneck, and white sneakers, sporting an unfettered Afro. He bows slightly to the audience, and asks, "How's everybody today? Good? I'm feeling great." The man is dressed exactly like Steve Jobs at his legendary product launch of the Apple iPhone in 2007.[1] In this moment, we're viewing the corporate rollout of a rather different product. An audience of mixed ethnicities cheers enthusiastically. A rapid montage of images plays on the screen behind the Black host, displaying a portrait of Jim Crow, KKK cross burnings, lynching scenes, and flying Nazi flags. The images are at first shown on the full screen, and then compressed to a screen within a screen—an iPhone—with more modern images of alt-right racists and more Nazi flags. The host, played by Terence Nance, cheerfully announces, "Random Acts of Flyness is so ecstatic to reveal to you today BITCH BETTER HAVE MY MONEY." The audience goes wild. "Bitch Better Have My Money is the very first proximity-based social app to use the now ubiquitous technology of genetic ancestry to locate the nearest white person that owes a black user money."

The provocative power of this sketch from *Random Acts of Flyness* (Terence Nance, 2018) is revealed by comments on the YouTube page where the sketch was posted by HBO.[2] For example, "HBO has lost there (sic) freaking minds . . .

---

[1] John Schroter, "Steve Jobs Introduces iPhone in 2007," *Macworld Keynote Address*. YouTube video, October 8, 2011, https://youtu.be/MnrJzXM7a6o (accessed January 2, 2021).
[2] HBO, "Random Acts of Flyness: Bitch Better Have My Money (Season 1 episode 3 Clip)," *Terence Nance*, YouTube video, August 17, 2018, https://www.youtube.com/watch?v=BjnqAFgUSP0 (accessed January 2, 2021).

Sorry HBO ... your racist bs is another reason I'm not buying anything labeled HBO:" and "Thank you Random Acts of Flyness. Glad to see my return on those centuries of genocide and colonization our white brothers and sisters invested in! Downloading the app as we speak;" and "I will never apologize for being white;" and more poetically, "oceans of white tears rising from the salty icebergs of the Arctic Circle melting in the bright sunrise of truth. Snowy white fragility pooling away..." Perhaps most appropriately: "What in the world did I just watch?"

The original "Bitch Better Have My Money" is a 2015 trap song by Rihanna et al.,[3] with the pertinent lyrics:

> Pay me what you owe me
> Don't act like you forgot
> Bitch better have my money

In *Random Acts of Flyness*, "Bitch Better Have My Money" is a corporate product launch where the high-tech product is reparations for both historic and ongoing depredations against the African American community. The reference to Apple Inc. is apropos, given that whiteness has been a defining style element in signature Apple products such as the iPod, iMac, iPhone, Airport Express, and AirPods. Apple's power supplies are still all white. At the time of this writing (early 2020), the Apple board of directors is almost entirely white: there are no African Americans on the board. Corporations represent one of the pinnacles of non-state institutional wealth and power in the present-day United States of America. Banks are another major non-state power and wealth center in the U.S. Pairing the case for reparations with the corporate product launch of a new banking loan resolution product constitutes a bold mixture of codes, and an artistic inversion of existing power relationships.

By addressing the topic of reparations, *Random Acts of Flyness* invites comparison with the highly influential 2014 article, "The Case for Reparations" by Ta-Nehisi Coates,[4] which served as the primary source of inspiration for "Bitch Better Have My Money."[5] Where Coates's 16,000-word article is didactic, comprehensive, reasonable, and patient, "Bitch Better Have My Money" is a

---

[3] Rihanna, "Bitch Better Have My Money," YouTube video, July 2018, https://youtu.be/ukW82Ico4U0 (accessed January 2, 2021).
[4] Ta-Nehisi Coates, "The Case for Reparations," *The Atlantic*, June 2014, https://www.theatlantic.com/magazine/archive/2014/06/the-case-for-reparations/361631/ (accessed January 2, 2021).
[5] Personal communication with Terence Nance. January 14, 2021.

multimedia onslaught that makes its case with speed, precision, vividness, humor, high-tech veneer, and a complete lack of compromise, all within the span of three minutes. The political agenda of *Random Acts of Flyness* draws historical connections from past to present, with a high modernist corporate gloss that turns remembering into a product. In doing so, it incorporates found footage, imagery, and media artifacts from various eras, which take on new meaning under the show's subversive gaze. This media assembly often results in jarring registral shifts that can disorient the viewer and invite shifts of consciousness.

Continuing the corporate rollout, Nance cheerfully announces, "We help Black people connect with the white folks who owe them that 40 acres and a mule." At this point, some of the white members of the audience appear considerably less enthusiastic, slow clapping at this new information. Across a montage of corporate-friendly biomorphic CGI and smiling app users, the host continues, "simply put, our technology allows our Black users whose ancestors have been enslaved to locate and match with the white families whose ancestors enslaved them. Now, once the match is made, we make it very simple for you. Payments can be made anonymously through the app." A smiling young white woman taps on her smart phone; cut to a Black man walking in an urban environment and smiling as the payment appears on his phone. (See figure 18.1) "Or, if the Black debtor and the white lender agree, they can meet up in person at the nearest ATM or notary public." Cut to two men, one white and one Black, smiling and shaking hands.

The faux-corporate proceedings are notably cheerful. There's no sense of anger or grimness, at least not at first. We are watching a fun, exciting rollout of a new technology, and, at least on stage, everyone is in a great mood. This attitude is in striking contrast to TV's traditional approach to historical crimes and their ramifications, which demand a certain level of gravitas. The series *American Crime*, which ran on ABC from 2015 to 2017, created by John Ridley, exemplifies this traditional approach, with its ascetic lack of a sense of humor and grim tone throughout. *American Crime* is gripping in its own way, with a single-minded focus on navigating the toxic matrix of race relations in present-day America. The double meaning of the generic-sounding title is worked out through investigation of an individual crime each season that gradually spirals outward toward an indictment of the society as a whole. The series received much critical praise, with multiple prestigious awards and nominations including Emmys and Critics Choice Television Awards. The traditional approach taken by *American Crimes*, along with its aesthetic excellence across many areas, working with,

**Figure 18.1** A white debtor repays an algorithmically selected Black lender through the app "Bitch Better Have My Money" from *Random Acts of Flyness* (Terence Nance, 2020).

rather than subverting established TV norms, likely contributed to this success. By contrast, *Random Acts of Flyness* moves in a more unstable aesthetic space that exercises problematic gestures in multiple registers, embracing disruption as a core operating principle. *Random Acts of Flyness* won just one award, the highly prestigious Peabody Award, which is open to recognizing small, local programming, and is particularly interested in exemplary and disruptive digital content.[6]

Shift consciousness.

Nance, again. "I know what you're thinking. Why a notary public? Everybody asks me that." That gets a chuckle from a few people of color in the audience. "Well, some of these reparations are so enormous that they require the signing over of assets—land, cars, intellectual property, etc., what have you." Not one of the white members of the audience is smiling at this point. "Our algorithm takes into account medical records, criminal records, real-estate records, land rights and treaties, in order to make extremely precise determinations about how much our Black users are owed, due to the contemporary injustices of red-lining, hiring discrimination, and of course, mass incarceration." This ends the expository part of the presentation. Next comes the obligatory Steve Jobs "magic trick" of this corporate product launch.

"I've got a surprise for you today. Everybody, take out your phones." The background music becomes an insistent drumbeat, underscoring the urgency of this moment when the audience finally gets to try out the new technology for themselves. "Every one of your phones has been updated with a new version of our OS, which includes version 1.0 of Bitch Better Have My Money." The host's voice takes on a tone of excitement at being able to share this wonderful new product. "And for you, our first group of Beta users, we've invited a random selection of Blacks to the launch, just to see what happens." At this point, the music turns ominous as a group of unsmiling young Black people dressed identically to the host march to the front of the stage. CGI black clouds appear over their heads (figure 18.2). Their phones start buzzing, and the product demo begins. The technicians disperse to assist the audience in the use of the "Bitch Better Have My Money" app. As multiple smart phones buzz, several white audience members walk to the exits. The doors are locked. We discover that we have been watching a captive audience. Welcome to *Random Acts of Flyness*.

---

[6] Jeffery P. Jones, "Message from the Executive Director," *Peabody awards*, http://www.peabodyawards.com/about (accessed January 2, 2021).

**Figure 18.2** *Random Acts of Flyness* (Terence Nance, 2020) technicians prepare to assist the audience in the use of "Bitch Better Have My Money."

Like several other networks in recent years, HBO has seen an uptick in its Black programming with shows such as *2 Dope Queens* (2015–2018), *Insecure* (2016– ), *A Black Lady Sketch Show* (2019– ), and *Wyatt Cenac's Problem Areas* (2018). This uptick mirrors a general increase in Black TV programming in the 2010s with shows such as *Scandal* (ABC, 2012–2018), *How to Get Away with Murder* (ABC, 2014–2020), *Empire* (Fox, 2015– ), *Atlanta* (FX, 2016– ), *Dear White People* (Netflix, 2017– ) and others. Issa Rae, showrunner for HBO's *Insecure* has suggested that cable TV has been influential in motivating these changes: "Where television in the early '90s and early 2000s was geared to capture the biggest audiences possible, which meant not addressing certain things or being tame with certain storylines, I think with the advent of cable and streaming, the desire to be ridiculously broad isn't really important. It's really about authenticity."[7]

Representation is now considered an important value in popular culture in a way that was not really true even a decade ago. Representation reflects cultural

---

[7] Henry Chu, "Issa Rae Talks About Her White Audience, Shonda Rhimes and Male Nudity," *Variety*, October 17, 2018, https://variety.com/2018/tv/news/issa-rae-insecure-hbo-shonda-rhimes-male-nudity-1202982674/ (accessed January 2, 2021).

change but is not necessarily subversive in design or intent. Shows like *2 Dope Queens* and *Insecure* reinforce the HBO aesthetic that espouses high production values and carefully curated excellence within established genres. *2 Dope Queens* is a comedy show, and *Insecure* is a 30-something dramedy about striving after personal and career goals in Los Angeles. *2 Dope Queens* is not interested in challenging its audience's politics, but rather in skillfully exercising the authority of Black curation of comedy in front of a supportive audience (the validating audiences in *2 Dope Queens* could not be more different from the problematized depictions of audiences presented in *Random Acts of Flyness*). *Insecure* does at times gently call out its white characters who are often blissfully unaware of their privilege, but the main focus of the show is on examining and celebrating the lives, challenges, and achievements of young Black people navigating the American social and professional world as it is, rather than expressing any desire to change the system.

## Against the Grain

At first viewing, *Random Acts of Flyness* doesn't even look like it belongs on HBO. Rather than a smooth surface, the show demonstrates a vast range of production values and visual aesthetics. This makes sense from a production standpoint since the first season was not filtered by HBO management, but rather presented as a *fait accompli* to HBO by its creators. HBO did not ask for any changes. The show is an assemblage of work by a Brooklyn-based Black artists' collective, curated by Terence Nance, while preserving the individual voices of the contributing artists.

Ordinarily, Prestige TV shows involve carefully managed surfaces with a unified and distinctive tone and style, enumerated in a "show bible." Any major deviations from the show bible are intended to be understood by the audience as meta-gestures deployed for dramatic effect. But generally, formal coherence is a design goal for prestige TV, in order to welcome the viewer back each week to an established and reliable environment that meets audience expectations for viewing pleasure.

Through its unpredictable, non-uniform surfaces, *Random Acts of Flyness* eschews formal coherence, leaning into unpredictability, rapid information blasts, an expectation of subaltern cultural and historical context, extremes of register, and a lack of affinity with moderation. In short, *Random Acts of Flyness* is an avant-garde show that has sneaked its way onto a mainstream platform, thereby raising its own outreach, and elevating the platform in the process. In this respect, it shares a strategy with the 2019 film *Joker*, which in the words of its director Todd Phillips, intended to "sneak a real movie" into the studio system.[8]

## Thematic Areas of *Random Acts of Flyness*

So, who is the intended audience? The technological mode of address can be a way into that question. In an interview with *The* Ringer in which he discusses the incorporation of a purpose-built game to *Random Acts of Flyness*, Nance stated:

> My goal is always to have a very integrated, relevant conversation, and especially with kids, with younger people. If you're not working on the stage that they're looking at, then you have no chance, you know? ... The power of games to simulate subjectivities outside of your own is what we're trying to suggest with the show, because everyone watching the show isn't a gender-nonconforming brown person. There's a way, though, to access some sort of subjectivity of different people if you understand the formal idea of gaming. Which, because of the world we live in now, almost everyone will or does.[9]

*Random Acts of Flyness* is obsessed with its audience. An audience bears witness, and thereby provides a platform for multiple subjectivities. Incorporating an audience to an artwork risks creating distance, such as the alienating effect of laugh tracks in TV comedies, or the way a voiceover can take the audience out of the sense of direct experience. Alternatively, when an audience is depicted on a TV show, the viewer can feel less alone, as if one is of a larger group witnessing the event. But *Random Acts of Flyness* does not use the onscreen audience to provide comfort for the actual viewing audience. The audience becomes a participant in the "Bitch Better Have My Money" skit, where white audiences and audiences of color enact very different reactions to a shared performance. The white witness serves as an audience, various sketches are interrupted by

---

[8] Omar Sanchez, "'Joker' Director Todd Phillips Rebuffs Criticism of Dark Tone: 'We Didn't Make the Movie to Push Buttons,'" *The Wrap*, September 25, 2019, https://www.thewrap.com/joker-director-todd-phillips-rebuffs-criticism-of-dark-tone-we-didnt-make-the-movie-to-push-buttons-exclusive/ (accessed January 2, 2021).
[9] Alison Herman, "Terence Nance is Indescribable," *The Ringer*, August 20, 2018, https://www.theringer.com/tv/2018/8/20/17757956/terence-nance-interview-random-acts-of-flyness (accessed January 2, 2021).

conversations about their audience, and perhaps most disturbingly in a sketch called "The Apology" in episode 4, a Black man apologizes for an incident of sexual assault in front of a Black audience that shifts between serious engagement, skepticism, and derisive laughter. The ongoing range of different modes of addressing the audience requires the viewer to recalibrate his or her engagement constantly, creating a disorientating and destabilizing effect.

## The Technologies of *Random Acts of Flyness*

Technology in film and TV can be a historical marker, or a speculative plot device. Putting a computer into a film will date it. It's impressive how precisely and effortlessly a real-life computer or a smart phone in a film dates a film. Think of the bulky mobile phones of the *X-Files* (Fox, 1993–2002) or the presence of a colored plastic 1990s iMac. Speculative technologies such as "beaming" and phaser in *Star Trek* solve staging problems (e.g., avoiding the expense of filming the Enterprise landing on a planet), and create plot options. In either case, the technology is expected to be part of the world of the show. *Random Acts of Flyness* finds other uses for technology: as a framing device, for its allusive qualities (such as the whiteness of Apple Inc., as previously discussed) or as a way of rapidly shifting from immersion in the show, to the making of the show itself.

In the first episode, Jon Hamm appears as himself in an ad for a product called "White Be Gone," a product based on "13th century nanotechnology" designed to cure white people of their white thoughts, which are given the medical term "Acute Viral Perceptive Albinitis." The presence of Jon Hamm is on-the-nose stunt casting, referring to Hamm's signature role as *Mad Men* (AMC, 2007–2015) character Don Draper, a great adman who could sell any product, and in the sketch, Hamm plays both himself as Jon the actor, and also an unnamed salesperson who appears to be channeling Don Draper. Deeper allusions beckon to the centering and problematizing of the white male in the *Mad Men* show itself (figure 18.3).

While critiquing latent racism in standard white discourse, the sketch simultaneously seems to be gently poking fun at the "woke" white person who might actually not be woke enough. For example, the Jon Hamm character says to himself, "none of this applies to me. I read Noam Chomsky. I'm not racist." He's immediately corrected by the Don Draper character: "Maybe not, Jon. But sadly, 'I'm not racist' spoken aloud is a classic white thought." This clip clearly touched

Figure 18.3 An advertisement for the "White Be Gone" nanotechnology skin-care product from *Random Acts of Flyness* (Terence Nance, 2020).

a nerve. At the time of this writing, the HBO posting of "White Thoughts" on YouTube shows 1.2K likes and 24K dislikes.[10]

The actor "Jon" struggles to deliver his speech as it reaches some of its more political conclusions about "white thoughts," at which point the fourth wall is broken, and the camera pulls back to reveal the studio in which the sketch is being filmed. After a tense conversation with the director, played with exasperated patience by Michael Potts, we are returned to the normative TV frame of the advertisement. "Don Draper" reads the list of symptoms associated with "white thoughts," and when he reaches the symptom that occurs while sitting at one's laptop, the frame shifts frames to "Jon" sitting at his laptop, wrestling with white thoughts. Further complexities ensue, including a CGI cloud over Jon's head, an intercut of Shaka King's short film LaZercism[11] (a film which describes a fictional medical solution to "racial glaucoma," a disease which prevents white people from seeing the contributions of people of color), and the introduction of turn-of-the-thirteenth-century nanotechnology, when the sketch is interrupted by a series of messages sent with the Apple Message app. We pull back to yet another technological frame, this time watching Terence Nance editing the "White

---

[10] HBO, "*Random Acts of Flyness*: Jon Hamm 'White Thoughts' (Season 1 episode 1 Clip)," Terence Nance, YouTube video, August 3, 2018, https://youtu.be/6m0oMrMUiWQ (accessed January 2, 2021).

[11] Shaka King, "LaZercism," *King Me Multimedia Inc.*, Vimeo, 2017, https://vimeo.com/218549048 (accessed January 2, 2021).

Thoughts" sketch on Premiere Pro, a digital video editing program. The video sketch is frozen on his screen, and Nance, apparently working on edits in his Brooklyn apartment, fields a critique in the format of instant messages, sent by *Random Acts of Flyness* assistant director Annalise Lockhart: "as ARTISTS we should be addressing whiteness less and affirming Blackness more." Nance agrees. The episode does indeed proceed to affirm Blackness, but in ways that seem deliberately exploitative, notably in a segment entitled, "The Sexual Proclivities of the Black Community," with its music and title lettering suggesting blaxploitation cinema of the 1970s. This deployment of blaxploitation, with its allusions to the 1970s Black power movement is destabilizing, perhaps especially so for white viewers. Are we being called out as voyeurs, or being challenged to dig deeper into Black historical and artistic codes? Over-the-top sexual display and innuendo are interrupted by an engaging interview with a bisexual man that shifts over to stop-action dolls, as he tells his story. These ongoing media shifts within a single sketch keep the viewer in a disoriented state, even after repeated viewings.

The smart phone is a technological frame of reference at many points throughout *Random Acts of Flyness*, frequently as a videographic frame. But the smart phone also serves as a space for zooming down into smart phone apps. In episode 5, which presents the most exasperated and angry political tone of the entire series, the final image shown is the icon of a drained iPhone battery, symbolizing Nance's (and perhaps the viewer's) affective state by the end of that episode. Gaming is a key technology for episode 2, inviting subjectivities for the more personal subjects of that episode.

In the "White Thoughts" sketch, and throughout much of *Random Acts of Flyness*, technologies provide frame shifts, somewhat like slipping into a Narnian wardrobe, and wandering into constantly varied environments and often ironic perspectives. The continuity is very much in line with Steven Shaviro's theory of "post-cinematic affect." Shaviro writes, "This is the world we live in: a world of hypermediacy and ubiquitous digital technologies, organized as a 'timeless time' and a 'space of flows' through which 'divergent series are endlessly tracing bifurcated paths.'"[12] Earlier in his book, Shaviro expresses concerns regarding the implications of post-cinematic affect for Afrofuturism. "Postmodern finance capital 'transgresses' the very possibility of 'transgression,' because it is always only transgressing itself ... Of course, all this has grave consequences for the

---

[12] Steven Shaviro, *Post Cinematic Affect* (London: Zero Books, 2010), 131.

Afrofuturist project. Without transgression, how can there be transformation or transcendence?" This concern seems to hit *Random Acts of Flyness* where it lives, given its tag line, "shift consciousness." We could argue back that in our networked world, there is plenty of room for transgression. As previously mentioned, the YouTube posting of "White Thoughts" received a 24:1 ratio of negative rating, with no shortage of unambiguously racist YouTube commentary. The YouTube space can be seen as one more technological frame that attaches itself to *Random Acts of Flyness*, and in that frame, transgression seems to function just fine. At the same time, critical response, another media pendant to *Random Acts of Flyness*, is overwhelmingly positive.

## Angels and Devils—A Jungian Perspective

As previously mentioned, episode 5 reaches the pinnacle of political anger and intensity, starting with an exploration of how Martin Luther King Jr.'s dream has become a nightmare. The episode then launches into an investigation into media, starting with a bidding war for a lifetime pass to use a derogatory word that arguably should never be used by white people. A fake Quentin Tarantino wins the bidding, perhaps in recognition of the repetitive and tasteless use of the word in his film *Pulp Fiction* (1994). The episode then invests heavily in the story of Joel (played by Paul Sparks), a writer/director/actor profoundly lacking in self-awareness who decides to make a white savior film. Despite all manner of poor and tasteless decisions, and incredible shallowness, Joel ultimately wins an Oscar for the film. African demons finally claim retribution, but not before Joel is privileged to bask in unearned adulation.

Late in episode 5, Nance delivers a media analysis of the "white devil" as a film and TV character. Nance's analysis starts by pointing out that *Birth of a Nation* (D. W. Griffith, 1915), a film which presents the Ku Klux Klan as a heroic force, was the first film screened in the White House. Next, Nance poses the question, "does [electronic bleep] get elected without the existence of Walter White? … Movies and television are propaganda. It's no coincidence the cold war gave rise to the action hero, or that Hollywood set the table for an affable, non-threatening Black president." Nance next discusses the Tyler Durden character from *Fight Club* (David Fincher, 1999) in the context of the 2017 racist riot in Charlottesville and asks if *Fight Club* could have inspired the alt-right. Nance then introduces the "white devil" as a TV character trope. A video montage shows clips of such "white devils" as Tony Soprano, Dexter Morgan, Hannibal Lecter, Jack Bauer, Nurse Jackie, Patty Hewes, Don Draper, Ray Donavan, and several others:

> The white devil is a white man or woman cast into a situation with other white people who lack their genius and skill. The wrinkle is that the white devil is his or her own antagonist. Often some embodiment of their Jungian shadow persistently invades their life and relationships. The power that the shadow generates is the key source of the white devil's technical genius ... The function of the white devil is to win in the battle with his or her shadow self ... Why them? Is it because white people are the best at everything? Not only the best cops but also the best drug dealers? I ask because Breaking Bad is a show about a guy who has the American dream. But when his mortality is threatened, he has to turn to a life of crime. In order to survive, he has to tap into a rage that's always been there, just below the surface. In doing so, he embraces his Jungian shadow, and goes from impotent to virile. Clearly, the surname "White" is symbolic of this archetypal pathology, but to what degree is Breaking Bad intentionally about white men's pathological fragility? To what degree is this show intentionally inspiring white men to re-establish their social and cultural dominance? Given the aforementioned, did Walter White intentionally inspire the popular resurgence of the alt-right?

It's unlikely that the *Breaking Bad* (AMC, 2008–2013) writers' room intended to inspire white racist violence, any more than Taylor Swift hoped to be embraced by the alt-right. But the fact that *Breaking Bad* became a surprise hit during its 2008–2013 run does speak to a latent cultural desire to embrace the psychopathic "white devil" as a relatable hero. And the racial and Jungian analysis is highly plausible. Most literally, Dexter Morgan, the lovable serial killer, refers to his compulsion to kill as his "Dark Passenger," a perfect match to the Jungian shadow (*Dexter*: Showtime, 2006–2013). Nance concludes his analysis thus, "The White Devil archetype in contemporary film and television functions as yet another display of white male dominance and social power, by framing the Jungian shadow as a source of white male cultural and social centrality ... The desired effect is for the melanated masses to hesitate before challenging white supremacy."

Nance's white devil analysis is complementary to Adam Kotsko's book *Why We Love Sociopaths*.[13] Kotsko cites many of the same antiheros (called sociopaths in the book) that Nance referred to as white devils. But interestingly, where

---

[13] Adam Kotsko. *Why We Love Sociopaths: A Guide to Late Capitalist Television* (London: Zero Books, 2012).

Nance identifies the white devil as an antagonist, Kotsko finds the sociopath relatable and desirable. He writes, "My hypothesis is that the sociopaths we watch on TV allow us to indulge in a kind of thought experiment, based on the question, 'What if I really and truly did not give a fuck about anyone?' And the answer they provide? 'Then I would be powerful and free.'"[14] Kotsko's analysis is compatible with Nance's to a degree. But it is notable that whiteness is never mentioned as a feature of the sociopath, even though nearly all of Kotsko's examples are white.

## Scenius and Genius

Brian Eno's neologism "scenius" provides a Bohemian social theory of high artistic creation. "I was encouraged to believe that there were a few great figures like Picasso and Kandinsky, Rembrandt and Giotto and so on who sort of appeared out of nowhere and produced artistic revolution. As I looked at art more and more, I discovered that that wasn't really a true picture. What really happened was that there was (sic) sometimes very fertile scenes involving lots and lots of people—some of them artists, some of them collectors, some of them curators, thinkers, theorists, people who were fashionable and knew what the hip things were—all sorts of people who created a kind of ecology of talent. And out of that ecology arose some wonderful work."[15] *Random Acts of Flyness* seems to be very much a work of "scenius," strategically located in Brooklyn.

Nance states, "I think that in general the thing that I was most excited about in making the show was finding a way for it to be a platform for our community of artists to make stuff at a scale and with a speed that we don't usually get to work at. I think that freedom, that opportunity to exercise that freedom, it was very central to the tonality in the room and how the room extrapolated out into the shoot. Basically, everybody in the room is also a director and directed on the show. I think it was really important to retain the level of intimacy within our community in terms of making things and being in conversation—that birthed all my work but also definitely birthed the pilot."[16]

---

[14] Ibid., 4.
[15] Brian Eno, "Luminous Sydney 2009, Introductory Speech," *More Dark Than Shark*, May 26, 2009, http://www.moredarkthanshark.org/feature_luminous2.html (accessed January 2, 2021).
[16] Alison Herman, "Terence Nance is Indescribable," *The Ringer*, August 20, 2018, https://www.theringer.com/tv/2018/8/20/17757956/terence-nance-interview-random-acts-of-flyness, (accessed January 2, 2021).

## The Sound of *Random Acts of Flyness*

Sound in *Random Acts of Flyness* follows the heterogenous texture of the series as whole, according to the contingencies of the moment. The music is credited to multiple creators including Terence Nance, and his brother Norvis Junior, recording under the estimable name Nelson Bandela. The music and sound design adapt fluently to the wide aesthetic range of the sketches. Coherence is not a goal here. But a digital sensibility is very much in play.

Each of the six episodes of *Random Acts of Flyness* features an after-show interview that provides information and perspectives from one of the creators. The first interview is with Nelson Bandela Nance. But rather than drawing a strict tonal divide between the actual show and the interview, *Random Acts of Flyness* uses the last sketch as a lead-in to the interview. This blurring of boundaries is a strategy used repeatedly throughout the series.

The final sketch returns to Terence Nance being chased by the white police officer, and when it appears he will be captured, Nance addresses the audience. "It might be curtains for your boy. Cop's after me. I don't know what's gonna happen, so before I get outta here, I'll put you up with some new music. My brother, Norvis Junior. 'Music in the Mountains.'" Then, in a perfect random act of flyness, Terence leaves the ground and flies away to freedom. A phrase invoking Sankofa is written across the sky, transitioning to three musicians walking in a jungle and performing intricate drumming patterns on djembes. The camera pans to Norvis Junior, and as the three musicians walk by him, Norvis starts up a funky electronic groove on his phone, which is patched into a serious stack of speakers. Norvis pulls down a 1940s-style vocal microphone (as if it is the most natural thing in the world to find a microphone hanging in the jungle) and sings a short R&B jam over the beat. At the conclusion of the song, Norvis leaves the scene, with just the microphone, iPhone, sound system, and his hat in the frame.

After the credits, Nelson Bandela is shown, working on a home digital audio studio. "I used to tell people I play computer. And they'll be like, what instrument do you play back? I play computer." Bandela's music plays through the segment, which itself is subjected to video stutters, blurring the distinction between *Random Acts of Flyness* proper, and the after-show interview. Bandela describes wanting to create R&B and ambient music, but also "beep with like a keyboard onstage and vocoder." Bandela concludes by observing that even if you're homeless, you can go to the library and make beats. The ubiquity and accessibility of digital sound technology underwrites Bandela's fluidity and range as a composer for the show.

The mercurial nature of the sound design is on full display at the opening of episode 5, "Tried to Tell My Therapist about My Dreams/MARTIN HAD A DREEEEAAAAM." The introductory HBO soundicon (an upward filter sweep followed by white noise cadencing into a C major chord) is directly extended with an accelerating tremolo effect imposed on the chord. The chord is deepened, and reverberance creates a sense of distance and "underwater-ness."[17] A Black man with close-cropped hair and mustache, in pajamas awakes, turns on the lights, and drops out of bed, falling directly into the ocean. A bass drop underwrites his fall. As he descends, water sounds are added to the mix. The texture gradually thickens and adds a beat low in the mix, as the man explores the environment. Diegetic sounds are added as he swims past organisms, and there is a quick swell as he is swallowed by a whale, connecting the biblical prophet Jonah to the twentieth-century prophet, Dr. Martin Luther King Jr. The sound and music aggressively follow the action in detail, creating an exquisite and allusive audiovisual underwater experience.

## Conclusion

*Random Acts of Flyness* commands attention through its particularity, cultural significance, technological fluency, subversive politics, and outstanding execution. For many prestige TV shows, technology is a means to an end—a way of creating effects, or thematic fetish objects. *Random Acts of Flyness* both employs and unpacks technology as part of its analysis of contemporary Black experience. Nance has described the show as "a portrait of a community's subconscious that has the potential to shift consciousness."[18] His theorization of media explains the purposefulness and personality of the show: "You know, what is a program? A set of instructions that will hopefully change how something behaves. What is a channel? What is channeling? Those concepts are really important to the idea of the show, which is to hopefully retool or be a means of

---

[17] This is the only example I know of that subjects the HBO soundicon to sonic extrapolations as the opening gambit in a series episode.
[18] Phoebe Unterman and Kye Ryssdal, "Can Television Shift Consciousness? A Look Inside the HBO Show That's Working on It," *Marketplace*, November 5, 2018, https://www.marketplace.org/2018/11/05/can-television-shift-consciousness-behind-hbo-thats-working-it/, (accessed January 2, 2021).

retooling their point of view. You know, television is kind of uniquely suited and has always been a means of propagandizing or shifting consciousness in a variety of ways."[19]

Nance's acknowledgment of *Random Acts of Flyness* as a form of propaganda is refreshing. The post-internet world has become a mediascape of immersive advertisement for pretty much everything: politics, ideologies, products, personalities, brands, nationalities, religions, etc. In this space, where attention management is a problem for everyone, two main choices beckon—either choose sides such that algorithmically curated, ideologically aligned communications are consumed as responsible reporting and everything else is dismissed as fake news, or simply accept that everything is propaganda. The logical outcome of the fluidity and radical proximity of the contemporary digital world is to resist binaries. But this is the opposite of what is happening in our increasingly polarized online world. Trolling is a normal mode of communication, and civil war is regularly discussed in the comment sections to news articles. Empathy for otherness is thin on the ground. *Random Acts of Flyness* can't solve these problems and can't break through the membrane of ideological tribalism. What it can offer is perhaps the most needed mental therapy for our time—destabilization.

## Bibliography

Chu, Henry. "Issa Rae Talks About Her White Audience, Shonda Rhimes and Male Nudity." *Variety*, October 17, 2018. https://variety.com/2018/tv/news/issa-rae-insecure-hbo-shonda-rhimes-male-nudity-1202982674/ (accessed January 2, 2021).

Coates, Ta-Nehisi. "The Case for Reparations." *The Atlantic,* June 2014. https://www.theatlantic.com/magazine/archive/2014/06/the-case-for-reparations/361631/. (accessed January 2, 2021).

Eno, Brian. "Luminous Sydney 2009, Introductory Speech." *More Dark Than Shark,* May 26, 2009. http://www.moredarkthanshark.org/feature_luminous2.html (accessed January 2, 2021).

HBO. "Random Acts of Flyness: Bitch Better Have My Money (Season 1 Episode 3 Clip)." *Terence Nance*, YouTube video, August 17, 2018. https://www.youtube.com/watch?v=BjnqAFgUSP0. (accessed January 2, 2021).

---

[19] Ibid.

HBO. "Random Acts of Flyness: Jon Hamm 'White Thoughts' (Season 1 Episode 1 Clip)." *Terence Nance*. YouTube video, August 3, 2018. https://youtu.be/6m0oMrMUiWQ. (accessed January 2, 2021).

Herman, Alison. "Terence Nance is Indescribable." *The Ringer*, August 20, 2018. https://www.theringer.com/tv/2018/8/20/17757956/terence-nance-interview-random-acts-of-flyness (accessed January 2, 2021).

Jones, Jeffery P. "Message from the Executive Director." *Peabody Awards*. http://www.peabodyawards.com/about (accessed January 2, 2021).

King, Shaka. "LaZercism." *King Me Multimedia Inc.*, Vimeo, 2017. https://vimeo.com/218549048. (accessed January 2, 2021).

Kotsko, Adam. *Why We Love Sociopaths: A Guide to Late Capitalist Television*. Winchester and Washington: Zero Books, 2012.

Rihanna. "Bitch Better Have My Money." YouTube video, July, 2018. https://youtu.be/ukW82Ico4U0 (accessed January 2, 2021).

Sanchez, Omar. "'Joker' Director Todd Phillips Rebuffs Criticism of Dark Tone: 'We Didn't Make the Movie to Push Buttons.'" *The Wrap*. September 25, 2019. https://www.thewrap.com/joker-director-todd-phillips-rebuffs-criticism-of-dark-tone-we-didnt-make-the-movie-to-push-buttons-exclusive/ (accessed January 2, 2021).

Schroter, John. "Steve Jobs introduces iPhone in 2007." *Macworld Keynote address*. YouTube video, October 8, 2011. https://youtu.be/MnrJzXM7a6o. (accessed January 2, 2021).

Shaviro, Steven. "Post Cinematic Affect." Washington and Winchester: Zero Books, 2010.

Unterman, Phoebe and Ryssdal, Kye. "Can Television Shift Consciousness? A Look Inside the HBO Show That's Working on It." *Marketplace*, November 5, 2018. https://www.marketplace.org/2018/11/05/can-television-shift-consciousness-behind-hbo-thats-working-it/ (accessed January 2, 2021).

19

# Expecting the Twist

## How Media Navigate the Intersections Among Multiple Sources of Prior Knowledge

Noah Fram

Can a comedian make a convincingly frightening horror film? Can a freewheeling, surrealist comedy veer into sci-fi horror and back without losing its balance? And if so, what are the complex cognitive processes that allow those things to happen, and make such works effective?

These processes rely on audiences' prior knowledge, which comes from a variety of sources, including other art works and cultural norms. Genre bridges between this prior knowledge and the cognitive rules, or expectations, that audience members formulate to make sense of multimedia works. Understanding this cognitive role for genre requires a transdisciplinary unifying theory. In this chapter, I will detail how two films—Jordan Peele's *Get Out* (2017) and Boots Riley's *Sorry to Bother You* (2018)—and a mini-series—Terence Nance's *Random Acts of Flyness* (2018)—make use of, and are enabled by, a combination of representation-building, genre formation, and hierarchical prediction.[1] We will see how these works retain coherence while drawing upon disparate and often contradictory sources of prior contextual knowledge. Along the way, I will touch on the cognitive nature of genre, its entanglement with other instantiations of categorization and metacognition, and the power of fiction to shape reality.

The approach to genre I advocate here is built, in part, on concepts from music scholarship, in which a genre is a hybrid between aesthetic and sociocultural rules typified by works that meet some socially agreed-upon set of criteria,[2] which can include both the aesthetic properties of the work itself and

---

[1] Jordan Peele, *Get Out*: Universal Pictures, 2017; Boots Riley, *Sorry to Bother You*: Annapurna Pictures, 2018; Terence Nance, *Random Acts of Flyness*: HBO, 2018.
[2] Franco Fabbri, "What Kind of Music?," *Popular Music* 2 (1982): 131–43.

any social, cultural, political, or economic elements.[3] Such views of genre are uniquely useful in my cognitive approach to the perception of artworks, as they imply that audience members synthesize various sources of prior knowledge into a single feature, which is used to determine what sort of rules should be applied to understand a work.

To illustrate this synthesis, I model the process by which people acquire prior knowledge about the aesthetic and sociocultural context of media with Arnold Sameroff's unified theory of child development,[4] which indicates that humans interact with the world through *representations*, or "encodings of experience ... [that bring] order to a variable world, producing a set of expectations of how things should fit together."[5] These representations are stored, modified, and accessed as in James McClelland's complementary learning systems theory (CLST) of interactions between long-term and short-term memory in the brain.[6] Genre mediates between these representations and the final interpretation of a multimedia work, which I model by extending Fred Lerdahl's theorization that composers and listeners each possess their own set of rules, called a grammar, for comprehending a musical work to include all forms of media.[7] My approach is similar to Annabel Cohen's Congruence Associationist Model with Working Narrative,[8] differing mostly in which part of the process we emphasize: Cohen emphasizes the interactions among multiple sensory modalities, while I focus on the construction and modification of representations in long-term memory.

Before an individual can make sense of a work, they must first determine what *kind* of work it is. Different forms of media have different norms, from rhythmic patterns and visual motifs to performance contexts. This can be

---

[3] David Brackett, *Categorizing Sound* (Oakland, CA: University of California Press, 2016).
[4] Arnold Sameroff, "A Unified Theory of Development: A Dialectic Integration of Nature and Nurture," *Child Development* 81, no. 1 (2010): 6–22.
[5] Ibid.
[6] James L. McClelland, Bruce L. McNaughton, and Randall C. O'Reilly, "Why There are Complementary Learning Systems in the Hippocampus and Neocortex: Insights from the Successes and Failures of Connectionist Models of Learning and Memory," *Psychological Review* 102, no. 3 (1995): 419–57; Dharshan Kumaran, Demis Hassabis, and James L. McClelland, "What Learning Systems Do Intelligent Agents Need? Complementary Learning Systems Theory Updated," *Trends in Cognitive Sciences* 20, no. 7 (2016): 512–34.
[7] Fred Lerdahl, "Cognitive Constraints on Compositional Systems," *Contemporary Music Review* 6, no. 2 (1992): 97–121.
[8] Annabel J. Cohen, "Congruence-Association Model of Music and Multimedia: Origin and Evolution," in *The Psychology of Music in Multimedia*, ed. Siu-Lan Tan et al. (Oxford, UK: Oxford University Press, 2013), 17–47; Annabel J. Cohen, "Scoring Music for Westworld Then and Now—A Cognitive Perspective," in *Cybermedia: Explorations in Science, Sound, and Vision*, ed. Carol Vernallis, Holly Rogers, Selmin Kara and Jonathan Leal (New York: Bloomsbury, 2021).

characterized as a hierarchical expectation in which individuals build predictions on multiple levels of analysis, with each level dependent on the others in a causal loop. Audience members predict what kind of multimedia work they are encountering given everything they have learned about the work up to that moment, what rules apply to works of that kind given other contextual information, what the broad shape of the work will likely be given those rules, and what will happen next given that shape. After assessing this last expectation, they update what they know about the work itself and the process starts again.[9] In the case of music, and many other aesthetic domains, these "kinds" are analogous to genres, and they are inferred based on all available information, including aesthetic properties of the piece itself (e.g. length, instrumentation, and form), its sociocultural context, prior knowledge about its creators or community of listeners, means of performance or communication, and references to other pieces or cultural products.

This process of building and assessing expectations is not unique to artistic works. Rather, it is grounded in core psychological principles derived from neurophysiological constraints on human cognition. Sameroff's theory of child development describes how children form representations of people and objects in their environments within the dynamic interplay between self and context indicated by cultural psychologists.[10] Other work has probed the learning of cultural norms through children's books,[11] music,[12] and even facial expressions.[13] McClelland's CLST and Russell Poldrack's and Karin Foerde's analysis of multiple memory systems (MMS) theory,[14] among other similar work, indicates a possible neurological basis for the formation and modification of representations. Since these representations, especially in Sameroff's framework, are derived from experience, they are dependent on the nature and content of those experiences, rendering them susceptible to phenomena such as naïve realism, or an

---

[9] David Huron, *Sweet Anticipation: Music and the Psychology of Expectation* (Cambridge, MA: The MIT Press, 2006).

[10] Hazel R. Markus and Shinobu Kitayama, "Cultures and Selves: A Cycle of Mutual Constitution," *Perspectives on Psychological Science* 5, no. 4 (2010): 420–30.

[11] Jeanne L. Tsai et al., "Learning What Feelings to Desire: Socialization of Ideal Affect Through Children's Storybooks," *Personality and Social Psychology Bulletin* 33, no. 1 (2007): 17–30.

[12] Dave Miranda et al., "Towards a Cultural-Developmental Psychology of Music in Adolescence," *Psychology of Music* 43, no. 2 (2015): 197–218.

[13] Jeanne L. Tsai et al., "Leaders' Smiles Reflect Cultural Differences in Ideal Affect," *Emotion* 16, no. 2 (2016): 183–95.

[14] Russell A. Poldrack and Karin Foerde, "Category Learning and the Memory Systems Debate," *Neuroscience & Biobehavioral Reviews* 32, no. 2 (2008): 197–205.

individual's mistaken belief that their context is typical of everyone's context and that they are an objective observer of their environment.[15]

These theories rely on inference, or making decisions about the environment based on limited information. There are two primary frameworks for modeling inference—classical and Bayesian—and these approaches reflect different ways of thinking about how we make sense of the world and develop expectations about what is likely to happen next.

Classical inference aims to choose between a limited set of proposals by picking the one that maximizes the probability of the observed data. Since we must be more certain than before to change our conclusion, classical inference can only get *more* sure about which hypothesis is best. Bayesian inference, on the other hand, considers all possible hypotheses by finding the probability of each hypothesis given the observed data. Representing this probabilistically clarifies this distinction. In classical inference, we compare the probability of the data given an experimental hypothesis, $P(\text{data} \mid H_1)$, and a null hypothesis, $P(\text{data} \mid H_0)$, while Bayesian inference computes the probability of a hypothesis given the data, $P(H \mid \text{data})$, for each possible hypothesis. These probabilities can be obtained using Bayes' Theorem:

$$P(H \mid \text{data}) = \frac{P(\text{data} \mid H) P(H)}{P(\text{data})}. \tag{1}$$

Crucially, the $P(H \mid \text{data})$ for all possible hypotheses must add to 1, because the observed data at any point in time are constant. This means that certainty in Bayesian inference is characterized by how the probability is distributed: when any hypothesis becomes more likely, other hypotheses become less likely. No matter how sure we are about the most likely hypothesis, if we observe data that supports an alternative, we will become *less* certain about which hypothesis is most likely. To a Bayesian, certainty is fleeting.

In addition, while classical inference emphasizes choosing between alternative hypotheses, Bayesian inference involves the continuous refinement of a hypothesis based on new observations, resulting in a constant cycle of assessing and adjusting models of the world (figure 19.1). This structure is reflected in Box's model of how the scientific method itself is an iterated process of

---

[15] Lee Ross, David Greene and Pamela House, "The 'False Consensus Effect': An Egocentric Bias in Social Perception and Attribution Processes," *Journal of Experimental Social Psychology* 13, no. 3 (1977): 279–301; Dale W. Griffin and Lee Ross, "Subjective Construal, Social Inference, and Human Misunderstanding," *Advances in Experimental Social Psychology* 24 (1991): 319–59.

## Classical inference

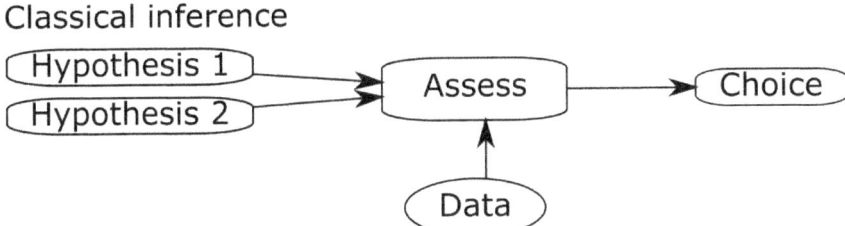

## Bayesian inference

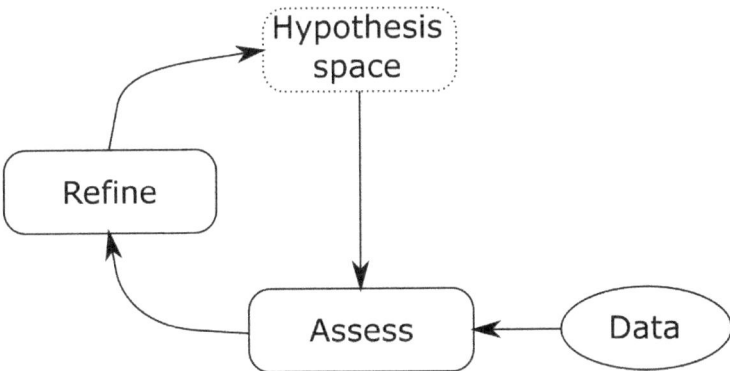

**Figure 19.1** Classical and Bayesian inference structures. In both cases, new data is gathered each time through the inference process. The classical model is shown to be a linear process of choosing the hypothesis that maximizes the probability of the observed data and must be repeated for each new set of observations, while the Bayesian model is a continuous process of refining the probability distribution over the entire hypothesis space in response to new observations.

inference.[16] The similarities between this structure and cognitive models such as CLST imply that Bayesian inference is a more appropriate model for cognitive learning and inference processes than its classical cousin, and its capacity for reducing certainty must also be a characteristic of making inferences about art.

Artists in all forms of media manipulate this uncertain certainty by constructing hierarchical expectations and then defying them in deliberate and informative ways. Because these expectations frequently rely on norms associated with specific genres, they are intrinsic in my model of media interpretation.

---

[16] George E. P. Box, "Science and Statistics," *Journal of the American Statistical Association* 71, no. 356 (December 1, 1976): 791–99.

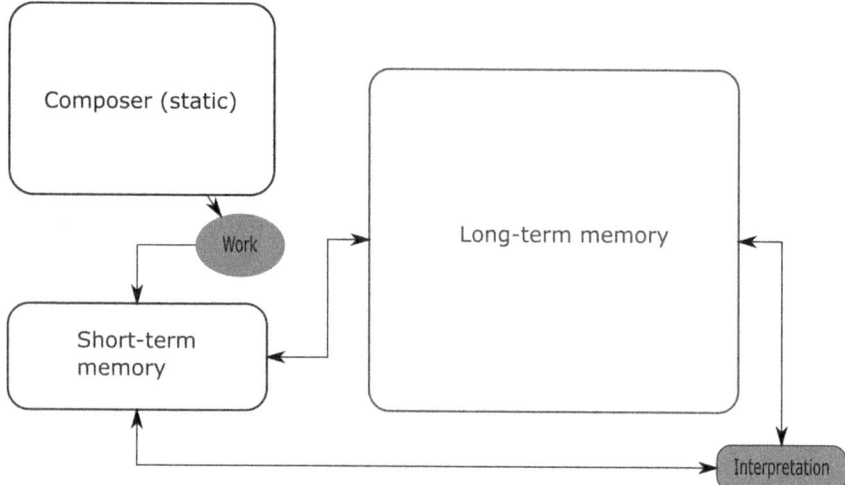

**Figure 19.2** The CLST model in the instance of a listener interpreting a work. Note the bidirectional connections among short-term memory, long-term memory, and the interpretation, and their isolation from the composer node.

At the most abstract level, my approach embeds CLST in the process of creating and interpreting a multimedia work by inserting the interactions between long-term and short-term memory (figure 19.2). This shows that once a work is made, its composer has no other direct influence on its interpretation, and that the interpretation is a stage in a constant cognitive cycle, not the final product of a linear process. The inability of a work's creator to directly influence its interpretation is akin to Roland Barthes's concept of the "death of the author," and allows for significant discrepancies between a work's intent and its reception.[17] Superimposing Lerdahl's grammars onto the existing sketch of the CLST framework and inserting genre between prior knowledge, or context, and the compositional and listening grammars demonstrates how the listener's inference process mirrors the composer's creative process and how a work's interpretation influences the listener's prior knowledge (figure 19.3). These feedback processes are crucial to interpreting works because works and their contexts are mutually constituted in a similar fashion to cultures and selves.

In addition, because the listener infers a work's genre from their own context, they may reach a different conclusion as to what genre a work belongs to, or even

---

[17] Roland Barthes, "The Death of the Author," in *Image, Music, Text*, trans. Stephen Heath (London: Fontana, 1977), 142–48.

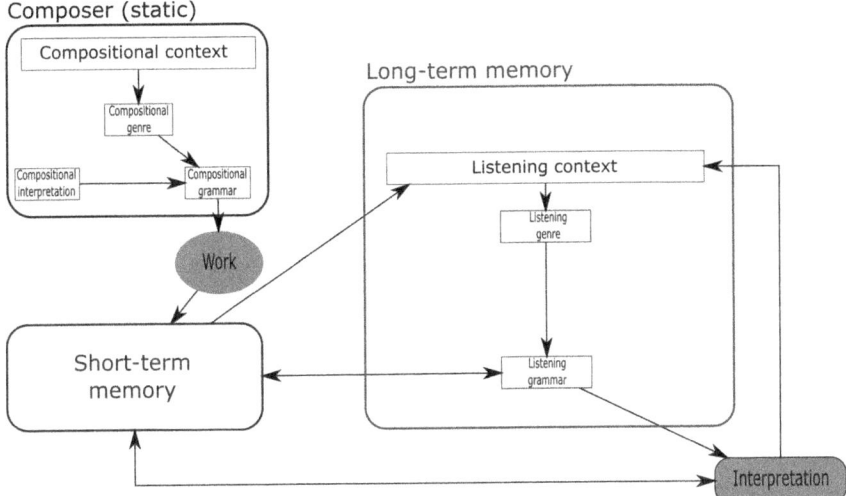

**Figure 19.3** Expansion of the composer and long-term memory nodes. In each case, genre mediates the link between the context and the grammar used for either composing or perceiving a work.

if it belongs to a genre at all: these conclusions may be different from what the creator intended.

Further detail in how context and short-term memory are modeled reveals specifics of how different components in the interpretive process impact different kinds of prior contextual knowledge (figure 19.4). Short-term memory includes the hippocampus, which stores short-term memories, and limbic system, particularly the amygdala, which process emotional responses to perceived events. I have separated the compositional and listening contexts into three components: prior aesthetic knowledge, prior sociocultural knowledge, and prior knowledge about the composer. This allows feedback loops to connect to specific types of prior knowledge: sensory and affective responses to the work itself contribute to the listening aesthetic context, while the final interpretation influences both the listener's inferences about the composer and the listener's prior knowledge of the sociocultural context.

Often, listeners encounter works by composers they do not know. In these cases, their prior knowledge about the composer is limited, and they are forced to build a representation of them almost entirely out of inferences. The dotted line surrounding the listener-inferred composer indicates that this does not need to include prior information. If the listener does possess prior knowledge

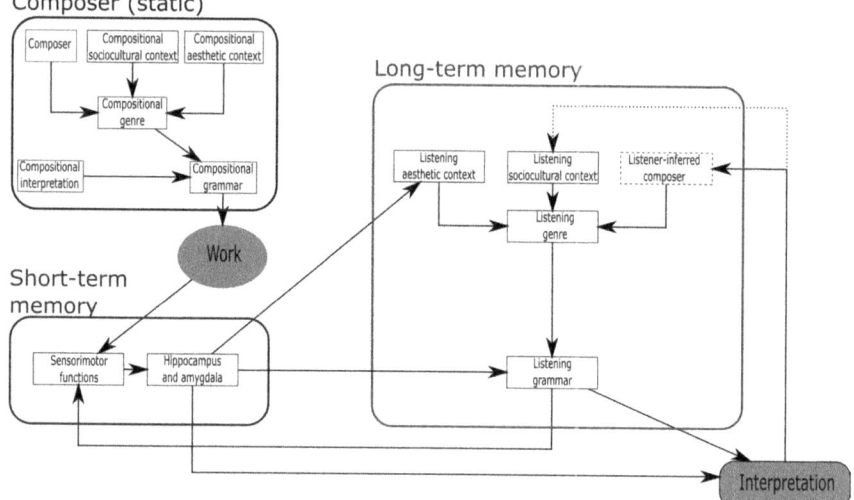

**Figure 19.4** The full model, with various sources of context and the short-term memory node expanded. The dotted line connecting the interpretation to the listening sociocultural context implies that this process often unfolds over multiple works, rather than one, and the dotted line surrounding the listener-imposed context indicates that this might not be based on prior knowledge.

about the composer, they use it in their interpretive analysis; if the listener does not, they use their inferences about the composer instead.

Throughout this model, I assume that all prior knowledge is stored as representations, which is consistent with Sameroff's theory and has been observed in the organization of neural structure.[18] Genre, then, is a representation derived from interactions between the perceived work and various forms of prior knowledge. It synthesizes multiple sources of context into a single phenomenon, mediating the relationship between prior information and the grammar used to make sense of a multimedia work.

Riley's *Sorry to Bother You* makes use of multiple sources of prior knowledge for dramatic and emotional effect and contends directly with the same real-world effects of prior knowledge that I engage in this chapter. Because of this, it can be analyzed both as an example of the cognitive process detailed above and as a metacognitive discussion of that process's implications in the real world. It is Riley's directorial debut, so unless the audience were familiar with his prior work

---

[18] Alex Martin, "The Representation of Object Concepts in the Brain," *Annual Review of Psychology* 58, no. 1 (December 6, 2006): 25–45.

with the hip-hop group The Coup, they would likely have little prior knowledge of him, his work, and his intentions. In this case, the film's interpretation will rely, at first, almost entirely on prior knowledge of its sociocultural and aesthetic contexts. However, if the audience is aware of The Coup or the film's advertising campaign, they will likely expect a politically or socially conscious bent to the film. In addition, The Coup released an album in 2012 titled *Sorry to Bother You*, which is directly inspired by Riley's original draft of the screenplay for the film and provides an even more direct source of prior knowledge.

Although *Sorry to Bother You*'s plot follows a relatively well-trodden path— the hero joins an organization; rises through the ranks, leaving their old life and friends behind; uncovers a scandal; and uses their newfound knowledge to help their old companions and bring down a corrupt leader—it draws heavily on genres that are less mainstream. Stefanie Dunning highlights Afro-futurism, Afro-Pessimism, and the Weird, a constellation of genres that combine in what she calls the Black Weird, as primary inspirations for Riley's film.[19] Although Riley himself has rejected the idea that *Sorry to Bother You* is a genre film, these prior representations are crucially important for audiences attempting to make sense of his creation.[20] In other words, although he may not have created it within a genre, audience members rely on genres to interpret it. Audience members who lack prior familiarity with Dunning's genres, in either an aesthetic or a sociocultural domain, would not be able to rely on a clear representation for *Sorry to Bother You*. Imagine, for instance, encountering a platypus for the first time: each of its parts indicates something completely different, so you might interpret it as an awkward fusion of radically disparate kinds of things, rather than an ideal example of its own kind of thing. This is evidenced by reviewers who observed that *Sorry to Bother You*'s abrupt shift between satirical comedy and horror, for instance, met with varied responses from their audiences.[21]

*Sorry to Bother You*'s overtly political tone, while in keeping with Riley's past work and other examples of the Black Weird Dunning references, further complicates its audiences' responses.[22] It was released amid a nation-wide

---

[19] Stefanie K. Dunning, "What Is the Future? Weirdness and Black Time in Sorry to Bother You," *Studies in the Fantastic* 9 (2020): 44–62.
[20] Michael T. Martin, Yalie Kamara, and Boots Riley, "Boots Riley on Sorry to Bother You and the Matter of the 'Good Fight,'" *Black Camera* 11, no. 2 (2020): 176–215.
[21] David Sims, "Sorry to Bother You Is Fizzy, Flawed, and Fascinating," *The Atlantic*, July 6, 2018, https://www.theatlantic.com/entertainment/archive/2018/07/sorry-to-bother-you-review/564263/ (accessed January 10, 2021).
[22] Dunning, "What Is the Future?" 45–46.

discussion of economic inequality, racial bias, and police brutality, which shaped the prior sociocultural knowledge and assumptions of its audience, and draws on tropes such as the universal dislike of telemarketers and high school nostalgia. However, the prior knowledge associated with the specific cultural moment, such as the political context of *Sorry to Bother You*'s release, is perhaps more susceptible to change than any other and is highly sensitive to real-world developments that contextualize a person's interpretative experience. The most impactful moments shift as the sociocultural context changes: portrayals of clashes between protestors and security or law enforcement at sporting events, for instance, may have gotten more attention if the film were released and reviewed in 2020. Even though the aesthetics remain unchanged, the specific events that capture the audience's attention change along with the broader sociocultural context. This impact of sociocultural context on the audience's experience of the work highlights how representations of multiple kinds of prior knowledge, stored in long-term memory and synthesized as genres, can impact short-term memory and sensory processes by directing the audience's attention to particular stimuli.

However, the most powerful connection between *Sorry to Bother You* and the expectation processes central to this chapter is the movie's dominant motif: code-switching. It is surprising when David Cross's voice (eternally linked to the hapless Tobias Fünke from *Arrested Development* [2003–2019] at least in my mind) comes from Lakeith Stanfield's mouth and when Omari Hardwick's eyepatched Mr. \_\_\_\_\_ is dubbed by Patton Oswalt. Riley uses these creative choices to emphasize the cognitive processes that shape how we experience and make sense of our social world, which are based on the same inference-making behaviors that underpin my model of genre and interpretation. The human brain is designed to make inferences from extremely limited information. When the only available information is someone's voice, as in phone conversations, then the mental representation of the person at the other end of the line will be constructed entirely from that stimulus. The people Cassius Green calls while using his white voice do not imagine themselves to be talking to someone who looks like Lakeith Stanfield—rather, they picture someone like David Cross. Given a voice, the customers select the most likely representation *of an entire person* as their conversational partner, in much the same way as listeners with no prior knowledge of a composer infer a representation of that composer based on the works they produce.

Unlike *Sorry to Bother You*, *Get Out* occupies a clearly delineated genre space, which makes its deviations from that genre's expectations even more apparent. One indication of this adherence is its musical score. While *Sorry to Bother You* draws heavily on The Coup, featuring sharp backbeats, overdriven guitars, and edgy vocal delivery, *Get Out* sounds exactly like a horror movie. Its soundtrack is redolent with precisely timed dynamic shifts, unsettling modalities, and rumbling bass heartbeats, encouraging the episodic build and release of tension characteristic of the horror style. It is also studded with jump-scares and unsettling interactions, such as Walter's and Georgina's too-wide smiles, Logan King's seizure, and Jeremy Armitage's somehow-intimidating ukulele playing. In other words, the aesthetics of *Get Out* point unambiguously to the horror genre.[23]

It differs from the typical mold of a horror movie with the overtness of its message. While horror films are often allegorical commentaries on social or political problems,[24] *Get Out*'s contention with slavery and racial fetishization is surprisingly direct; Alison Landsberg refers to this aesthetic as "horror *vérité*."[25] This situates it at a point of tension between horror and commentary,[26] and brings it uncomfortably close to absurdist satire for a movie as serious and frightening as it is. Sonically, the engagement with African and African American musical and lyrical themes, and their juxtaposition with tracks like Bill Medley's and Jennifer Warnes's "(I've Had) The Time of My Life" (1987) and Flanagan and Allen's "Run Rabbit Run" (1939) highlights the central themes of the movie while deviating from the soundscape more typical of a horror film.[27] This puts sociocultural context—epitomized by the opening scene, which directly recalls the shooting of Trayvon Martin,[28] and one character's insistence that he is not racist while actively participating in literally dehumanizing violence against a Black man—in conflict with aesthetic context, which encourages a more abstract engagement with real-world issues.

Since the aesthetic context lies firmly within the horror-film space, the unexpected ways in which *Get Out* contends with its sociocultural context operate

---

[23] Anthony Carew, "American Horror: Genre and the Post-Racial Myth in Get Out," *Screen Education* 94 (2019): 14–21.
[24] Robin Wood, "An Introduction to the American Horror Film," in *The American Nightmare: Essays on the Horror Film*, ed. Robin Wood and Richard Lippe (Toronto: Festival of Festivals, 1979), 7–28.
[25] Alison Landsberg, "Horror Vérité: Politics and History in Jordan Peele's Get Out (2017)," *Continuum* 32, no. 5 (September 3, 2018): 629–42.
[26] Isabel Pinedo, "Get Out: Moral Monsters at the Intersection of Racism and the Horror Film," in *Final Girls, Feminism and Popular Culture*, ed. Katarzyna Paszkiewicz and Stacy Rusnak (Cham: Springer International Publishing, 2020), 95–114.
[27] Charles Pulliam-Moore, "The Hidden Swahili Message in Get Out the Country Needs to Hear," *Splinter*, March 1, 2017, https://splinternews.com/the-hidden-swahili-message-in-get-out-the-country-needs-1793858917 (accessed January 10, 2021).
[28] Manohla Dargis, "Review: In Get Out, Guess Who's Coming to Dinner? (Bad Idea!)," *The New York Times*, February 23, 2017, https://www.nytimes.com/2017/02/23/movies/get-out-review-jordan-peele.html (accessed January 10, 2021).

more as surprises that capture attention than confusions of the film's genre. Unlike *Sorry to Bother You*, which is perfectly comfortable as a genre chameleon, *Get Out*'s adherence to the aesthetic rules of horror and its resultingly unambiguous membership in the horror genre work to amplify its social commentary. As one of my old theater instructors was fond of saying: the eye goes to what is different. This causes audience members to have quite different reactions to *Get Out* than to other horror movies and may have long-term ripple effects on the norms governing the genre itself that cannot yet be effectively observed.

*Random Acts of Flyness* (*RAoF*) moves in the opposite direction. As indicated by Elizabeth Reich elsewhere in this volume, *RAoF* relies heavily on sampled textures.[29] Sampling has driven hip-hop from its earliest days, with some scholars identifying it as both a dominant aesthetic within the genre and as a culturally-situated practice: since then sampling has been foundational in more recent "nostalgic" genres such as chillwave, vaporwave, and hypnagogic pop.[30] Unlike *Get Out*, which uses well-established genre conventions and associations to highlight its message, or *Sorry to Bother You*, which draws on genres strongly affiliated with specific cultural backgrounds, *RAoF* flips between references to genres as disparate as documentary, satire, surrealism and the movie musical, including direct samples such as footage from police dash cameras and newsreels, Frances Bodomo's "Ripa the Reaper" sketch from *Collective: Unconscious* (2016), the arcade game *Dance Dance Revolution* (1998– ), and drag performer Moi Renee's single "Miss Honey" (1992). It also makes liberal use of remixing: the "Nuncaland" sequence, a bilingual, samba-infused retelling of *Peter Pan*, stands out in its use of music, dance, staging, cinematography, and script to "remix" a classic story in a novel way. This variability prevents the audience from settling on a specific interpretation. As a result, reviewers have assessed the series using language such as "overwhelming,"[31] "uncategorizable,"[32] and "a mass-hallucination

---

[29] Elizabeth Reich, "The Gift of Black Sonics: Interface and Ontology in Sorry to Bother You and Random Acts of Flyness," in *Cybermedia: Explorations in Science, Sound, and Vision*, ed. Carol Vernallis et al. (Bloomsbury, n.d.).

[30] Georgina Born and Christopher Haworth, "From Microsound to Vaporwave: Internet-Mediated Musics, Online Methods, and Genre," *Music and Letters* 98, no. 4 (November 1, 2017): 601–47; James B. Peterson, *Hip-Hop Headphones: A Scholar's Critical Playlist* (New York: Bloomsbury Academic, 2016).

[31] Caroline Framke, "TV Review: HBO's 'Random Acts of Flyness,'" *Variety*, August 4, 2018, https://variety.com/2018/tv/news/acts-of-flyness-hbo-review-terence-nance-1202894706/?sub_action=logged_in (accessed January 10, 2021).

[32] James Poniewozik, "Review: 'Random Acts of Flyness' Is a Striking Dream Vision of Race," *The New York Times*, August 1, 2018, https://www.nytimes.com/2018/08/01/arts/television/review-random-acts-of-flyness-terence-nance.html (accessed January 10, 2021).

experiment."³³ This is itself indicative of the neurophysiological constraints that require representation-forming in the first place: the world is too complex for our mental machinery to make sense of without representations.

Nance and his collaborators' refusal to force their work into a predetermined niche—it was once described as "a kaleidoscopic, nearly unclassifiable variety show"³⁴—pressures the audience to find a stable frame of reference. Some audiences default to viewing it as part of a generic "Black art" category, a tendency which has met with considerable and justified critique. Racquel Gates and Michael Gillespie push back against the impulse to interpret Black artworks as aesthetic novelties when they are simply drawing on aesthetic predecessors that are not culturally available to existing criticism.³⁵ In a similar vein, Eric Forthun highlights how prestige television companies such as HBO (on which *RAoF* appeared) implicitly interpret artworks that draw on culturally distinct sources as, in his words, "quality."³⁶ Both articles cite *RAoF* as a work where its formal innovation and connection to artistic inspirations outside the cultural mainstream of art criticism are obscured by such perceptual essentialization.

These habits can be explained, although not excused, by my model. While *Get Out*'s content deviated from the clear expectations of its genre, and *Sorry to Bother You*'s genre itself veered wildly with the third-act twist, *RAoF* is six episodes of near-constant surprises that exists within an artistic and cultural context characterized by shows such as FOX's *In Living Color* (1990–1994) and BET's *Uncut* (2000–2006), engages with Black artistic practices such as hip-hop sampling and Afro-surrealism, and consistently contends with the contemporary Black experience.³⁷ Although the existence of this context is impossible to miss, its nuances are not available to many audiences.

When audiences lack experience with a culture's specific modes of artistic expression, their representations for interpreting those expressions will be

---

[33] Evan Narcisse, "'Random Acts of Flyness' Doesn't Give a F–k If White People Get It," *Rolling Stone*, September 7, 2018, https://www.rollingstone.com/tv/tv-features/random-acts-of-flyness-719221/ (accessed January 10, 2021).
[34] Reggie Ugwu, "Is America Ready for the Mind of Terence Nance?" *The New York Times*, July 26, 2018, https://www.nytimes.com/2018/07/26/arts/television/terence-nance-random-acts-of-flyness-hbo.html (accessed January 10, 2021).
[35] Racquel J. Gates and Michael Boyce Gillespie, "Reclaiming Black Film and Media Studies," *Film Quarterly* 72, no. 3 (March 1, 2019): 13–15.
[36] Eric Forthun, "'We Should Be Addressing Whiteness Less, and Affirming Blackness More': Random Acts of Flyness, Afrosurrealism, and Quality Programming," *Communication, Culture and Critique*, no. tcaa013 (June 17, 2020).
[37] Alison Herman, "Terence Nance Is Indescribable," *The Ringer*, August 20, 2017, https://www.theringer.com/tv/2018/8/20/17757956/terence-nance-interview-random-acts-of-flyness (accessed January 10, 2021).

comparatively broad because they have never been called upon to interpret such art before. To such an audience, those modes of expression have no meaning independent of their cultural context, so their cultural context substitutes for a more nuanced understanding of how a work operates. While my cognitive approach to genre and expectation requires that every work belong to, or participate in, some genre, it does not require that the *perceived* genre and the *intended* genre be the same. Rather, it implies that the more a creator's and a listener's prior knowledge and experience overlap, the likelier it is that they will share a representative framework for the creator's output. And since a work's interpretation influences the inferences an audience draws about that work's creator, any lack of specificity in the genre used to interpret a work will be reflected in how its creators themselves are regarded. This connection holds regardless of the medium and its relationship with reality.

It is, I hope, clear that the responses of audiences and critics are reflective of underlying cognitive processes, and that this cognition relies heavily on the representation of the work being perceived. But what exactly counts as a "work" in this analysis? While this chapter has focused on films and other forms of artistic expression, since the neurocognitive structures underpinning the interpretation process are applicable to perception more generally, a "work" could more accurately be defined as *any object that is created by one individual or community and perceived by another*. Therefore, this model can be applied to all forms of communication, with the analysis of artistic media acting as more of a case study of the broader mechanism.

Notably, this model does not require much, if any, information in any contextual variable for an inference to be made. Take the code-switching that plays such a dominant role in *Sorry to Bother You* as an example. Code-switching occurs in everyday life, and in these instances, a speech act would operate as a "work," and would interact with prior knowledge about the sociocultural context, aesthetic (or linguistic) context, and the type of speech (e.g., public or private, persuasive or informative, and so on). Unless the people talking had met before, the listener would have no other knowledge about the speaker, but will nevertheless make inferences about the speaker themselves, even with scant or flawed prior information about them (see figure 19.4). In those cases, they will rely almost entirely on context to fill in the gaps; inaccuracies or biases in this prior contextual knowledge may be a primary source of unconscious of implicit biases. The importance of this contextual information to making accurate or useful inferences, especially about other people, raises an immediate question: where do we get this context?

According to naïve realism, we all believe that we are reasonable observers of the world, and that the things we observe are typical of the things other people would observe.[38] However, each of our experiences is constrained by our surroundings, cultural background, and economic situation. As a result, not only are our representations of phenomena we have experienced directly more detailed than others, our representations of things outside our personal experience comes from our consumption of media, such as the news, movies, television, books, and music. The consequences of this are depicted, to both humorous and sinister effect, in each of the works analyzed here. In other words, this model and these case studies give a cognitive explanation and demonstration of how fiction does not merely reflect reality, but in fact helps to shape it.

## Bibliography

Box, George E. P. "Science and Statistics." *Journal of the American Statistical Association* 71, no. 356 (December 1, 1976): 791–99.
Brackett, David. *Categorizing Sound*. Oakland, CA: University of California Press, 2016.
Carew, Anthony. "American Horror: Genre and the Post-Racial Myth in Get Out." *Screen Education* 94 (2019): 14–21.
Cohen, Annabel J. "Congruence-Association Model of Music and Multimedia: Origin and Evolution." In *The Psychology of Music in Multimedia*, edited by Siu-Lan Tan, Annabel J. Cohen, Scott D. Lipscomb, and Roger A. Kendall, 17–47. Oxford, UK: Oxford University Press, 2013.
Cohen, Annabel J. "Scoring Music for Westworld Then and Now—A Cognitive Perspective." In *Cybermedia: Explorations in Science, Sound, and Vision*, edited by Carol Vernallis, Selmin Kara, Jonathan Leal, and Holly Rogers. Bloomsbury, 2021.
Dargis, Manohla. "Review: In Get Out, Guess Who's Coming to Dinner? (Bad Idea!)." *The New York Times*, February 23, 2017. https://www.nytimes.com/2017/02/23/movies/get-out-review-jordan-peele.html. (accessed December 30, 2020).
Dunning, Stefanie K. "What Is the Future? Weirdness and Black Time in Sorry to Bother You." *Studies in the Fantastic* 9 (2020): 44–62.
Fabbri, Franco. "What Kind of Music?" *Popular Music* 2 (1982): 131–43.
Forthun, Eric. "'We Should Be Addressing Whiteness Less, and Affirming Blackness More': Random Acts of Flyness, Afrosurrealism, and Quality Programming." *Communication, Culture and Critique*, no. tcaa013 (June 17, 2020): n.pag.

---

[38] Lee Ross and Andrew Ward, "Naïve Realism: Implications for Social Conflict and Misunderstanding," in *Values and Knowledge*, ed. Edward S. Reed, Elliot Turiel, and Terrance Brown (Mahwah, NJ: Lawrence Erlbaum Associates, Inc., 1996), 103–35.

Framke, Caroline. "TV Review: HBO's 'Random Acts of Flyness.'" *Variety*, August 4, 2018. https://variety.com/2018/tv/news/acts-of-flyness-hbo-review-terence-nance-1202894706/?sub_action=logged_in. (accessed December 31, 2020).

Gates, Racquel J., and Michael Boyce Gillespie. "Reclaiming Black Film and Media Studies." *Film Quarterly* 72, no. 3 (March 1, 2019): 13–15.

Griffin, Dale W., and Lee Ross. "Subjective Construal, Social Inference, and Human Misunderstanding." *Advances in Experimental Social Psychology* 24 (1991): 319–59.

Herman, Alison. "Terence Nance Is Indescribable." *The Ringer*, August 20, 2017. https://www.theringer.com/tv/2018/8/20/17757956/terence-nance-interview-random-acts-of-flyness. (accessed December 31, 2020).

Huron, David. *Sweet Anticipation: Music and the Psychology of Expectation*. Cambridge, MA: MIT Press, 2006.

Kumaran, Dharshan, Demis Hassabis, and James L. McClelland. "What Learning Systems Do Intelligent Agents Need? Complementary Learning Systems Theory Updated." *Trends in Cognitive Sciences* 20, no. 7 (2016): 512–34.

Landsberg, Alison. "Horror Vérité: Politics and History in Jordan Peele's Get Out (2017)." *Continuum* 32, no. 5 (September 3, 2018): 629–42.

Lerdahl, Fred. "Cognitive Constraints on Compositional Systems." *Contemporary Music Review* 6, no. 2 (1992): 97–121.

Markus, Hazel R., and Shinobu Kitayama. "Cultures and Selves: A Cycle of Mutual Constitution." *Perspectives on Psychological Science* 5, no. 4 (2010): 420–30.

Martin, Alex. "The Representation of Object Concepts in the Brain." *Annual Review of Psychology* 58, no. 1 (December 6, 2006): 25–45.

Martin, Michael T., Yalie Kamara, and Boots Riley. "Boots Riley on Sorry to Bother You and the Matter of the 'Good Fight.'" *Black Camera* 11, no. 2 (2020): 176–215.

McClelland, James L., Bruce L. McNaughton, and Randall C. O'Reilly. "Why There Are Complementary Learning Systems in the Hippocampus and Neocortex: Insights from the Successes and Failures of Connectionist Models of Learning and Memory." *Psychological Review* 102, no. 3 (1995): 419–57.

Miranda, Dave, Camille Blais-Rochette, Karole Vaugon, Muna Osman, and Melisa Arias-Valenzuela. "Towards a Cultural-Developmental Psychology of Music in Adolescence." *Psychology of Music* 43, no. 2 (2015): 197–218.

Narcisse, Evan. "'Random Acts of Flyness' Doesn't Give a F-k If White People Get It." *Rolling Stone*, September 7, 2018. https://www.rollingstone.com/tv/tv-features/random-acts-of-flyness-719221/. (accessed December 31, 2020).

Peterson, James B. *Hip-Hop Headphones: A Scholar's Critical Playlist*. New York: Bloomsbury Academic, 2016.

Pinedo, Isabel. "Get Out: Moral Monsters at the Intersection of Racism and the Horror Film." In *Final Girls, Feminism and Popular Culture*, edited by Katarzyna Paszkiewicz and Stacy Rusnak, 95–114. Cham: Springer International Publishing, 2020.

Poldrack, Russell A., and Karin Foerde. "Category Learning and the Memory Systems Debate." *Neuroscience & Biobehavioral Reviews* 32, no. 2 (2008): 197–205.

Poniewozik, James. "Review: 'Random Acts of Flyness' Is a Striking Dream Vision of Race." *The New York Times*, August 1, 2018. https://www.nytimes.com/2018/08/01/arts/television/review-random-acts-of-flyness-terence-nance.html. (accessed December 31, 2020).

Pulliam-Moore, Charles. "The Hidden Swahili Message in Get Out the Country Needs to Hear." *Splinter*, March 1, 2017. https://splinternews.com/the-hidden-swahili-message-in-get-out-the-country-needs-1793858917. (accessed December 30, 2020).

Reich, Elizabeth. "The Gift of Black Sonics: Interface and Ontology in Sorry to Bother You and Random Acts of Flyness." In *Cybermedia: Explorations in Science, Sound, and Vision*, edited by Carol Vernallis, Holly Rogers, Jonathan Leal and Selmin Kara. New York: Bloomsbury, 2021.

Ross, Lee, David Greene, and Pamela House. "The 'False Consensus Effect': An Egocentric Bias in Social Perception and Attribution Processes." *Journal of Experimental Social Psychology* 13, no. 3 (1977): 279–301.

Ross, Lee, and Andrew Ward. "Naïve Realism: Implications for Social Conflict and Misunderstanding." In *Values and Knowledge*, edited by Edward S. Reed, Elliot Turiel, and Terrance Brown, 103–35. Mahwah, NJ: Lawrence Erlbaum Associates, Inc., 1996.

Sameroff, Arnold. "A Unified Theory of Development: A Dialectic Integration of Nature and Nurture." *Child Development* 81, no. 1 (2010): 6–22.

Schloss, Joseph G. *Making Beats: The Art of Sample-Based Hip-Hop*. Middletown, CT: Wesleyan University Press, 2004.

Sims, David. "Sorry to Bother You Is Fizzy, Flawed, and Fascinating." *The Atlantic*, July 6, 2018. https://www.theatlantic.com/entertainment/archive/2018/07/sorry-to-bother-you-review/564263/. (accessed December 30, 2020).

Tsai, Jeanne L., Jen Ying Zhen Ang, Elizabeth Blevins, Julia Goernandt, Helene H. Fung, Da Jiang, Julian Elliott, et al., "Leaders' Smiles Reflect Cultural Differences in Ideal Affect." *Emotion* 16, no. 2 (2016): 183–95.

Tsai, Jeanne L., Jennifer Y. Louie, Eva E. Chen, and Yukiko Uchida. "Learning What Feelings to Desire: Socialization of Ideal Affect Through Children's Storybooks." *Personality and Social Psychology Bulletin* 33, no. 1 (2007): 17–30.

Ugwu, Reggie. "Is America Ready for the Mind of Terence Nance?" *The New York Times*, July 26, 2018. https://www.nytimes.com/2018/07/26/arts/television/terence-nance-random-acts-of-flyness-hbo.html. (accessed December 31, 2020).

Wood, Robin. "An Introduction to the American Horror Film." In *The American Nightmare: Essays on the Horror Film*, edited by Robin Wood and Richard Lippe, 73–110. Toronto: Festival of Festivals, 1979.

# 20

# Face Color

Bevil R. Conway

The first bowel movement of our newborn daughter was met with cheers from the staff at the neonatal intensive care unit in Ottawa. I rose blurry eyed from a night on the lazy-boy chair in the corner of the bay to inspect the source of celebration: a pinhead-sized speck of black on the diaper. It didn't seem that impressive. But I was told that it, and its color meant that her inside bits were working. Our daughter was naked except for the diaper. She weighed less than three pounds, the embodiment of vulnerability. My overwhelming urge was to clothe her, but the nurses said doing so would make their jobs harder. A day later, I learned why. Hours after our daughter was born, both her lungs collapsed. A short, terrifying, emergency surgery was done to save her life. In the hours that followed, I looked intensely at her through the plexiglass incubator walls, as if my gaze would sustain her recovery. On the second day of our daughter's life outside the womb, a nurse asked me if I wanted to hold her. Shirtless and shaking, I sat with outstretched arms as the nurse disconnected all the monitoring equipment. The reassuring chirp from the heartrate monitor went silent. The glow of the green trace reporting the ebb and flow of her respiration went dim. In the quiet, my daughter was placed on my chest. I asked the nurse why on Earth she had turned off all the vital monitoring equipment—such irresponsibility. The nurse rested a warm hand on my shoulder, then casually tipped forward the head of my infant girl. Without the strength to right her airway, my daughter's lips, face, and body went a startling color. She looked as if she had been spray painted with a grayish blue pigment. I reflexively reached over, lifted her head, and straightened her neck. My reaction seemed preconscious, instantaneous, tapping into an ancient fight-or-flight response. Within seconds, her cheeks flushed with bright red blood. The nurse smiled. Patting my shoulder, she said "your response to looking at her will be more decisive than your response to a change in some artificial beep or a flicker of the digital display." Clothes obstruct the view.

## The Color Problem

Pinning down the role of color in behavior, and how color vision works, has been a surprisingly contentious business, in culture and in science. As early as the sixteenth century, in writings by Giorgio Vasari and Lodovico Dolce, the debate pitted *Disegno* against *Colore*.[1] The argument seems to tip in favor of design— the relative merit of colorless form over color. This idea was promulgated through the eighteenth century by Johann Joachim Winckelmann, who famously argued that "color should have a minor part in the consideration of beauty, because it is not [color] but the structure that constitutes its essence."[2] Winckelmann's contention was that ancient Greek sculptures were the pinnacle of beauty, and that they achieved beauty despite—or because—they were colorless,[3] perpetuating the myth "that the lofty ancient Greeks were too sophisticated to color their art."[4] The view held by Winckelmann is ironic, since the sculptures were originally colored, and are devoid of color because of the inadvertent effects of exposure to the elements not aesthetic choices of their makers. Any persistent difficulty in accepting the color of ancient sculptures may reflect the mendacity of a false logic that perverts the ancient Greeks: since ancient sculpture is the epitome of beauty; and ancient sculpture is colorless; the epitome of beauty must be colorless.[5] According to the racist logic, the white race is conceived as colorless (like ancient Greek sculpture) and is the default against which other races are seen as colored.[6] This abhorrent reasoning may underwrite the under-representation of Black slaves in sculpture.[7] Attempts to reconstruct the color of classical sculptures are still met with amusement, if not distaste: the color is seen

---

[1] Giorgio Vasari, *The Lives of the Artists* (1550; repr. Oxford: Oxford University Press, 2008); Lodovico Dolce, *Aretin: Or, A Dialogue on Painting From the Italian of Lodovico Dolce* (1557; repr. London: Forgotten Books, 2018).
[2] Matthew Gurewitsch, "True Colors," *Smithsonian Magazine*, July 2008, https://www.smithsonianmag.com/arts-culture/true-colors-17888/ (accessed January 15, 2021).
[3] Johann Joachim Winckelmann, *History of the Art of Antiquity*, Trans. Harry Francis Mallgrave (Los Angeles: Getty Publications, 2006).
[4] The characterization of Winckelmann's argument by Nell Irvin Painter, *The History of White People* (New York: W. W. Norton, 2010), quoted by Sarah Bond in "Why we need to start seeing the classical world in color," *Hyperallergic*, June 7, 2017, https://hyperallergic.com/383776/why-we-need-to-start-seeing-the-classical-world-in-color/ (accessed January 15, 2021).
[5] Colleen Flaherty, "Threats for What She Didn't Say," *InsideHigherEd.com*, June 19, 2017, https://www.insidehighered.com/news/2017/06/19/classicist-finds-herself-target-online-threats-after-article-ancient-statues (accessed January 15, 2021).
[6] Richard Dyer, *White: Essays on Race and Culture* (New York: Routledge, 1997).
[7] Kirk Savage, *Standing Soldiers, Kneeling Slaves* (Princeton: Princeton University Press, 1997).

as obscuring the essence of the Greek forms, garish and distracting.[8] That these sculptures were originally colored has been known for centuries, yet they persist in our collective imagination as white, *pure* white, as fantastic as dinosaurs, which remain naked in our mind's eye even though they were almost certainly clothed in feathers. Design trumps color. Or so it seems.

Information conveyed by color is not simply difficult to quantify, but difficult to defend. Respected newspapers resisted printing in color long after printing technology readily enabled color reproduction.[9] The late start cannot be chalked up to a technical challenge, but rather to a chromophobia arising from the conviction that grayscale images imply seriousness, and color is frivolous.[10] But this is too glib. My argument is that under the covers of the debate between *Disegno* against *Colore*, or between black-and-white photographs and color photographs, hides a mystery: color and form give rise to fundamentally different forms of knowledge. They cannot compete in the same ring. They are apples and oranges. Design (or shape, or form), tells us about the identities of stuff in the world; color colors how we feel about the stuff.[11] Design and color are not symmetric properties. At any given instant, a given shape has one color or pattern of colors. But a given color can simultaneously relate to many shapes. The banana changes color as it ripens, and with that change comes a change not in its shape or identity—it is always a banana—but whether we care about it. Adjudicating feelings runs against the grain of institutions, like newspapers, that defend objective reporting. As David Batchelor writes, color "is regarded as alien and therefore dangerous; ... merely as a secondary quality of experience, and thus unworthy of serious consideration. Color is dangerous, or it is trivial, or it is both. ... Either way, colour is routinely excluded from the higher concerns of the Mind."[12]

The history of modern color science, like the Western cultural history that preceded it, is founded on a sequence of inflamed debates, rife with *ad hominem* attacks: Goethe versus Newton in the first half of the nineteenth

---

[8] Margaret Talbot, "The Myth of Whiteness in Classical Sculpture," *The New Yorker*, October 22, 2018, https://www.newyorker.com/magazine/2018/10/29/the-myth-of-whiteness-in-classical-sculpture (accessed January 15, 2021).
[9] The front page of *The New York Times* printed its first color image only in 1997, as recounted by Will Higginbotham, "When the Gray Lady Started Wearing Color," *The New York Times*, October 4, 2018, https://www.nytimes.com/2018/10/04/insider/history-times-color-photos.html (accessed January 15, 2021).
[10] David Batchelor, *Chromophobia* (London: Reaktion Books, 2000).
[11] I first made this argument explicit in Bevil R. Conway, "The Organization and Operation of Inferior Temporal Cortex," *Annual Review of Vision Science* 4, no.1 (2018): 381–402.
[12] Batchelor, *Chromophobia*, 23.

century;[13] Hering versus Helmholtz in the second half of the nineteenth century.[14] These debates ostensibly were about biological mechanisms of color vision, but the subtext was really the role of color in behavior. Is color a cue, understandable in simple mechanistic terms, that assists in the visual recognition of objects (Newton, Helmholtz)? Or is it a metaphysical, almost divine, subjective phenomenon that relates to experience, emotion, and affect (Goethe, Hering)? Contemporary textbooks in psychology skirt the debate, asserting that color assists object recognition while overlooking the obvious fact that almost every quantitative faculty related to object vision can be achieved without color.[15] People, places, things, and actions: they are all easily seen in black-and-white movies. What do we get from color?

## What is Color For?

I was taught the standard textbook account that color is "for the detection of ripe fruit."[16] This conception relegates color to a supporting role—color assisting design. And yet people who are genetically colorblind show little difficulty in object recognition and are generally unimpaired in discriminating ripeness (smell and touch are ultimately more important than color, even for bananas). Genetic colorblindness happens when one of the three color-coding genes is lost when the chromosomes replicate in sexual reproduction.[17] Genetic colorblindness is quite common, affecting about 1 in 12 men and 1 in 200 women. Many colorblind people do not become aware of their visual difference until adulthood, following an entire childhood of ripe-fruit detection and consumption. What's more, people with colorblindness use appropriate color terms far more often than they should, given their retinal deficiency.[18] Colorblind people are not distinguished from those with genetically normal vision in terms of body weight, education, ability to attract partners, or fertility. It is hard to defend the position

---

[13] Dennis L. Sepper, *Goethe Contra Newton: Polemics and the Project for a New Science of Color* (Cambridge, UK: Cambridge University Press, 1988).
[14] Ian P. Howard, "The Helmholtz–Hering Debate in Retrospect," *Perception* 28, no.5 (1999): 543–9.
[15] For example, E. Bruce Goldstein, *Sensation and Perception (10th Edition)* (Boston: Cengage Learning, 2016), 195–6.
[16] Ibid., 196.
[17] Joseph Carroll and Bevil Conway, "Color Vision," in *Handbook of Clinical Neurology* Vol. 178, eds. J. Barton and A. Leff (Amsterdam: Elsevier, 2021).
[18] Dorothea Jameson and Leo M. Hurvich "Dichromatic Color Language: "Reds" and "Greens" Don't Look Alike but Their Colors Do," *Sensory Processes* 2, no. 2 (1978): 146–55.

that the role of color is to enhance fitness in pursuit of better object vision. A skeptic might counter that this evidence is suspect because it is from humans, a species that is not simply subject to selective pressures, but actively changes them. So, what about other primate species? Irresistibly cute, if smelly, New World monkeys provide a natural experiment. These creatures, which include capuchins, squirrel monkeys, and tamarins, are common to the tropical forests of Central and South America. They naturally come in dichromatic and trichromatic versions, comingled in each population. The trichromatic version has the full complement of three color-coding genes, just like people with normal color vision. The dichromatic animals, meanwhile, are missing one of the genes, and they are colorblind in the same way as most genetically colorblind humans. Despite their genetic and physiological impairment, colorblind members in the monkey populations are no slower in finding ripe fruit.[19] Clearly primate color vision has not evolved simply to identify food sources.

Similar conclusions are obtained from experiments in humans. In an influential paper, Biederman and Ju reported that the reaction time taken to recognize objects is no shorter for objects in color photographs than for line drawings.[20] They observed no advantage of color even for objects that have characteristic colors, such as bananas and forks. Spurred by these provocative results, a cottage industry emerged to quantify a role of color in behavior. Gegenfurtner and Rieger showed that color can provide a subtle boost that facilitates encoding and retrieval of visual information.[21] Oliva and Schyns showed that color can help mediate scene recognition when the colors are sensible.[22] Yip and Sinha showed that when images are degraded, for example by blurring them, color confers a modest benefit for parsing objects like faces.[23] And Therriault and colleagues showed that when objects are presented in their

---

[19] Nathaniel J. Dominy, Paul A. Garber, Julio César Bicca-Marques and Maria Aparedcida De O. Azevedo-Lopes, "Do Female Tamarins Use Visual Cues to Detect Fruit Rewards More Successfully than do Males?" *Animal Behaviour* 66, no. 5 (November 2003): 829–37; Erin R. Vogel, Maureen Neitz, and Nathaniel J. Dominy, "Effect of Color Vision Phenotype on the Foraging of Wild White-faced Capuchins, Cebus Capucinus," *Behaviour Ecology* 18, no. 2 (2007): 292–7.

[20] Irving Biederman and Ginny Ju, "Surface Versus Edge-based Determinants of Visual Recognition," *Cognitive Psychology* 20, no.1 (1988): 38–64.

[21] Karl R. Gegenfurtner and Jochem Rieger, "Sensory and Cognitive Contributions of Color to the Recognition of Natural Scenes," *Current Biology* 10, no. 13 (2000): 805–8.

[22] Aude Oliva and Philippe G. Schyns, "Diagnostic Colors Mediate Scene Recognition," *Cognitive Psychology* 41 (2000): 176–210.

[23] Andrew W. Yip and Pawan Sinha, "Contribution of Color to Face Recognition," *Perception* 31, no.8 (August 2002): 995–1003.

normal colors, the time it takes to name them can be slightly reduced.[24] But these studies seem to miss what any parent recognizes: color tells us about each other, about our emotional and physical welfare. As evident across art history,[25] color carries meaning about the stuff in the world, meaning that exists in a dimension that is independent of object recognition. Color is the language we turn to as metaphor for other ineffable subjective experiences. Consider sound color. And whatever one thinks about the relative advantages of design versus color, there is near universal agreement that people like color. We ask each other about favorite colors, and only rarely about favorite shapes. Color may even contribute to the addictiveness of smart phones.[26]

The insatiable appetite for color has inspired many technological advances, from the extraction of natural pigments used in cave painting over ten thousand years ago, to the invention of the first synthetic pigments in the eighteenth century, the creation of new structural pigments such as Vanta Black, and the development in the twenty-first century of expanded color-gamut displays that use more than three primaries. Cumulatively over history, humans have expended enormous effort and resources on obtaining, creating, retaining, and experiencing color. None of this enterprise is in pursuit of better object recognition. Meanwhile, the desire for color has underwritten some dark chapters in human history. The desire for color—specifically indigo blue pigment—fueled the slave trade.[27] This history uncovers the Jekyll and Hyde of the human obsession with color: the distinction of slave and slaveholder reflects an insidious preference of one color (of skin); while the abhorrent business of slavery was financed by the transcendent pleasure of a specific color (of pigment). Contemporary neuroscience has yet to determine whether color is a primary reward, but it need not be to explain the desire for color. Color is often predictive of the reward value of an object: the yellowness of the banana reflects its sugar content. Color may become rewarding by proxy, in the same way that clickers become rewarding when used to train animals. The dolphin learns the association of a click with a piece of fish and will eventually learn tricks for click reward.

[24] David J. Therriault, Richard H. Yaxley, and Rolf A. Zwaan, "The Role of Color Diagnosticity in Object Recognition and Representation," *Cognitive Processing* 10, no.4 (June 2009): 335–42.
[25] For example see John Gage, *Color and Meaning: Art, Science, and Symbolism* (Berkeley and Los Angeles: University of California Press, 1999).
[26] Nellie Bowles, "Is the Answer to Phone Addiction a Worse Phone?" *The New York Times*, January 12, 2018, https://www.nytimes.com/2018/01/12/technology/grayscale-phone.html (accessed January 15, 2021).
[27] Catherine E. McKinley, *Indigo: In Search of the Color That Seduced the World* (London: Bloomsbury, 2011); Andrea Feeser, *Red, White and Black Make Blue: Indigo in the Fabric of Colonial South Caroline Life* (Athens: University of Georgia Press, 2013).

The neuroscience of color has historically focused heavily on the front end: mechanisms for encoding color found in the eye and primary visual cortex.[28] Dogma is that these chromatic signals are multiplexed with signals about form to provide a unified experience of the visual world.[29] As in the cultural debate of *Disegno* versus *Colore*, the standard neuroscientific account seats color as second fiddle. By this account, color is back-up music that can, if absolutely needed, help segment the objects in the scene when achromatic form vision is compromised. It is this context that made the discovery of an extensive network of color-processing regions within the cerebral cortex a surprise.[30] The discovery was the culmination of a series of experiments by my colleagues and me, aimed at figuring out how the brain interprets color signals encoded by the retina.[31] In these experiments, we used brain imaging to measure responses of the cerebral cortex of participants while they were shown a battery of images and moving images, in color and black-and-white. The results show that large parts of the cerebral cortex are much more strongly engaged by color than by black-and-white. Moreover, these color-biased regions are located within the large swath of cerebral cortex called the ventral visual pathway, which is implicated in high-level object vision. This cortical real estate runs for many centimeters along the temporal lobes from the back of the brain along the ventral belly of the brain. It is tissue that is typically thought to enable face recognition, object categorization, and scene perception— challenging tasks, that understandably require lots of brain power to execute. The discovery that the same cortical territory also houses an extensive and largely independent network for processing color compelled a re-evaluation, at least among neuroscientists, about the role of color in behavior. If color engages so much cortical terrain, and within cortical regions known to be essential for high-level cognition, color must be more than a supporting player.

One of my flatmates during graduate school was obsessed with the stock market. He could not fathom why I wasn't. As he pointed out, one can make relatively safe

---

[28] Gregory D. Horwitz, "Signals Related to Color in the Early Visual Cortex," *Annual Review of Vision Science* 6, no. 1 (September 2020): 287–311.
[29] Karl Gegenfurtner, "Cortical Mechanisms of Colour Vision," *Nature Reviews | Neuroscience* 4 (2003): 563–72.
[30] Bevil R. Conway, "The Organization and Operation of Inferior Temporal Cortex," *Annual Review of Vision Science* 4 (2018): 381–402.
[31] Rosa Lafer-Sousa and Bevil R. Conway, "Parallel, Multi-stage Processing of Colors, Faces and Shapes in Macaque Inferior Temporal Cortex," *Nature Neurosci*ence 16, no. 12 (October 2013): 1870–8; Rosa Lafer-Sousa, Bevil R. Conway, and Nancy G. Kanwisher, "Color-Biased Regions of the Ventral Visual Pathway Lie between Face- and Place-Selective Regions in Humans, as in Macaques," *The Journal of Neuroscience* 36, no.5 (February 2016): 1682–97.

investments in established, stable, well-recognized corporations. Blue chips. I was much more interested in why the stocks were "blue." The choice of a logo color reflects more than aesthetic preferences. Blue relays reliability (IBM, NIH); orange connotes warmth and friendliness (Dunkin' Donuts, UPS); red conjures attention and excitement (Harvard, MIT); gray implies neutrality (The New York Times, known as The Gray Lady). Logo-color conventions tempt us to draw the conclusion that color meaning is reflexive, hardwired: that there is a lookup table that translates color to meaning. But this is wrong. A hundred years ago in Western culture, pink was the color preference for little boys. Today, pink is the choice for girls. A brief dip into the anthropology of color teaches us that the meanings of colors emerge as a result of social pressures and communicative goals[32] in the same way that common businesses coalesce in certain cities, or on certain streets. A new financial services company would be well-served to open offices in London and New York City, where customers seeking those services already congregate. Similarly, a new restaurant would do well to open next door to a current one, where it can steal hungry customers at the threshold. An organization that seeks to project stability and reliability can communicate that information using a blue logo because other organizations have taught us to make this association. But context is essential to decoding color meaning. In Europe, red is the color of socialists; in America, red is the color of capitalists. And everywhere, blue sometimes implies sadness.

## Skin Color

Perhaps the most significant context of human behavior concerns our well-being. Complexion is one of the first signs evaluated by nurses and paramedics to evaluate health status. When the hemoglobin of blood is oxygenated, it turns from dusky cyan to fire-engine red. This transition is most evident in the lips and cheeks, where the external layers of skin are thinnest, and the vasculature is close to the surface. Skin color is, of course, also impacted by the amount of melanin and carotinoids, which are largely determined by genetics and diet. Carrot consumption turns the skin orange. But of the three main factors that determine

---

[32] Stephen C. Levinson, "Yéli Dnye and the Theory of Basic Color Terms," *Journal of Linguistic Anthropology* 10, no.1 (2000): 3–55; Bevil R. Conway, Edward Gibson, Richard Futrell, Julian Jara-Ettinger, Kyle Mahowald, Leon Bergen, Sivalogeswaran Ratnasingam, Mitchell Gibson and Steven T. Piantadosi, "Color Naming Across Languages Reflects Color Use," *Proceedings of the National Academy of Sciences of the United States of America* 114, no. 40 (2017): 10785–90.

skin color—blood, genetics, diet—it is the first that can change most rapidly, and that signals dynamic changes in health and emotion. The meaning of bluish green in the context of faces is universal, cross-racial, reflected in emojis of sickness.

A simple experiment by Maryam Hasantash, Rosa Lafer-Sousa, Arash Afraz and me provided clear evidence of the importance of face color in human behavior.[33] Our experiment made use of low-pressure sodium (LPS) light, which was in widespread use as streetlighting in the 1980s because of its energy efficiency. But unlike natural lights such as the sun, which consist of many wavelengths, LPS light consists of a single dominant wavelength. The light has an eerie yellowness. Because the light is monochromatic, retinal mechanisms for encoding color are completely impaired—there is no objective way that the visual system can compute color information under LPS light. Looking at things under LPS light is bizarre. One has no difficulty recognizing objects, but their color is entirely bleached. Everything appears the same colorless brownish gray. Strawberries and bananas are easily distinguished. Faces can also be recognized. But the fruit looks unappealing. And faces, regardless of race, look surprisingly green. This is curious because there is nothing objectively green in the stimulus. The light is monochromatic yellow, not green. The paradoxical greenness seems to be a kind of error signal—the brain reporting an alarming situation: the face does not appear a normal color. The experiment provides a powerful example of the impact on perception of cognition. In this case, knowledge of the healthy face color influences the color appearance of faces under objectively colorless viewing conditions. The fact that object-color knowledge influences only the color appearance of faces and no other objects suggests that color plays a fundamental role in social interactions. The experiment reinforces the conclusion that color tells us about the state of things, not about what they are. Color is not for the cool analytic task of recognizing objects or faces, but rather for the subjective task of determining meaning, especially when that information is unavailable to language. My newborn daughter was years away from speaking; evolution had equipped us with another medium, her complexion, for communicating pressing information.

Red blood cells live for about 120 days. When they die, the hemoglobin is broken down to toxic bilirubin that is then converted by the liver into a chemical form that can be excreted by the gall bladder into the gastrointestinal tract. The liver of premature babies is not yet able to do its job, and the skin yellows as

---

[33] Arash Afraz, Bevil R. Conway, Maryam Hasantash and Rosa Lafer-Sousa, "Paradoxical Impact of Memory on Color Appearance of Faces," *Nature Communications* 10 (July 2019): 3010.

bilirubin accumulates. The jaundice can be treated by exposing the skin to blue light, which photo-oxidizes the bilirubin to make it water soluble. The toxin can then be expelled by the kidneys in urine. About a week after birth, our daughter looked as if she had been bruised all over. I felt physically ill looking at her tiny yellow body, and I was filled with existential panic. We signed a release form, and the nurses placed her under blue lights. Though reassured that she was now being treated, I felt strangely disconnected from her. Like LPS illumination, the blue light in the NICU prevented my retinal mechanisms from encoding color. The vital color signals relayed by my daughter's skin were eclipsed by a peculiar colorlessness that enveloped her and the sheets she lay on. I could still see her chest rise and fall with the respirator; I could still see her face asleep without distress. But I was cut off from the most primal form of communication about her health that I had come to depend upon: the color of her skin was hidden under a cloth of blue light. She seemed to exist in another world, and I once again turned to the digital displays, the beeps, and the graphs, for insight into her welfare, now acutely aware of how impoverished these devices were.

The importance of skin color is evident not only in the parental care it excites, but also in the odiousness of racism. Like color itself, that lives a double life both dangerous and trivial, skin color exerts effects in contradictory ways, vital and deadly. The social and cultural construction that underlies racism, in its most primitive form, is built on associations linked to appearance, especially levels of melanin in the skin.[34] These levels may fluctuate with changes in sun exposure, but for the most part they are determined by genetic factors beyond our control. Contemporary culture continues to use skin color to tag identity[35]—this behavior is inextricably linked to our status as a social species, a manifestation of the same forces that endow us with the ability to read emotions and health from color changes in faces. Indeed, a person's race, evident in shape features alone, impacts the perception of face color: Black faces are perceived as darker than white faces even in photographs that have been adjusted to have the same lightness.[36] Of

---

[34] There is a long history of attempts to quantify skin tones, dating to at least as early as Paul Broca, "*Instructions générales pour les récherches anthropologiques à faire sur le vivant* (Paris: Masson, 1879). The implicit racism that influenced anthropology in the nineteenth century, as it relates to measurements of skull and brain but not skin tone, is discussed in Stephen J. Gould, *The Mismeasure of Man* (New York: Norton, 1981).

[35] Evident in the plethora of options in makeup; see also Byron Kim's art installation *Synecdoche* (Washington DC: National Gallery of Art, 1991-present), https://www.nga.gov/collection/art-object-page.142289.html (accessed January 15, 2021).

[36] Mahzarin R. Banaji, and Daniel T. Levin, "Distortions in the Perceived Lightness of Faces: the Role of Race Categories," *Journal of Experimental Psychology: General* 135, no.4 (2006): 501–12; Afraz et al. "Paradoxical Impact of Memory on Color Appearance of Faces," 3010.

course, humans use color to project identity in domains beyond race, including sports and politics (Red Sox shirts and MAGA caps). The use of color to signal team membership reverberates in cultural discourse, such as in the debate about the infamous #thedress where people assigned themselves to team "white-and-gold" versus team "blue-and-black?"[37]

## Semiotics of Color in Film

Judgments we make from color seem reflexive. Yet they are not unlearned. And that we experience them as automatic reveals the profound importance of color to human behavior. It follows that color use in visual media presents a wealth of data on culture. As Dorothy walks from an achromatic world into Oz, we watch the scene become flushed with color, and we come to understand what Dorothy is feeling through our own experience of the movie becoming "alive." In *The Wizard of Oz* (Victor Fleming, 1939), color is used as a blunt instrument to signal a richly animated fantasy world. With subsequent advances in color technology that has expanded the options of visual moving media from film to digital and beyond, makers of images have been equipped to exploit color in potent ways, able to control the local color of a scene and to use color correction during editing. Color can be used to direct attention, establish credibility, conjure nostalgia, disgust, pleasure, apathy, hopelessness, euphoria, or psychosis—indeed the full range of human emotion. As such, color offers a lens trained on the subjective forces that impact how culture works, and it does so precisely because the associations of color and meaning are fungible. We are invited to tease apart the impact of color choices, intentional or not, and to explore the cultural context that makes these choices successful or not.

As you look around, from one glance to the next, light levels can vary from only a few photons to many billions. Moreover, the colors of objects appear relatively consistent—this is true even when the lighting conditions change from one kind of light to another, such as from daylight to indoor light. The efficacy of vision derives from sophisticated adaptive mechanisms that accommodate the huge dynamic range of light levels and the changes in the spectral properties of

---

[37] Bevil R. Conway, "Why Do We Care About the Colour of the Dress?" *The Guardian*, February 27, 2015, www.theguardian.com/commentisfree/2015/feb/27/colour-dress-optical-illusion-social media#:~:text=Colour%20helps%20us%20to%20recognise,for%20indigo%2C%20which%20fuelled%20slavery (accessed January 15, 2021).

lighting to achieve color constancy.[38] By contrast, light capturing technologies are impoverished. For example, film stock can only be sensitive to a restricted range of light levels—and historically film stock has been made so that it is most sensitive to Caucasian faces, a reverberation of institutional racism.[39] It remains inescapable, even with digital technology, that every color image must involve light and color correction because the technology works within material parameters that differ from those of the visual system.

Advances in technology will continue to expand the efficacy of the algorithms involved in producing the colors reproduced in images. Some of these algorithms are hardwired into the technology, with settings chosen automatically or with user input about the illumination. For example, one might select different settings on a digital camera for natural light, fluorescent light, or incandescent light, which will correct for the peculiar spectral bias associated with each light. Other image-correction algorithms are accessible during editing. Hard-wiring of the algorithms makes cameras easier to use but obscures prejudices about what looks good, which can have devastating consequences, as evident in Dyer's history of film stock which was manufactured to prioritize Caucasian faces.[40] Digital technology has made it easier to expose these algorithms and to expand the creative potential of image making, providing ample opportunity to shape emotional valence not only in what is captured, but in how images are processed.

In the remainder of this essay, I will consider color choices in *Random Acts of Flyness* (HBO 2008), one of the television shows taken up by other essays in this volume.[41] Nuotama Frances Bodomo, a director of the series, describes her objective to identify "the core of our emotional world that we don't necessarily have a language for."[42] Terence Nance, the creator of the show, affirms that "color is a big part of how emotion is conveyed visually."[43] The collaborative team worked with the colorist Elias Nousiopoulos of The Mill to color grade the film. Here I will attempt to unpack the color grading, which, according to Nance,

---

[38] Bevil R. Conway, "Color Vision, Cones, and Color-coding in the Cortex," *Neuroscientist* 15, no.3 (May 2009): 274–90.
[39] Dyer, *White*, 82–145.
[40] Ibid.
[41] Elizabeth Reich, "The Gift of Black Sonics: Interface and Ontology in Sorry to Bother You and Random Acts of Flyness" in *Cybermedia* eds. Carol Vernallis, Holly Rogers, Selmin Kara, and Jonathan Leal (New York: Bloomsbury, 2021).
[42] Interview with Bodomo following episode 4.
[43] The Mill, "Random Acts of Flyness Q&A With Terence Nance," *The Mill*, September 28, 2018. http://archive.themill.com/millchannel/1786/random-acts-of-flyness-q%26a-with-terence-nance- (accessed January 23, 2021).

**Figure 20.1** Screenshots of four vignettes from episode 1, *Random Acts of Flyness*, directed by Terence Nance, Frances Bodomo, and Shaka King (first aired, August 4, 2018). The upper left scene was shot in black and white; the other scenes made instrumental use of local color and color editing as described in the text (see original clips for color).

posed a challenge "on a project like this where there is no chromatic 'home base' and the creative and emotional intent is constantly in flux."[44]

One of the first vignettes in episode 1 of the first season illustrates number 473/1000 of worries that a Black person should not have to worry about. A Black man mistakenly gets into the wrong blue jalopy. The owner of the automobile, a white woman, calls the police. The narrator, speaking to the Black man, says "it is likely that you, friend, have found the turn of events humorous, but I assure you it is not." The sequence is shot in black-and-white, communicating that the set of events is being reported objectively, neutrally, *The New York Times*. We, the viewers, are left to add the inescapable color: the blue of the cars, the humor, the injustice.

Another vignette showcases Ripa the Reaper, the host of a TV show that teaches Black children about the futility of life, "Everybody dies!" The show adopts the aged-yellow color production associated with 1970s sitcoms, evoking a dated subplot in need of revision. The juxtaposition with Ripa's deadly message underscores the exasperation and resignation of contemporary Blacks faced with the same old story.

[44] Ibid.

A subsequent sketch features Jon Hamm, the actor we associate with the powerful advertising executive Don Draper. Here, Hamm is the star for an infomercial, "White be Gone." The sequence repositions whiteness not as the default colorless condition implied by contemporary experience of ancient Greek sculptures, but as a color that must be confronted and can be removed. Hamm is uncertain about his task. To reassure him, a director is filmed saying "You are here because the people who call themselves white, those victims, for whatever reason, they trust you and that beautiful beige face of yours. You see, drunk with whiteness, stumbling in their stupor, you have what it takes to sober them, with that bullish sincerity in your spirit. They need you. Help them." The filmmakers manipulate the images to align with the message, using a color correction that casts a bluish hue spanning the entire sequence of images. This color-editing choice creates cohesion that links the disparate settings of a stage, film set, and a bedroom. But together with ubiquitous blue local color of the computer displays, clothing, armchair, cushions, and bedspread, the color editing reassures the viewer: you can trust this guy and his message. It works because this color-editing choice is a convention often used in documentaries. The clip ends by complicating the narrative with an unsettling scene. A blue lightning bolt striking Hamm's alter ego as he dashes to escape the realization of his own white privilege. The episode is unequivocally directed at white viewers, and here we see our (white) trust betraying ourselves.

The penultimate vignette in the first episode aims to remedy the "invisibility [in film] of the bisexual Black man." The clip recounts through interview the experiences of a person confronting gender stereotypes, illustrating the man's intimate story through playful, colorful Claymation marked by a sunny yellow backdrop. The interview is shot in thick saturated color—purples and yellowish greens—showcasing a prominent bromeliad flower: proud, erect, and red with giggly excitement. The exuberant color choices throughout this sequence accentuate this message. The final scene returns us to the opening. We see director Nance filming himself while on his bicycle. The footage is shot casually, using a smart phone whose color correction bias we won't be able to recognize until the tectonic plates that govern the meanings of color choices shift and we can look back with some objectivity. Taken together, the set of vignettes that constitute the body of the episode shows a striking color contrast, made startling because the color lies beneath the surface of the articulated narrative. The film makers use color to create cohesion within each vignette, holding each vignette together in a cohesive emotional key, while using distinct color choices for each vignette to unearth the dynamic range of emotional engagement. *Random Acts* provides compelling evidence that by moving beyond the binary of design versus color, we can recognize the complexity of color and its multifaceted role in visual cognition. Just as the brain does not process color in a single computational step, but rather through a series of interacting stages, so is color embedded in a network of many functions and a variety of cultural processes.

# Acknowledgements

I thank Sylvia Chong and Alexander Rehding for helpful discussions and comments on the manuscript.

# Bibliography

Afraz, Arash, Bevil R. Conway, Maryam Hasantash and Rosa Lafer-Sousa. "Paradoxical Impact of Memory on Color Appearance of Faces." *Nature Communications* 10 (July 2019): 3010.

Banaji, Mahzarin R. and Daniel T. Levin. "Distortions in the Perceived Lightness of Faces: the Role of Race Categories." *Journal of Experimental Psychology: General* 135, no. 4 (2006): 501–12.

Batchelor, David. *Chromophobia* (London: Reaktion Books, 2000).

Biederman Irving and Ginny Ju. "Surface Versus Edge-based Determinants of Visual Recognition." *Cognitive Psychology* 20, no. 1(1988): 38–64.

Bond, Sarah. "Why We Need to Start Seeing the Classical World in Color." *Hyperallergic,* June 7, 2017. https://hyperallergic.com/383776/why-we-need-to-start-seeing-the-classical-world-in-color/. (accessed January 15, 2021).

Bowles, Nellie. "Is the Answer to Phone Addiction a Worse Phone?" *The New York Times*, January 12, 2018. https://www.nytimes.com/2018/01/12/technology/grayscale-phone.html. (accessed January 15, 2021).

Broca, Paul. *Instructions Générales pour les Récherches Anthropologiques à Faire sur le Vivant* (Paris: Masson, 1879).

Carroll, Joseph and Bevil R. Conway. "Color Vision." *Handbook of Clinical Neurology* 178 (2021).

Conway, Bevil R. "Color Vision, Cones, and Color-coding in the Cortex." *Neuroscientist* 15, no. 3 (May 2009): 274–90.

Conway, Bevil R. "The Organization and Operation of Inferior Temporal Cortex." *Annual Review of Vision Science* 4, no. 1 (2018): 381–402.

Conway, Bevil R. "Why Do We Care About the Colour of the Dress?" *The Guardian*, February 27, 2015. www.theguardian.com/commentisfree/2015/feb/27/colour-dress-optical-illusion-social media#:~:text=Colour percent20helps percent20us percent20to percent20recognise,for percent20indigo percent2C percent20which percent20fuelled percent20slavery. (accessed January 15, 2021).

Conway, Bevil R., Edward Gibson, Richard Futrell, Julian Jara-Ettinger, Kyle Mahowald, Leon Bergen, Sivalogeswaran Ratnasingam, Mitchell Gibson and Steven T. Piantadosi. "Color Naming Across Languages Reflects Color Use." *Proceedings of the National Academy of Sciences of the United States of America* 114, no. 40 (2017): 10785–90.

Dolce, Lodovico. *Aretin: Or, A Dialogue on Painting From the Italian of Lodovico Dolce* (1557; repr. London: Forgotten Books, 20018).

Dominy, Nathaniel, Paul A. Garber, Julio César Bicca-Marques and Maria Aparecida de O. Azevedo-Lopes. "Do Female Tamarins Use Visual Cues to Detect Fruit Rewards More Successfully than do Males?" *Animal Behaviour* 66, no. 5 (November 2003): 829–37.

Dyer, Richard. *White: Essays on Race and Culture* (New York: Routledge, 1997).

Feeser, Andrea. *Red, White and Black Make Blue: Indigo in the Fabric of Colonial South Carolina Life* (Athens: University of Georgia Press, 2013).

Flaherty, Colleen. "Threats for What She Didn't Say." *InsideHigherEd.com*, June 19, 2017. https://www.insidehighered.com/news/2017/06/19/classicist-finds-herself-target-online-threats-after-article-ancient-statues. (accessed January 15, 2021).

Gage, John. *Color and Meaning: Art, Science, and Symbolism* (Berkeley and Los Angeles: University of California Press 1999).

Gegenfurtner, Karl. "Cortical Mechanisms of Colour Vision." *Nature Reviews / Neuroscience* 4 (2003): 563–72.

Gegenfurtner, Karl and Jochem Rieger. "Sensory and cognitive contributions of color to the recognition of natural scenes." *Current Biology* 10, no. 13 (2000): 805–8.

Goldstein, Bruce. *Sensation and Perception (10th Edition)*, 195–6. Boston: Cengage Learning, 2016.

Gould, Stephen J. *The Mismeasure of Man* (New York: Norton, 1981).

Gurewitsch, Matthew. "True Colors." *Smithsonian Magazine*, July 2008. https://www.smithsonianmag.com/arts-culture/true-colors-17888/. (accessed January 15, 2021).

Higginbotham, Will. "When the Gray Lady Started Wearing Color." *The New York Times*, October 4, 2018. https://www.nytimes.com/2018/10/04/insider/history-times-color-photos.html (accessed January 15, 2021).

Horwitz, Gregory D. "Signals Related to Color in the Early Visual Cortex." *Annual Review of Visual Science* 6, no.1 (September 2020): 287–311.

Howard, Ian P. "The Helmholtz–Hering Debate in Retrospect." *Perception* 28, no.5 (May 1999): 543–9.

Jameson, Dorothea and Leo M. Hurvich. "Dichromatic Color Language: 'Reds' and 'Greens' Don't Look Alike But Their Colors Do." *Sensory Processes* 2, no. 2 (1978): 146–55.

Kim, Byron. *Synecdoche*. Installation, Washington DC: National Gallery of Art, 1991-present. https://www.nga.gov/collection/art-object-page.142289.html. (accessed January 15, 2021).

Lafer-Sousa, Rosa and Bevil R. Conway. "Parallel, Multi-stage Processing of Colors, Faces and Shapes in Macaque Inferior Temporal Cortex." *Nature Neuroscience* 16, no. 12 (October 2013): 1870–8.

Lafer-Sousa, Rosa, Bevil R. Conway, and Nancy G. Kanwisher. "Color-Biased Regions of the Ventral Visual Pathway Lie between Face- and Place-Selective Regions in Humans, as in Macaques." *The Journal of Neuroscience* 36, no. 5 (Febuary 2016): 1682–97.

McKinley, Catherine E. *Indigo: In Search of the Color That Seduced the World* (London: Bloomsbury, 2011).

Nance, Terence. "Random Acts of Flyness Q&A With Terence Nance", in *The Mill* (September 28, 2018) at http://archive.themill.com/millchannel/1786/random-acts-of-flyness-q percent26a-with-terence-nance- (accessed January 23, 2021).

Oliva, Aude and Philippe G. Schyns. "Diagnostic Colors Mediate Scene Recognition." *Cognitive Psychology* 41, no. 2 (September 2000): 176–210.

Painter, Nell Irvin. *The History of White People* (New York: W. W. Norton, 2010).

Reich, Elizabeth. "The Gift of Black Sonics: Interface and Ontology in Sorry to Bother You and Random Acts of Flyness." In *Cybermedia*. Edited by Carol Vernallis, Holly Rogers, Selmin Kara, and Jonathan Leal (New York: Bloomsbury, 2021).

Savage, Kirk. *Standing Soldiers, Kneeling Slaves: Race, War, and Monument in Nineteenth-Century America* (Princeton: Princeton University Press, 1998).

Sepper, Dennis L. *Goethe Contra Newton: Polemics and the Project for a New Science of Color* (Cambridge, UK: Cambridge University Press, 1988).

Stephen C. Levinson. "Yélî Dnye and the Theory of Basic Color Terms." *Journal of Linguistic Anthropology* 10, no. 1 (2000): 3–55.

Talbot, Margaret. "The Myth of Whiteness in Classical Sculpture." *The New Yorker*, October 22, 2018. https://www.newyorker.com/magazine/2018/10/29/the-myth-of-whiteness-in-classical-sculpture. (accessed January 15, 2021).

Therriault, David J., Richard H. Yaxley, and Rolf A. Zwaan. "The Role of Color Diagnosticity in Object Recognition and Representation." *Cognitive Processing* 10, no. 4 (June 2009): 335–42.

Vasari, Giorgio. *The Lives of the Artists* (1550; repr. Oxford: Oxford University Press, 2008).

Vogel, Erin, Maureen Neitz, and Nathaniel J. Dominy. "Effect of Color Vision Phenotype on the Foraging of Wild White-faced Capuchins, Cebus Capucinus." *Behavioral Ecology* 18, no. 2 (March 2007): 292–7.

Winckelmann, Johann Joachim. *History of the Art of Antiquity*. Translated by Harry Francis Mallgrave (Los Angeles: Getty Publications, 2006).

Yip, W. Andrew and Pawan Sinha. "Contribution of Color to Face Recognition." *Perception* 31, no. 8 (August 2002): 995–1003.

Part Six

# Productive Neuropathologies

21

# Digital Vitalism

Marta Figlerowicz

The digital world is human-made, of course. But in its immersive quality it has become like a second nature to many of us, with quasi-natural laws and processes in which we are swept up with an apparent inevitability. Our digital environments are, moreover, often much more uncontrollably persistent and proliferating than we might like them to be. Data replicates, varies, and spreads within them with apparent aimlessness; once present online, it often becomes stubbornly difficult to eradicate.

This chapter considers two films, Jonathan Glazer's *Under the Skin* (2013) and Lars von Trier's *Nymphomaniac* (2014), that reflect on these qualities of our new media environments. Comparing them to natural ecosystems, they represent these environments as frighteningly vibrant and proliferating; as a new nature that does not only match, but at times appears to exceed, the reproductive vitality of the nature that gave rise to our species.

On an immediate political level, both films are explorations of a femininity liberated by contraception and the ostensible loosening of social norms that came with it. They are also films about working women whose control over their own bodies and fates often appears to threaten the men around them. These women treat family relationships not as a necessity but as a choice, one they can modify or withdraw from. They are also not above instrumentalizing their temporary ties with the men they meet, for profit, pleasure, or simple amusement.[1] Glazer's film treats this vision of liberated femininity with sympathy but also skepticism, as less readily accommodated by our social norms—and more

[1] This chapter is the revised version of Marta Figlerowicz, "Inanimism: *Nymphomaniac, Under the Skin*, and *Capitalist Late Style*," Camera Obscura 33, no.2 (98) (2018): 41–67. Copyright, 2018, Duke University Press. All rights reserved. Republished by permission of the publisher. www.dukeupress. edu. For feminist readings of these films along similar lines—as representing or participating in the late capitalist exploitation and objectification of women—see, for example, Ara Osterweil, "Under the Skin: The Perils of Becoming Female," *Film Quarterly* 67, no. 4 (2014): 44–51; Linda Williams, "Cinema's Sex Acts," *Film Quarterly* 67, no.4 (2013): 9–25.

threatened by violent backlash—than it might seem in theory. Von Trier casts upon it the half-horrified, half identificatory gaze he tends to cast onto his female protagonists: turning the central character into an occasionally relatable, but generally fearsome, contraceptively enhanced *vagina dentata*.

But in the midst of these politics, both films are also marked by a more ontological undercurrent. Their protagonists' controlled reproductive sexuality is continually contrasted against, and often overwhelmed or exceeded by, a different kind of fertility or generativity that seems indiscriminately both organic and inorganic, material and affective. The social environments in which they exist and the bodies they inhabit and interact with, appear constantly to produce and reproduce expansions, copies, or continuations of themselves. Indeed, the very form of both films represents them as teeming with more footage, and more recorded or technologically produced aliveness, than can even be contained in a feature film (even one which, like *Nymphomaniac*, is four hours long). There is, to these depictions of teeming life, an unstructured quality, and a schizophrenic one in Deleuze's sense of the term: rather than compose themselves into a symbol of, say, patriarchy or motherhood, they continue to erupt in scattered fashion that also does not recognize clear boundaries between background and foreground, the diegetic and the metatextual.[2] Aligned with these women, the viewer finds herself or himself overwhelmed by this sense of uncontrolled generativity, while at the same time being shown that the forms of self-control for which these characters have struggled are somehow not sufficient—that what they appear to dream of is an escape not only from sexual reproduction in particular, but from embodied existence itself. The point is not just that these people cannot remain isolated and unbounded, but that there is something excessive and difficult mentally to span even about the extent of this reproducibility. A great outward vitality—one that eventually comes to seem more technological than organic—invades the characters' ordered lives, setting rules of its own, suggesting a wealth of instinctual and bodily contingencies that a person might not be able to predict or control in the way she tries to make her body do her bidding. This vitality is ambiguously both non-human and manmade; a floodgate we might have ourselves created, but by which we are now ourselves swept up in and governed by, as by an inborn bodily instinct.

---

[2] Gilles Deleuze and Félix Guattari, *Anti-Oedipus: Capitalism and Schizophrenia* (New York: Bloomsbury, 1972).

At first blush, *Nymphomaniac* and *Under the Skin* do not have much in common, besides their release dates. *Nymphomaniac* is the third installment of von Trier's so-called "Depression Trilogy." Seligman, an asexual bachelor, lets a wounded, bruised woman named Joe recover in his bedroom, where she gradually tells him her life story. Glazer's *Under the Skin* is a surreal sci-fi thriller. Scarlett Johansson plays an alien dressed in a woman's skin who drives a van around Scotland trying to seduce and capture lonely men. Despite these superficial differences, both films follow versions of the same narrative arc: a woman (or humanoid being) is swept up in pursuits of non-reproductive (or very controlledly productive) sex, only to find that the productivity and reproductivity of the bodies and environments around them—and in the case of *Nymphomaniac*, also that of the protagonist's own body—is out of control. These protagonists' femininity also places these films within the very long cultural history of representing women as sexualized machines: a history from which they ultimately diverge, but which both films share as a point of departure. In ostensible accordance with these conventions, *Nymphomaniac* and *Under the Skin* at first focus on the degree of agency these women are able to wrest over the men and women around them, and on the ethics of these acts of control and overpowering. They question whether their protagonists' intense sexual exploits are turning them into subjects or into objects; whether these labors are making them freer or more enslaved to the communities around them. But as the world around these characters seems endlessly reproductive and generative no matter what they themselves try to do, their insistence on the non-teleological nature of their actions begins to seem ever more futile—it no longer appears to matter whether their particular bodies are organically reproductive or not, and to what extent they engage in or try to prevent that. Both films also deploy the mechanics of film recording to produce a similar effect of overwhelming generativity for their viewers—and to explicitly involve their medium as an agent of the intense air of teeming life that they create. *Under the Skin* and *Nymphomaniac* share a preoccupation with images of light and fire—which are also conventional metaphors for cinema—and they suggest that these phenomena are generative of the human figures one might at first have supposed them merely to illuminate.

One way to describe the narrative arc of *Nymphomaniac* is as a gradual transition from a finite to an infinite notion of sexual desires and of the acts they can give rise to. Borrowing from Marquis de Sade's careful sexual mathematics in *120 Days of Sodom* (1785) and elsewhere, von Trier's protagonist Joe is initially obsessed with counting aspects of her sexual experience. She describes her first sexual encounter by detailing the number of thrusts—vaginal and then anal—that she receives from her partner. She then competes with a friend to seduce as many men as possible on a crowded train, winning a climactic victory when she cajoles into oral sex a man who was saving his sperm for his wife's narrow fertility window. Gradually, these sexual exploits become more complicated and more open-ended, involving not just arithmetic but also combinatorics—as when she dates three men at once on a complicated, randomized schedule, using a pair of dice to determine how nice she will be to each of them during any particular date. The breaking point of these mathematics comes when she tries, and fails, to narrow down these forking paths, and settle down with her husband: only to find herself attracted to ever more people and things, to the point where—in a comically overwhelmed attempt to control her urges—she wraps every corner of her furniture in gauze, to prevent herself from masturbating on it. Eventually, she capitalizes on her knowledge of pain and pleasure by becoming a professional extortionist—a job in which she always finds herself potentially discovering a new obsession or fetish, some combination of potent contexts, subjects, and objects that had not occurred to her before. Where in Sade, we are awed by how exhaustive are the sexual exploits he depicts, here they remain always insufficient, just barely past the mark, artificially constrained—and the constraint, not the generality, is what she ultimately seems to be after and cannot get, like the long-lost father whom she tries to cling to. In his conversations with Joe, Seligman highlights her experience of the world as endlessly productive, and reproductive, of sexuality, by likening her sex acts to things like hunting and fishing in plentiful environments and waters—catching a small bit of life from within a sea of plenty.

*Nymphomaniac*'s codified sexual practices closely resemble the repetitive rituals of seduction in *Under the Skin*. In an interview with Hank Sartin, Glazer says that a large point of the film is to make the alien's aggressive flirtations resemble a precisely executed "job."[3] As Amy Herzog puts it, "she is an instrument, a vehicle that performs her tasks in tandem with other operatives."[4] At times, it is suggested that this protagonist is a murderous cyborg: in a scene midway through the film, when she meets with one of her fellow aliens, he carefully

---

[3] "Glazer: I think routine is what we wanted to show. Her job, her at work. Really the first half of the film is that. It's watching her go about her job if you like. If you make her presence equivalent to ours, it's like she's in a routine and a job she hates and she begins to lose her focus and leaves her job." Jonathan Glazer and Hank Sartin, "Jonathan Glazer Talks about His New Film *Under the Skin*," *RogerEbert.com*, April 10, 2014, http://www.rogerebert.com/interviews/jonathan-glazer-talks-about-under-the-skin (accessed April 22, 2015).

[4] Amy Herzog, "Star Vehicle: Labor and Corporeal Traffic in *Under the Skin*," *Jump Cut* 57, 2016, https://www.ejumpcut.org/archive/jc57.2016/-HerzogSkin/index.html (accessed October 22, 2016).

inspects her: as Glazer puts it in his interview for *Rolling Stone*, he is looking "for something ... like a hairline fracture or a crack in the wing of an airplane."[5] Behind the scenes, she presumably collects the bodily fluids of her victims, as well as the skins from which these fluids gradually separate; indeed, it appears that what we see her wearing is the skin of one of such victims who was captured earlier, and which she dons in the film's opening sequence. All of this gives an air reminiscent of *Invasion of the Body Snatchers (1978)* or other older horror films about alien abduction. It creates the impression that what we will be following, as the plot of the film progresses, is an increasingly overt and escalating human-alien conflict.

But that is not what actually happens; instead, the alien we follow is gradually overwhelmed by, and swallowed up in, the mass of human beings from whom she at first tries to pluck out her singular victims, and whose teeming numbers are almost comically indifferent to, and unchanged by, her violent seduction efforts. Driving around in her van ceaselessly, smiling and talking to a succession of men no matter what time of day it is, the alien takes on an aura less reminiscent of an efficient fem-bot than of a hunter in a virgin forest whose constant activities barely make a dent in the opulent landscape.[6] But this collection seems oddly small and inefficient when compared to the masses of people around her. Glazer's shots insistently take us into crowded cities where nobody seems to notice the missing men; we are reminded of the masses of other human beings whose existence, and capacity to keep multiplying, the alien cannot touch. No police force or detective is put on her tracks; at most, one of the deaths she witnesses (but does not herself orchestrate) makes it onto the regional evening news. During shots in crowded discos and in city streets, there always seem to be many more human beings than the camera can capture. To bring out this sense of indifference and overabundance even further, Glazer shot most of the film by capturing Johansson's actual (endlessly repeated, and then drastically edited) improvisatory efforts to invite random passers-by into her car and get them to

---

[5] "Glazer: I think the explanation of things like her cohorts, these alien entities as bikers—they're performing a function. We see him clearing up after her, inspecting her at one point. There's a sense in that scene that there's something not quite right with her that he's detecting, like a hairline fracture or a crack in the wing of an airplane. He's satisfied that there isn't and carries on with his day." Jonathan Glazer and Sam Adams, "Space Oddity: Jonathan Glazer on *Under the Skin*," *Rolling Stone*, April 4, 2014, http://www.rollingstone.com/movies/news/space-oddity-jonathan-glazer-on-under-the-skin-20140404 (accessed April 22, 2015).

[6] To invoke the central example defining what, in *The Managed Heart*, Arlie Russell Hochschild was the first to term "emotional labor." Arlie Russell Hochschild, *The Managed Heart: Commercialization of Human Feeling* (Berkeley, CA: U of California P, 1983).

flirt with her in ways that might be used in the film's storyline.[7] To these men, she is only one woman among many; for every solitary male she gets into her car, many more easily walk away. Her violent sexuality is barely making a dent here; people are reproducing and coexisting to a degree that appears to make them generally indifferent to her occasional successes.

This sense of indifferent, unstoppable generativity is tempting: in the course of the film, the alien appears ineluctably drawn to these human beings, to the point of wanting something like a human family and lifestyle.[8] In the course of her gruesome harvests, and apparently because of them, she develops a (deeply ironized) fantasy of becoming human. She spares the life of one of her potential victims, and leaves him to roam around naked in the countryside. She then abandons her van and takes to crowded streets and buses, and eventually also tries to actually have sex with a kind stranger—to be left puzzled by what seems to either be the inoperability of her vagina, or the sheer fact of having one.

In the midst of these labors, both films' protagonists falter. Between volumes one and two of *Nymphomaniac*, Joe finds herself unable to reach orgasm. For the remaining two hours of volume two, she makes a grueling, increasingly mechanical attempt to restore her initial, now unattainable, capacities for pleasure. Joe's body becomes weary and battered, and she tells Seligman that she feels guilt-ridden. *Under the Skin* involves a similar turn: the alien starts to relate to her body as an encumbrance that she does not know how to feed, protect, or satisfy. After she runs away from her co-workers and tries to live among human beings, she realizes that she cannot digest their food and that she does not understand much about human sexuality in practice.[9] And as she wanders away from this would-be lover, in her puzzlement, she is pursued and hunted down by another human male in a forest where he first tries to have sex with her and then sets her on fire, apparently using up the remnants of her body's resources without issue.

---

[7] J.D. Connor offers a Jamesonian reading of the way these narrative strategies reflect the film's own mode of production in "Independence and the Consent of the Governed: the Systems and Scales of *Under the Skin*," *Jump Cut* 57, 2016, http://www.ejumpcut.org/trialsite/-ConnorSkin/index.html (accessed October 22, 2016).

[8] As Osterweil puts it, "In discovering empathy, the alien discovers herself anew. Suddenly catching sight of her face in a tarnished mirror, she is startled by some new form of recognition." Osterweil, "The Perils of Becoming Female," 44–51.

[9] It's also possible to claim, with Elena Gorfinkel, that the alien discovers that she does not have a vagina at all—which only adds to the dark comedy of this scene, and reinforces my later point about the alien's reduction to mere matter. Elena Gorfinkel, "Sex, Sensation, and Nonhuman Interiority in *Under the Skin*," *Jump Cut* 57, 2016, http://www.ejumpcut.org/trialsite/-GorfinkelSkin/index.html (accessed October 22, 2016).

In this regard, both films superficially appear to follow the patterns of what Lauren Berlant has called cruel optimism.[10] They also seem to take quite seriously the systems of control, surveillance, and expectation that induce their protagonists to such strenuous, self-destructive effort.[11] But rather than simply identify with these women as suicidally overwhelmed, *Under the Skin* and *Nymphomaniac* also partly dissociate themselves from their expressively performed despair, panic, and self-pity (about whose sincerity, especially in the alien's case, we can never be sure anyway). In his review of von Trier's *Melancholia* (2011)—the second part of the "Depression Trilogy" of which *Nymphomaniac* is the last installment—Christopher Peterson describes it as embracing a utopian hope of survival beyond the catastrophes that will inevitably befall us. "The world is gone," Peterson paraphrases the film's ending, "yet as long we survive, which is to say as long we say 'yes' to life, we must carry this world into a future that survives the total destruction to which we can never bear witness."[12] I suggest that von Trier and Glazer fill us with a certainty about our world's persistence and reproduction in which our sense of ourselves becomes detached from what Peterson calls actively "saying 'yes' to life."

Both films open up this counterintuitive dimension of their plots through the way they allow their films' more fantastical, aestheticized elements—and the ones particularly generated by, and reminiscent of, digital technologies—to make the intense productivity and reproductivity their protagonists shun seem inevitable. Refusing the grand sublimity of Terrence Malick's *Tree of Life* (2011) or Roland Emmerich's *The Day After Tomorrow* (2004), or even of von Trier's earlier *Melancholia*, these films depict human efforts to reproduce as surprisingly easy, especially as facilitated by the many technical and physical means at our disposal.[13] At once excessive and sterile, ostensibly courting pathos but then skirting it, the two films constantly invite comparison to melodrama yet offer

---

[10] Cruel optimism, as Berlant describes it, is a destructive attachment to "conventional good-life fantasies—say, of enduring reciprocity in couples, families, political systems, institutions, markets, and at work—when the evidence of their instability, fragility, and dear cost abound." Lauren Berlant, *Cruel Optimism* (Durham, NC: Duke University Press, 2011), 2. Achille Mbembe's notion of "necropolitics" is, of course, an important background to both these arguments: Mbembe, "Necropolitics," *Public Culture* 15, no. 1 (2003): 11–40.
[11] As these films wear on, it thus also becomes increasingly tempting—but incorrect, I would argue—to see them as instances of what Nitzan Lebovic calls "biopolitical film." Nitzan Lebovic, "The Biopolitical Film (A Nietzschean Paradigm)," *Postmodern Culture* 23, no.1 (2012).
[12] Christopher Peterson, "The Magic Cave of Allegory: Lars von Trier's *Melancholia*," *Discourse* 35, no. 3 (2013): 419.
[13] See David Sterritt, "Days of Heaven and Waco: Terrence Malick's *Tree of Life*," *Film Quarterly* 65, no. 1 (2011): 52–7. I make a similar point in Marta Figlerowicz, "Comedy of Abandon: Lars von Trier's *Melancholia*," *Film Quarterly* 65, no. 4 (2012): 21–6.

their viewer only its dry, hollowed shell.[14] Both films do lean on tropes of female objectification to depict their two protagonists' demise. Their gender draws attention to human bodies' need to control their reproductivity in a way that registers as quite deliberate. Yet *Nymphomaniac*'s and *Under the Skin*'s uses of these feminist tropes are ultimately decontextualized and instrumental: they function less as a point about sexual freedom, and more as a point about other forms of reproducibility and productivity that we cannot control in our lives quite as well, and by which we find ourselves being bound at least as efficiently.[15] In this sense, the directors' choice of female rather than male protagonists seems further motivated by a wish to forestall an alternative reading of their aesthetics as machismo, or as what Moira Weigel terms "sadomodernism."[16] We are not witnessing, here, the apotheosis of human beings in pure, affect-less formal structures. Instead, human bodies reduce themselves to clusters of multiplying matter whose organic or non-organic quality ceases to matter, and which are most remarkable for the uncontrolled ways in which, and levels on which, they continue to multiply.

On a most basic and most comic level, the two films disrupt their protagonists' anxiety about reproduction by depicting them as apparently resurrectable (in *Nymphomaniac*) or as themselves incapable of understanding the deathbound exhaustion to which they eventually fall prone. *Nymphomaniac* thus begins with Joe's apparent rise from the dead: supine and bloodied on the pavement when Seligman stoops over her, she suddenly wakes up and proceeds to assert her unwillingness to follow him to his room. It then ends with her capacity to incite desire in someone who had considered himself asexual, and once again to defend herself from this person's assault—making this interaction go her way—despite

---

[14] I follow Linda Williams' definition of melodrama, as she articulates it in Linda Williams, *Playing the Race Card: Melodramas of Black and White from Uncle Tom to O.J. Simpson* (Princeton, NJ: Princeton UP, 2001), 10–44. See also Peter Brooks, *The Melodramatic Imagination: Balzac, Henry James, Melodrama, and the Mode of Excess* (New Haven: Yale UP, 1976) and Lauren Berlant, *The Queen of America Goes to Washington City* (Durham, NC: Duke UP, 1997). Williams notes these potential connections between feminist critique and melodrama in *Playing the Race Card*; see also Julianne Pidduck, "The Times of *The Hours*: Queer Melodrama and the Dilemma of Marriage," *Camera Obscura* 28 (2013): 37–67; Rebecca Wanzo, "Precarious-Girl Comedy: Issa Rae, Lena Dunham, and Abjection Aesthetics," *Camera Obscura* 31 (2016): 27–39.
[15] Ara Osterweil, "Under the Skin: The Perils of Becoming Female," 44–51, who gives a much more optimistically feminist reading of this film; and *see* also Marc Francis, whose reading, while ambivalent, also attributes to this film a much more vibrant political stance. Marc Francis, "Splitting the Difference: on the Queer-Feminist Divide in Scarlett Johansson's Recent Body Politics," *Jump Cut* 57, 2016, (accessed October 22, 2016). http://www.ejumpcut.org/trialsite/-FrancisSkin/index.html (accessed October 22, 2016).
[16] Moira Weigel, "Sadomodernism," *n+1* 16, 2013, https://nplusonemag.com/issue-16/essays/sadomodernism/ (accessed November 4, 2016).

her prior exhaustion and apparent defenselessness. In *Under the Skin*, the alien appears not to understand exhaustion and reproduction at all as concepts, which becomes ever more apparent the longer she hunts humans and then escapes them; her inability to sustain herself without her strange hunts and harvests starts to seem all the more puzzling, as we are made unable to understand how she actually eats, or otherwise sustains herself. She starts to seem merely parasitical on these other forms of life whose accessibility to her she has taken for granted.

The two films also surround the characters with tropes of an infinitely abundant, organic nature on the one hand, and with utopian but obviously dated ideals of encyclopedic containment on the other. *Nymphomaniac* is a bedroom dialogue that closely follows the style and rhetoric of de Sade's *Philosophy in the Bedroom* (1795). Its two main interlocutors, Joe and Seligman, refer to texts that David Denby lists as including "*Angler*, Burton's *Anatomy of Melancholy*, and ... such eighteenth-century libertine texts as the Marquis d'Argens's 'Thérèse Philosophe.'"[17] *Under the Skin* re-enacts sentimentalist narratives and paintings through a sequence of *tableaux vivants*. The man with neurofibromatosis whom the alien sets free, naked, into an empty field, echoes back—through David Lynch and Werner Herzog, the Elephant Man and Kaspar Hauser—all the way to Jean-Jacques Rousseau's and Denis Diderot's natural man. The film's natural landscapes—with their hills, fogs, wave spray, and intense play of light and shade—are clichéd versions of the paintings of Caspar David Friedrich and J. M. W. Turner (figure 21.1). In the former case, the introduction of these tropes produces the impression of an aspiration for a form of containment that is not actually available or forthcoming; confronting the speaker with notions of a kind of controllable, listable variability, and hopes for it, that are constantly subverted. In the latter, earth is depicted as a teeming natural sphere into which the alien's attempts to wrest control of it enter awkwardly and foolishly, their mechanics inherently suspicious. In both cases, we are made to feel that these protagonists do not apply the right lens to these landscapes, or that their lenses have ceased to fit these landscapes and become dated.

This stylistic turn is accompanied by a metatextual one: these films highlight the capture and storage capacities of their recording devices, as well as the meta-control these devices exercise over these characters' production and reproduction.

---

[17] David Denby, "The Story of Joe," *The New Yorker*, March 24, 2014, http://www.newyorker.com/magazine/2014/03/24/the-story-of-joe (accessed December 27, 2020).

**Figure 21.1** An image out of Caspar David Friedrich from *Under the Skin* (Jonathan Glazer, 2013).

They represent the camera as a means of capturing, but also of adding to and participating in, the human multitudinousness we see. *Nymphomaniac* and *Under the Skin* ultimately reimagine their protagonists as clusters of projected light and shadow. They represent the process of putting the film together frame after frame not as a means of capturing life, but as one of creating a site at which human bodies are revealed to persist beyond and outside their own immediate aliveness, acquiring an ever greater amount of permanence and vividness in this new medium—a medium which, ambiguously, appears at once to generate, and to render for us as viewers with heightened realism, something about these characters' worlds that they themselves cannot catch up to.

It is a hallmark of von Trier's Dogme 95-derived aesthetic that he undercuts the naturalness of his narratives by foregrounding the camera's vehement, clumsy efforts to infuse his films' settings and characters with animation. In his earlier *Dancer in the Dark* (2000), von Trier famously achieves this effect by filming his actors with no fewer than one hundred cameras at once, each of them focused more or less at random, out of whose disparate, baroquely redundant perspectives every scene is then stitched together (in an interview included in the DVD release of this film, von Trier wishes that he could have filmed *Dancer* from one thousand or even ten thousand such random, partly duplicated viewpoints).[18]

[18] *100 Cameras: Capturing Lars Von Trier's Vision*, Directed by Vincent Paterson (2000), New Line Home Video, DVD.

By means of similarly excessive, purposefully disjointed recording techniques, *Nymphomaniac* visually dissociates Joe's behaviors from her particular bodily presence, or from any articulable physical impulse or need. The film highlights how many actors, cameras, and special effects are condensed in representations of its protagonist's desires. The younger and the older Joe are played by two different actors, close enough in age and different enough in appearance that no viewer could actually mistake them for each other. Rather than viewing the development of a single female body, we witness the relay of two bodies whose combined presence represents the sexual energies of the film's central character. As both women's movements are multiplied on several parallel panels running simultaneously, or fast-forwarded, or rewound at high speed, our attention is drawn to the condensed, multiplied hyper-presence Joe is made to enjoy onscreen. At the end of the film, when Joe shoots Seligman after he tries to rape her, the screen goes black and Joe's presence is transposed onto a metallic sound, literalizing this sense of dependence on something that the film does not even, at this point, pretend to be a living creature. At the same time, the gun serves as a metaphoric phallus allowing her to regain control of the situation, an appendage that is more reliable than the one with which the man beside her tried to rape her. It controls not only the man's life, but also the light and darkness onscreen.

Rob White's review of *Antichrist* (2009), the first of von Trier's "Depression" films and one that uses similar visual effects, describes these effects astutely as embodying a tension between "videogame aesthetics" and "Old Master iconography." "Changeability and artificiality," White shows, "define the machine-made look of the film's environment; its monsters owe their existence to CGI and animatronics."[19] White's review is entitled "Against Nature," echoing the English translation of Joris-Karl Huysmans's *A rebours*, and he argues that these techniques allow von Trier's represented world to become purely artificial. While some of the animatronics White describes in *Antichrist*—such as the famous talking fox—do not reappear in *Nymphomaniac*, their structure and effect are similar. Even in relatively realist scenes, von Trier's successive cuts tend to be just slightly discontinuous, making us aware of some lost space and time in which these characters' bodies persisted: a space and time condensed within the emotional turmoil we are witnessing, as if there was more available to us, in every second of the film, than reality otherwise affords. One example of such obvious, alienating cuts is in the scene where Mrs. H. (Uma Thurman) invites

---

[19] Rob White, "Against Nature," *Film Quarterly* 63, no.1 (2009): 5.

herself into Joe's apartment to let her children see "the whoring bed." As it catches Mrs. H's face, von Trier's camera is shaky and not quite centered. Rather than smoothing out passages from one shot to another, von Trier lets us notice moments when our perspective onto her changes slightly. Mrs. H is suddenly reframed at a slightly different angle; her breathing is cut short halfway through a loud gasp, then recommences at a slightly different distance and rhythm. These obvious edits make the intensity of Mrs. H.'s anger seem condensed out of a plethora of further material; it makes her seem present in a way that could potentially stretch out considerably, even as her eruption onstage increases the stakes and apparent productivity of what was otherwise, for Joe, a relatively boring and routine encounter.

In *Under the Skin*, characters are reconceived as inorganically reproduced entities in a way that is even more abstract and geometrical. There is an aimlessness and lack of focus or intentionality both to the characters Glazer represents, and to his editing. People walk in and out of film frames in ways the camera does not follow; motorcycles speed along strips of unlit highways that do not have any known start or end as if they were being caught at random, as if there were always more potentially seeping into the picture. This is largely because, as Glazer recounts for *The Dissolve*, the film was shot "covertly" by cameras "[built] into the dashboard in her car or [hidden] in the street furniture," creating an eerie absence of intentional tracking and smoothness that echoes von Trier's editing. The soundtrack of *Under the Skin*, composed by Mica Levi of Micachu & the Shapes, is synthesized from "all the sonic chaos of the world that we tune out ... The role of this was to use all those things and have those things somehow becoming symphonic, just bubbling away in the background. All of the things you would normally cut out of the soundtrack for being noisy would be the things we used and pushed to the foreground."[20] All this gives us both the sense of endless footage being spliced and fused together, to ever more condensed effect, but also of a represented world whose plentifulness could only fully be captured by a camera that no longer needs to be limited and intentional; by a camera that shows that humans will keep randomly appearing within it no matter where it is turned.

---

[20] Jonathan Glazer and Scott Tobias, "Director Jonathan Glazer on *Under the Skin*'s Complex Honesty," *The Dissolve*, April 4, 2014, https://thedissolve.com/features/interview/496-director-jonathan-glazer-on-under-the-skins-comple/ (accessed April 22, 2015). For a more complete account of how the soundtrack for *Under the Skin* was made, see also Larry Fitzmaurice's *Pitchfork* interview with Glazer and Levi: "Interviews: Jonathan Glazer and Mica Levi," *Pitchfork.com*, March 31, 2014, http://pitchfork.com/features/interviews/9366-under-the-skins-jonathan-glazer-and-mica-levi/ (accessed April 22, 2015).

The effect of these experimental techniques, as Jonathan Romney aptly puts it, is that "the streets themselves become strange ... images of Earth women are superimposed on each other in a dense kaleidoscopic layering."[21] The dim light makes human figures look schematic. With no discernable purpose or meaning, their movements become as gravitational and accidental as those of the white sphere—at once an eye and an alien planet—in the film's opening shot. As they follow each other step by step, usually at some distance, the attraction the alien is able to create between herself and those whom she seduces looks literally rather than just metaphorically magnetic. Once they are finally caught by the alien, these moving human figures fold and shrink like burst balloons. Whatever organs we may have presumed them to have are nowhere to be seen. Rather than highlight their humanity, the alien's exploitation of her victims appears to merely confirm the ease of their fusions and transpositions.[22] And they easily add to a growing mass of energy to whose circulation they add but which they never take control or even complete stock of.

Perhaps most stunning, in this regard, is the film's representation of the three drowning figures: the dog and the couple, who all follow each other slowly into the water. The film tamps down the sound of their voices; all we hear is the movement of the waves on which these figures' heads bob up and down before disappearing (figure 21.2). They are transformed into floating shapes, obeying laws of physics that have nothing to do with their capacity to feel or move on their own. In the couple's absence, their baby no longer budges from the rocks among which they set him down. When a Czech surfer catches the drowning man and pulls him ashore, the man already resembles the watery skins that the alien submerges in her vat of mysterious fluid.

There is a sense, of course, in which the long history of film theory has partly presaged this odd, increasingly inorganic sense of permanence. As Stanley Cavell argues, quoting a long line of his predecessors, "Erwin Panofsky puts it this way: 'The medium of the movies is physical reality as such.'" André Bazin emphasizes essentially this idea many times and many ways: at one point he says, 'Cinema is committed to communicate only by way of what is real'; and then, 'The cinema [is] of its essence a dramaturgy of Nature.' ... That it is reality that we have to deal with, or some mode of depicting it, finds surprising confirmation in the way

---

[21] Jonathan Romney, "Film of the Week: *Under the Skin*," *Film Comment*, April 3, 2014, http://www.filmcomment.com/entry/under-the-skin-jonathan-glazer-review. (accessed December 27, 2020).
[22] See also Elena Gorfinkel's reading of these scenes in Gorfinkel, "Sex, Sensation, and Nonhuman Interiority."

Figure 21.2 Human bodies reduced to pieces of matter in *Under the Skin* (Jonathan Glazer, 2013).

movies are remembered and misremembered." But Cavell, Bazin, and Panofsky still understand film as a reality separate from and unthreatening to the viewer; the films I examine tantalizingly suggest that the reality to which they refer is our own.[23]

In *The World Viewed*, Cavell ironically invokes accounts of "those primitives [who] ... are terrorized upon seeing pictures of themselves. We are told that they fear their souls have been captured—and we laugh, pleased with that respect for our power. But it may be that they see their bodies being given a foreign animation. Is that wrong?—We are too used to what happens to us."[24] For Cavell, this is true only of photography: in film, "the photographic subject is released again" into a sense of control, if not over the world around her, at least over her own mind and body.[25] Borrowing Cavell's terms—and the Levi-Straussian anthropological contexts on which he draws—one might say that these films enact a similar dispossession, creating an alternative reality in which our previously held beliefs about who we are seem superstitious. Jacques Rancière describes contemporary images and spectacles as culturally productive to the

---

[23] Stanley Cavell, *The World Viewed: Reflections on the Ontology of Film* (Cambridge, Mass.:Harvard University Press, 1979), 16; Erwin Panofsky, "Style and Medium in the Moving Picture," *Film*, ed. David Talbot (New York, NY: Simon and Schuster, 1959), 31; André Bazin, *What is Cinema?*, trans. Hugh Gray (Berkeley, CA: U of California P, 1967), 110.
[24] Cavell, *The World Viewed*, 119.
[25] Ibid., 119.

extent that they "repel[] the big sleep of indifferent triteness or the great communal intoxication of bodies."[26] Art needs to wake us up, time and again, to the unknowability and incomplete representability of the people and environments around us. By contrast, the aesthetic and ontological paradoxes with which these films grapple concern not the other, but one's own being; they concern not the pretense of depicting the unrepresentable, but the fear that we have found a way of making ourselves more persuasively present in separation from our minds' and bodies' needs and urges, than we might ever have wished or expected—and that we have thereby rendered irrelevant our attempts to curb or control our bodily selves.

Against the backdrop of this synthetic quasi-naturalism, *Nymphomaniac* and *Under the Skin* forge alternative connections between human bodies and the inanimate physical environments that they increasingly come to resemble. Most often, they create such ties through the manipulation of light, by changing the saturation of the bodies in the frame or altering the degree of light or darkness in a given scene. In *Nymphomaniac*, light fills the screen during Joe's childhood walks with her father, and—most climactically—during her dramatic orgasms. Joe also, not just metaphorically but literally, sets many objects on fire. *Under the Skin* starts with a giant beam of light emitted out of something like a sun or a glowing eyeball. This beam of light is repeated in the film's many sunrises and sunsets, and—finally—in the way that the alien's body itself is set ablaze (figure 21.3). When the alien is stripped of her human skin and falls into the snow, she becomes a black, coal-like silhouette on which this fire indefinitely lingers without dying out in our view.

Such alchemical transformations from matter into light and back again provide these films with an alternative set of metaphors by which their characters' physical presence can be defined and tested. They invite the viewer to think of this light as an energy that one might, at first, mistake for exhaustive embodiment, but that is not, in fact, coterminous with it. If the camera captures these characters' presence as light, what does that mean their bodies are actually made of? The light beams from which these characters are constituted—and which they sometimes also exude or create—make their bodies seem inexhaustible both in the kinds and in the amounts of energy that can appear to radiate from them, and which they appear to reflect back onto ourselves.

---

[26] Jacques Rancière, *The Future of the Image*, trans. Gregory Elliott (London and New York: Verso, 2009), 46.

**Figure 21.3** A gruesome return to the flash of light from the opening of *Under the Skin* (Jonathan Glazer, 2013).

It is useful to juxtapose these images against an older idea that they clearly echo: Plato's allegory of the cave. The analogy between "reality" as seen in the cave—as shadows projected onto an illuminated wall—and the way film is created out of light and shadow, has long been remarked on by both philosophers and film theorists.[27] Critics who take up this trope, and turn it around to defend film as a medium, usually invoke film's capacity to help us face the ephemerality of our perceptions and our existences. If the shadows projected onscreen make us dream of some robust ideal forms that could stabilize and explain them, that is merely because—it is argued—they reveal to us the extent of our subjective fragmentation.

As the films I examine represent them, such plays of light and shadow are instructive and frightening not because of their ephemerality, but because they suggest that we might no longer be able to imagine any more robust or convincingly persistent iteration of our minds and bodies than the one found in them. The objects they illuminate impose upon us a standard of vividness and reproducibility before which—in our easily exhausted aliveness—we stand like Plato's disciples when they were confronted not with shadows but with ideas, from which the fleetingness of life has already been eradicated. The nightmare these films peddle is one in which we can no longer imagine anything more essential to ourselves than the media by which we maintain and enhance our

---

[27] A limited review of this trope is provided in Nathan Andersen, *Shadow Philosophy: Plato's Cave and Cinema* (New York, NY: Routledge, 2014).

experience—and rather than pity these media for their imperfections and impermanence, we have come to envy them for it, as our new touchstones of what reproduction of the self is like.

The ties these two films forge between light, cinema, and late capitalist technology, are not without precedent. On one level, one might see *Nymphomaniac* and *Under the Skin* as continuing a conversation opened by Quentin Tarantino's *Inglourious Basterds* (2009), which pits the fabled ahistorical permanence of digital cinema against celluloid's supposedly more immediate connection to (and inevitable decay alongside) the historical period during which it was shot. Tarantino's film is a counterfactual rampage in which an American lieutenant and eight Jewish-American soldiers, aided by the interracial couple Marcel and Shoshana, end the Second World War by killing Adolf Hitler, Hermann Göring, and Joseph Goebbels in a crowded movie theater. They accomplish this mass assassination by setting on fire 350 reels of celluloid from this cinema's archive while the Nazi leadership is watching a film about the heroic exploits of a German sniper. "Nitrate film burns more quickly than paper," says Samuel L. Jackson in a voiceover, as the shot offers an extra-diegetic demonstration of this chemical fact. This voiceover explains both why celluloid film will become an effective weapon against the Nazis, and why the history it was supposed to document will be so easy to erase and rewrite on Tarantino's own digitized screen. *Inglourious Basterds* is a brutal, nihilistic self-affirmation of the way its narrative is unbound from history both in physical terms, and in ideological or sentimental ones. By the end, Tarantino's film is literally illuminated by burning footage of the older kind whose air of fragile, archival historicity it has rejected. *Nymphomaniac* and *Under the Skin* are inspired by a similar—if less dramatically articulated—loss of commitment to the ephemeral historicity of human bodies as meaningful lenses onto the course of our lives or their future fates. The matter-of-factness with which they take such attitudes for granted arguably makes an even more forceful point about them than do Tarantino's provocations—and make the obstinacy with which they make their characters seem reborn and resurrected not once, but many times over, seem odd and fanciful.

On another level, these films respond to productions such as Terrence Malick's *Tree of Life*. In *Tree of Life*, fire and light function as part of an implicit argument for contemporary technology and culture as reliable, generous vessels for the long history of human and non-human life. Emerging and intensifying to the rhythm of a voice or an orchestral soundtrack, light beams and planet-sized balls of lava are represented as benevolent, godlike witnesses to the long narrative of our creation. The CGI technology that creates them is a means of imagining the prehistories of human beings in a way that makes these distant primal events seem continuous with, and unthreatening to, our bodily life. In a way that recalls D. N. Rodowick's description of *Jurassic Park* (1993) as forcing "physical reality entirely [to yield] to the imagination," Malick pretends to reconstruct the Ur-moment when a living creature first takes pity on another: a CGI allosaurus removes its foot from the face of a wounded herbivore.[28] The predator and its near-victim then share a moment of cognitive dissonance.[29] Jim Emerson reports that, in a 2007 draft, Malick scripts this encounter as follows:

> Reptiles emerge from the amphibians, and dinosaurs in turn from the reptiles. Among the dinosaurs we discover the first signs of maternal love, as the creatures learn to care for each other.
> Is not love, too, a work of the creation? What should we have been without it? How had things been then?
> Silent as a shadow, consciousness has slipped into the world.[30]

How could the light of such an event ever reach our eyes? Malick's film conveys the hope that, with new technologies and production budgets at our disposal, such otherwise unattainable breakthroughs could finally be recaptured. *Nymphomaniac* and *Under the Skin* take up similar intuitions about our contemporary capacity to enhance, and eventually surpass, our minds and bodies with their avatars, even while rejecting a perspective like Malick's as excessively hopeful. Malick still sees such artificially made worlds as merely an extension of human beings: they allow us to adopt the viewpoint of our supposed maker, and to imagine having answers to questions about our past that only someone who precedes us could possibly give. By contrast, the films I examine fantasize that these alternative, manufactured environments have reoriented both our troubles and our wishes to the point of making the epistemic and imaginative satisfactions in which Malick indulges irrelevant to our sense of self.

In response to Tarantino's melodrama and Malick's sublime optimism, these films might superficially seem to embrace the aspirations of object-oriented-ontologists: finding out what it is like to be a gust of air, a piece of wood, or a

---

[28] D.N. Rodowick, *The Virtual Life of Film* (Cambridge, MA: Harvard UP, 2007), 28.
[29] See an extended discussion of this scene, including Malick's own comments about it, on *RogerEbert.com*. Jim Emerson, "Tree of Life: The Missing Link Discovered!" *RogerEbert.com*, April 11, 2012, http://www.rogerebert.com/scanners/tree-of-life-the-missing-link-discovered (accessed December 27, 2020).
[30] As quoted in Malick, "Tree of Life."

stone. However, this conclusion does not hold up to closer scrutiny. Though they might echo new materialist and object-oriented rhetoric, these films pursue it to an extreme that is much more ironic and much more obviously anthropocentric.[31] These films' aesthetic is concerned with inorganic matter as a means of understanding not our surrounding objects, but ourselves.[32] Nor do *Nymphomaniac* and *Under the Skin* have much to do—though again they seemingly resonate—with the recent nihilistic tendencies of accelerationism.[33] If accelerationists such as Alex Williams or Nick Srnicek claim that, by speeding up capitalism, we might destabilize and destroy it, these films imagine, as the ultimate, fast-forwarded fate of their societies, a state in which we can no longer imagine ourselves as ever having been separate from the objects we create even on a most basic existential level.

In their combined interest in objects and in our futures, these films instead posit themselves as instances of what Theodor Adorno—and Edward Said following him—might have described as capitalist late style. "For Adorno," Said explains, "*lateness* . . . includes the idea that one cannot really go beyond lateness at all, cannot transcend or lift oneself out of lateness, but can only deepen the lateness."[34] Ludwig van Beethoven and Franz Schubert—who are Adorno's prime examples of late style—thus shake their impulse to mourn the decay of their surrounding culture and reconstitute their art around the premise that this culture's hopes and landmarks will be definitively absent.[35] As Adorno vividly describes it, "he who crosses the threshold between the years of Beethoven's death and Schubert's will shiver, like someone emerging into the painfully diaphanous light from a rumbling, newly formed crater frozen in motion, as he

---

[31] As compared, for example, to Graham Harman, *Towards Speculative Realism: Essays and Lectures* (New York, NY: Zero Books, 2010), or Ian Bogost, *Alien Phenomenology, or What It's Like to Be a Thing* (Minneapolis: U of Minnesota P, 2012). One person who reads *Under the Skin* as a new materialist film is Gorfinkel, in "Sex, Sensation and Nonhuman Interiority."

[32] That is admittedly not far from the anthropocentrism that Andrew Cole attributes to object-oriented ontology, new materialism, and actor-network theory alike. See Andrew Cole, "The Call of Things: A Critique of Object-Oriented Ontologies," *The Minnesota Review* 80 (2013): 106–18.

[33] As described, for instance, in Robin Mackay, *#Accelerate: The Accelerationist Reader*, ed. Robin Mackay and Armen Avanessian (New York: Urbanomic, 2014).

[34] Edward W. Said, *On Late Style: Music and Literature Against the Grain* (New York: Pantheon Books, 2006), 13.

[35] Said points to the following passage from Thomas Mann's *Doctor Faustus*, for which Mann repeatedly consulted Adorno, as an apt paraphrase of Adorno's essay: "An ego painfully isolated in the absolute, isolated too from sense by the loss of his hearing, [Beethoven is] a lonely prince of a realm of spirits, from whom now only a chilling breath issued to terrify his most willing contemporaries, standing as they did aghast at these communications of which only at moments, only by exception, they could understand anything at all." Thomas Mann, *Doctor Faustus*, trans. H. T. Lowe-Porter (New York: Vintage, 2002), 52. Quoted from Said, *On Late Style*, 8.

becomes aware of skeletal shadows of vegetation among lava shapes in these wide, exposed peaks, and finally catches sight of those clouds drifting near the mountain, yet so high above his head. He steps out from the chasm into the landscape of immense depth bounded by an overwhelming quiet at its horizon, absorbing the light that earlier had been seared by blazing magma."[36] For both Adorno and Said, the power of late style lies in its capacity to imagine decadence not simply as a catastrophe, but as a new given: a set of premises to which alternatives have become unthinkable. *Nymphomaniac* and *Under the Skin* similarly treat their implied environments and cultures as both a point of departure and a point of no return. Instead of trying to imagine our final degradation into automata or commodities, they confront us with the no less eerie spectacle of the aftermath of this degradation as a continuity so smooth that one can no longer conceive of its precedent. The possibility of this spectacle—and its potential appeal to these directors' implied exhausted, pessimistic ideal audience—is depicted as a form of discovery in and of itself.

# Bibliography

Adorno, Theodor. "Schubert (1928)." Translated by Jonathan Dunsby and Beate Perrey. In *19th-Century Music* 29, no. 1 (2005[1928]): 3–14.

Andersen, Nathan. *Shadow Philosophy: Plato's Cave and Cinema.* New York: Routledge, 2014.

Bazin, André. *What is Cinema?* Translated by Hugh Gray. Berkeley, CA: University of California Press, 1967.

Berlant, Lauren. *The Queen of America Goes to Washington City.* Durham, NC: Duke University Press, 1997.

Berlant, Lauren. *Cruel Optimism.* Durham, NC: Duke University Press, 2011.

Bogost, Ian. *Alien Phenomenology, or What It's Like to Be a Thing.* Minneapolis: University of Minnesota Press, 2012.

Brooks, Peter. *The Melodramatic Imagination: Balzac, Henry James, Melodrama, and the Mode of Excess.* New Haven, CT: Yale University Press, 1976.

Cavell, Stanley. *The World Viewed: Reflections on the Ontology of Film.* Cambridge, MA: Harvard University Press, 1979.

Cole, Andrew. "The Call of Things: A Critique of Object-Oriented Ontologies." *The Minnesota Review* 80 (2013): 106–18.

---

[36] Theodor Adorno, "Schubert," trans. Jonathan Dunsby and Beate Perrey, *19th-Century Music* 29, no. 1 (2005[1928]): 7.

Connor, J. D. "Independence and the consent of the governed: the systems and scales of *Under the Skin*." *Jump Cut* 57 (Fall 2016). https://www.ejumpcut.org/archive/jc57.2016/-ConnorSkin/index.html. (accessed January 6, 2021).

Deleuze, Charles and Félix Guattari, *Anti-Oedipus: Capitalism and Schizophrenia*. New York: Bloomsbury, 1972.

Denby, David. "The Story of Joe." *New Yorker*, March 17, 2014. https://www.newyorker.com/magazine/2014/03/24/the-story-of-joe. (accessed January 6, 2021).

Emerson, Jim. "Tree of Life: the Missing Link Discovered!" *RogerEbert.com*, April 11, 2012. http://www.rogerebert.com/scanners/tree-of-life-the-missing-link-discovered. (accessed January 6, 2021).

Figlerowicz, Marta. "Comedy of Abandon: Lars von Trier's *Melancholia*." *Film Quarterly* 65, no. 4 (2012): 21–6.

Francis, Marc. "Splitting the Difference: on the Queer-Feminist Divide in Scarlett Johansson's Recent Body Politics." *Jump Cut* 57, 2016. https://www.ejumpcut.org/archive/jc57.2016/-FrancisSkin/index.html. (accessed January 6, 2021).

Glazer, Jonathan, director. *Under the Skin*. New York: A24, 2013. Amazon Prime Video.

Glazer, Jonathan and Hank Sartin. "Jonathan Glazer Talks about His New Film *Under the Skin*." *RogerEbert.com*, April 10, 2014. http://www.rogerebert.com/interviews/jonathan-glazer-talks-about-under-the-skin.

Glazer, Jonathan and Sam Adams. "Space Oddity: Jonathan Glazer on *Under the Skin*." *Rolling Stone*, April 4, 2014. http://www.rollingstone.com/movies/news/space-oddity-jonathan-glazer-on-under-the-skin-20140404. (accessed January 6, 2021).

Gorfinkel, Elena. "Sex, Sensation, and Nonhuman Interiority in *Under the Skin*." *Jump Cut* 57 (2016). https://www.ejumpcut.org/archive/jc57.2016/-GorfinkelSkin/index.html. (accessed January 6, 2021).

Harman, Graham. *Towards Speculative Realism: Essays and Lectures*. New York: Zero Books, 2010.

Herzog, Amy. "Star Vehicle: Labor and Corporeal Traffic in *Under the Skin*." *Jump Cut* 57 (2016). https://www.ejumpcut.org/archive/jc57.2016/-HerzogSkin/index.html. (accessed January 6, 2021).

Lebovic, Nitzan. "The Biopolitical Film (A Nietzschean Paradigm)." *Postmodern Culture* 23, no.1 (2012).

Mackay, Robin and Armen Avanessian, eds. *#Accelerate: The Accelerationist Reader*. New York: Urbanomic, 2014.

Mann, Thomas. *Doctor Faustus*. Translated by H. T. Lowe-Porter. New York: Vintage, 2002.

Mbembe, Achille. "Necropolitics." Translated by Libby Meintjes. *Public Culture* 15, no. 1 (2003): 11–40.

Osterweil, Ara. "Under the Skin: The Perils of Becoming Female." *Film Quarterly* 67, no. 4 (2014): 44–51.

Panofsky, Erwin. "Style and Medium in the Moving Picture." In *Film: An Anthology*, edited by David Talbot, 15–32. New York: Simon and Schuster, 1959.

Paterson, Vincent, director. *100 Cameras: Capturing Lars Von Trier's Vision*. Los Angeles: New Line Home Video, 2000. DVD.

Peterson, Christopher. "The Magic Cave of Allegory: Lars von Trier's *Melancholia*." *Discourse* 35, no. 3 (2013): 400–22.

Pidduck, Julianne. "The Times of *The Hours*: Queer Melodrama and the Dilemma of Marriage." *Camera Obscura* 28 (2013): 37–67.

Rancière, Jacques. *The Future of the Image*. Translated by Gregory Elliott. London: Verso, 2009.

Rodowick, D. N. *The Virtual Life of Film*. Cambridge, MA: Harvard University Press, 2007.

Romney, Jonathan. "Film of the Week: *Under the Skin*." *Film Comment*, April 3, 2014. http://www.filmcomment.com/entry/under-the-skin-jonathan-glazer-review. (accessed January 6, 2021).

Russell Hochschild, Arlie Russell. *The Managed Heart: Commercialization of Human Feeling*. Berkeley, CA: University of California Press, 1983.

Said, Edward W. *On Late Style: Music and Literature Against the Grain*. New York: Pantheon Books, 2006.

Sterritt, David. "Days of Heaven and Waco: Terrence Malick's *Tree of Life*." *Film Quarterly* 65, no. 1 (2011): 52–7.

Wanzo, Rebecca. "Precarious-Girl Comedy: Issa Rae, Lena Dunham, and Abjection Aesthetics." *Camera Obscura* 31 (2016): 27–39.

Weigel, Moira. "Sadomodernism." *n+1* 16 (2013). https://nplusonemag.com/issue-16/essays/sadomodernism. (accessed January 6, 2021).

White, Rob. "Against Nature." *Film Quarterly* 63, no.1 (2009): 4–5.

Williams, Linda. "Cinema's Sex Acts." *Film Quarterly* 67, no. 4 (2013): 9–25.

Williams, Linda. *Playing the Race Card: Melodramas of Black and White from Uncle Tom to O.J. Simpson*. Princeton, NJ: Princeton University Press, 2001.

22

# Neuroplasticity

## From Experience to Healing

Sara Ferrando Colomer

Experiences create original memories leading to new states of mind, which engender new self-reflections and behaviors. This cycle endlessly flows throughout our lives. Memories can be retrieved or repressed; in all cases, they shape us. Distressing events, as depicted in the television series *Mr. Robot* (Sam Esmail, 2015–2019) or *Random Acts of Flyness* (*RAoF*: Terence Nance, 2018), may disrupt patterns of memory and behavior. Alterations of the mind, misconceptions of identity, and a need for restoration can follow. If we so desire, how can we facilitate healing? Can we resolve trauma through sleep, contemplation, or art and imagination? Could art heal an altered mind?

## Experiences

"Might as well do nothing" says Elliot Alderson, "You've brought up this issue before. This issue of not feeling like you are in control. Do you remember?" replies Krista Gordon, his therapist. From the beginnings of *Mr. Robot*, a critically acclaimed drama series, it is plausible to imagine that the state of mind of protagonist Elliot Alderson (Rami Malek), a professional hacker, evolves from trauma experienced as a child.[1] His distressing childhood events remain withheld from the audience and Elliot himself, until Fernando Vera (Elliot Villar), a drug dealer, pushes him to remember them—to retrieve hidden memories. *Mr. Robot* deploys a story of traumatic experiences that unleash mental disorders, and

---

[1] Sam Esmail, director, "Mr. Robot," HBO, 2015, Amazon Prime https://www.amazon.com/gp/video/detail/B00YBX664Q/ref=atv_dp_season_select_s1 (accessed February 2, 2021).

**Figure 22.1** *Random Acts of Flyness* directed by Terence Nance (2018–present). Season 1 episode 6. Un-wakened state of mind as a consequence of transgenerational trauma. A mother is rocking her daughter while pleading to her to stay asleep.

ultimately healing. Through filmic devices, spectators, too, participate in Elliot's cognitive and emotional processes.

Our mental state owes to more than just our own first experiences. Sometimes, an experience's aftermath can stretch across multiple generations. A fictional depiction of a transgenerational experience is staged at the opening of episode 6 of Terence Nance's *RAoF* (figure 22.1):[2] "[A]n unending multitude of terribly white smiles," as Elizabeth Reich describes in her chapter "The Gift of Black Sonics" elsewhere in this book, reflects the multidimensional breadth of an historical trauma. A Black woman, by means of her own experiences and those of the ones who lived and live, instigates her daughter to stay asleep, to elude the illogical reality that persecutes them. As Reich explains, this scene fairly represents the Du Boisian concepts of "the veil" and of "double consciousness." As Stuart Hall notes, for W. E. B. Du Bois, "the veil has biblical associations; double consciousness, philosophical ones." He argued that racism and the practices of segregation excluded Blacks from mainstream American life—it "shut them out of their world by a vast veil."[3] Exiled within, a stranger in his own

[2] https://play.hbomax.com/season/urn:hbo:season:GWzqKDAX83MPCDwEAAAAI?icid=hbo_streamingoverlay_max&hbo_source=hbo.com&hbo_medium=referral&hbo_campaign=hbomax_signinlink_hbomax_button_20200701&hbo_content=62&hbo_term=hbo_streamsignin&_ga=2.155376059.1657704859.1607471538-2115596285.1603681485

[3] Stuart Hall, "Tearing Down the Veil," *The Guardian*, February 22, 2003, https://www.theguardian.com/books/2003/feb/22/featuresreviews.guardianreview30 (accessed January 1, 2021).

home, always looking at himself through the eyes of another race, and being both African and American, "the Negro" was destined to have a double self, a divided soul, the bearer of what Du Bois refers to as a "double consciousness ... One ever feels his two-ness ... two souls, two thoughts, two unreconciled strivings, two warring ideals in one dark body."[4]

Double consciousness shows how so many Black people are subject to trauma: they have to both see themselves as they are and wish to be, as well as deal with the images that whites project onto them. I don't wish to co-opt Du Bois' terms, and others would be better, but "the veil," and "double consciousness" might be helpful for thinking about those who suffer from trauma, and must try to function in society while accessing themselves beyond surface, more socially structured and compliant, identities. PTSD survivors, for example, can also experience hurdles which need to be overcome in order to undergo a process of healing. Through a revolutionary representation of American society, *RAoF* seeks to re-form our feelings and thoughts and awaken our minds. Although the meaning of consciousness remains elusive to philosophers, psychologists, and neuroscientists, an "awake state of experiences" could be a useful definition. "Shift consciousness," the main tagline of *RAoF*, implies an awakeness to experiences. There is a need to be awake in order to be free; free from the past experiences that define us.

## Neuroplasticity and Self-Identity

Memories emerge from experiences, either the ones we accrue or the ones we inherit. In theory, experiences are not inheritable; yet scientists speculate about this possibility. Epigenetic inheritance studies in worms (*Caenorhabditis elegans*) and mice suggest that a genetic imprint from ancestral trauma passes down future generations.[5,6] Skepticism about transgenerationally inherited memories in humans predominates. Apart from epigenetics,[7] memories could be carried

---

[4] W.E.B. Du Bois quoted in Hall, "Tearing Down the Veil."
[5] Oded Rechavi, Gregory Minevich, and Oliver Hobert. "Transgenerational Inheritance of an Acquired Small RNA-based Antiviral Response in C. Elegans," *Cell* 147, no. 6 (2011): 1248–56.
[6] Brian G. Dias and Kerry J. Ressler, "Parental Olfactory Experience Influences Behavior and Neural Structure in Subsequent Generations," *Nature Neuroscience* 17, no. 1 (2014): 89–96.
[7] Epigenetics: study of changes in gene function that are mitotically and/or meiotically heritable and that do not entail a change in DNA sequence.

through next generations by a psychological, and, needless to say, metaphorical approach. Ancestors (parents or even societies) with unhealed traumatic experiences could pass them on to their children by their altered behavior. In any case, our experiences create episodic memories. Episodic or factual memory requires consciousness to be processed.[8] Our conscious brain encounters a datum of information (e.g., in a college course, we learn that the nervous system is made up of neurons and glial cells), and the central nervous system encodes this information into a memory. Although memory is integrated across several brain regions, the hippocampus has been shown to be crucial for episodic memory formation, as the studies of the famous, brain-damaged patient H.M. revealed.[9] This fascinating brain area, located in the temporal lobe and named after its resemblance to a seahorse, allows us to convert our experiences into short-term memories and then consolidate them into long-term memories stored across the brain based on category (e.g., faces are stored in fusiform gyrus). The physiological mechanism underlying both short-term and long-term memory formation is known as neuroplasticity.[10] Neuroplasticity refers to the ability of synapses, or connections between neurons, to undergo structural and functional change as a result of experience. Memories are formed by lasting changes in the synapses between specific populations of neurons distributed across multiple brain areas, often referred to as "engrams" or "memory traces."[11] Furthermore, the hippocampi, one in each cerebral hemisphere, belong functionally to the limbic system, also known as the "emotional" brain. The hippocampus forms part of a system able to imbue our experiences, and therefore, our factual memories, with emotional value. The likelihood of conversion from short- to long-term memory depends on emotional valence.

Neuroplasticity allows our brain to wire according to our experiences. Memories help us recollect our past and define who we are, how we feel, behave, and identify ourselves. A clear instantiation of the impact of childhood and transgenerational experiences on self-reflection during adulthood is explained by a young Black woman in episode 2 of *RAoF*: "You're raised with dark skin,

---

[8] M. Wheeler, "Episodic and Autobiographical Memory: Psychological and Neural Aspects," In *International Encyclopedia of the Social & Behavioral Sciences* (2001): 4714–17.

[9] Wilder Penfield, and Brenda Milner "Memory Deficit Produced by Bilateral Lesions in the Hippocampal Zone," *Archives of Neurology and Psychiatry* 79, no. 5 (1958): 475.

[10] T. V. P. Bliss, and G. L. Collingridge, "A Synaptic Model of Memory: Long-term Potentiation in the Hippocampus," *Nature* 361, no. 6407 (1993): 31–9.

[11] Richard Semon, "Die Mneme Als Erhaltendes Prinzip Im Wechsel Des Organischen Geschehens," *Nature* 73, no. 1893 (1906).

you're not beautiful. You learn shame early on." Her learned experiences shaped her brain, mind, and the paths of her self-reflection. As an antidote, Nance calls for a consciously shifting mental state, one that moves across vast terrains and finally modifies self-reflection. From a neuroscientific point of view, a new re-wiring of the brain must occur to shift self-awareness. Experiences free of trauma are crucial for this healing. But as a trans woman states in episode 2 of *RAoF*, how "to love ourselves in a world that fundamentally hates us?" How to change a brain that has been, and still is, freighted with traumatic experiences?

## Alterations of the Mind, Memory Recollection, and Pattern Completion

Emotional trauma is the seed of distressed states of mind and altered physiological mechanisms that spread across the human body. Altered mental scenarios are created to escape from the ruminations of a tormented mind. Unspeakable events, harmful words, unreasoning mindsets, and repressive societies enable mental disorders. Distorted, damaging views of ourselves often manifest physically as depression, social anxiety, addiction, dissociative identity disorder, and many more alterations of the mind. We witness, as if from an inner perspective, Elliot's mental scenarios in *Mr. Robot*. Elliot's brain, traumatized since childhood, exhibits altered neuroplasticity, presented as impaired memory formation and retrieval, and pathological diseases, which might include anxiety disorder, split personality, and schizophrenia.[12] In episode 8 of season 1, Elliot hacks himself. "Nothing. No identity" he soliloquizes. Next, Elliot opens his CD folder to find a blank disc that stores his blocked memories of traumatic childhood experiences. Elliot's conscience is unaware of his past. To cope with traumatic experiences, on certain occasions, a new self emerges from the unconscious mind, a new ego constructed to protect us, and, in the case of *Mr. Robot*, to protect Elliot from the unbearable mind state. In clinical terms, this is diagnosed as "dissociative identity disorder." "Do you think it's possible Mr. Robot is the reason why you can't remember?" asks Krista Gordon to Elliot. "He already told you he can't," replies Mr. Robot (Elliot's brain's manufactured

---

[12] Mazen A Kheirbek, Kristen C Klemenhagen, Amar Sahay, and René Hen, "Neurogenesis and Generalization: A New Approach to Stratify and Treat Anxiety Disorders," *Nature Neuroscience* 15, no. 12 (2012): 1613–20.

**Figure 22.2** *Mr. Robot* directed by Sam Esmail (2015–2019). Season 4 episode 7. Unlocking the memory of a traumatic childhood event, cracking the raison d'être of Elliot's dissociative identity disorder. The hacker breaks after the drug dealer's ambition (Fernando Vera) and his shrink's help (Krista Gordon).

second self), and "Remember" continues Elliot. The *raison d'être* of this second identity comes to light in the last season, episode 7 (figure 22.2). Mr. Robot represents a brain mechanism that blocks memory retrieval.

The neurobiological basis of memory retrieval or "ecphory" remains poorly understood; nevertheless, it is known that it involves the interaction between triggering cues and stored engrams. Through optogenetics and rodent behavior, neuroscientists have demonstrated that even in the presence of retrieval cues, the silencing of specific engrams, at the level of neuronal ensembles, prevents memory recovery. In contrast, the direct activation of certain engrams can induce ecphory.[13] The relationship between a retrieval cue and the probability of retrieving the original memory is called encoding specificity.[14] During the formation of an engram, two factors are involved: one external (or environmental), and one internal (or emotional). The external component refers to the environmental framework in which the memory representation was first encoded, and we understand the internal factor as the state of mind associated with the original experience. The hippocampal connectivity, through afferent and efferent pathways (neurons carrying signals either to or away from sources), enables the integration, and therefore, association, of these two components. As the initial experience unfolds, a specific cell ensemble of pyramidal neurons

---

[13] Paul W. Frankland, Sheena A. Josselyn, and Stefan Köhler, "The Neurobiological Foundation of Memory Retrieval," *Nature Neuroscience* 22, no. 10 (2019): 1576–85.
[14] Endel Tulving, and Donald M. Thomson, "Retrieval Processes in Recognition Memory: Effects of Associative Context," *Journal of Experimental Psychology* 87, no. 1 (1971): 116–24.

from the hippocampal CA3 subregion alters their firing pattern and coactivates with specific sensory brain regions.[15] This encoding association is the basis of the memory trace or engram. These engrams can be recalled by triggering cues to evoke the original memory. A retrieval cue is effective in episodic memory recollection to the extent that it matches the external and internal components of the original memory (e.g., context and emotion). Specific brain waves generated at the hippocampus, more concretely theta oscillation, play a decisive role in encoding specificity by a neuronal mechanism called pattern completion.[16] A partial input associated with the initial experience activates the memory trace in the hippocampus, and recollection is completed when sensory details are reinstated in neocortex.[17] Some theories of episodic memory retrieval, such as the representational account, limit the hippocampal engagement to specific memory content, such as associative memories.[18]

When trauma is experienced, a mark is etched on our brain. Neural connections are altered, which create a conscious or unconscious effect on our self-reflection and behavior. Memories may seem suppressed, thanks to the natural, protective ability of the brain to forget (as Elliot from *Mr. Robot* demonstrates), but still that mark chains us to the distressing event and the corollary alterations of the mind. The neural mark is and will be present in our brain unless a healing process occurs. After extensive therapy, Krista Gordon, Elliot's shrink, intuits the origin of his altered mental state. In a powerful and captivating scene, Krista's words guide Elliot's mind, bringing him to the emotional state linked to his traumatic memory, which reinstates the original engram. During this process of memory recollection, the memory trace is restored, and the original hippocampal firing pattern re-established. Krista's work reignites Elliot's traumatic experience and associated emotions. By gaining access to the blocked memories and facing the traumatic experience, Elliot's

---

[15] Stefan Leutgeb, and Jill K. Leutgeb, "Pattern Separation, Pattern Completion, and New Neuronal Codes Within a Continuous CA3 Map,"*Learning & Memory* 14, no. 11 (2007): 745–57.

[16] Bernhard P. Staresina, Sebastian Michelmann, Mathilde Bonnefond, Ole Jensen, Nikolai Axmacher, and Juergen Fell, "Hippocampal Pattern Completion Is Linked to Gamma Power Increases and Alpha Power Decreases During Recollection," *eLife* 5 (2016): e17397.

[17] Edmund T. Rolls, "The Mechanisms for Pattern Completion and Pattern Separation in the Hippocampus,"*Frontiers in Systems Neuroscience* 7 (2013). Benjamin J. Griffiths, George Parish, Frederic Roux, Sebastian Michelmann, Mircea Van Der Plas, Luca D. Kolibius, Ramesh Chelvarajah, et al., "Directional Coupling of Slow and Fast Hippocampal Gamma with Neocortical Alpha/Beta Oscillations in Human Episodic Memory," *Proceedings of the National Academy of Sciences* 116, no. 43 (2019): 21834–42.

[18] David A. Ross, Patrick Sadil, D. Merika Wilson and Rosemary A. Cowell."Hippocampal Engagement During Recall Depends on Memory Content," *Cerebral Cortex* 28, no. 8 (2018): 2685–98.

mind releases the burden of the Mr. Robot alter ego. His brain will need to rewire the original engrams carved out by the traumatic experiences. This is an excavation: a painful, arduous process of freeing his brain pathways. Healing commences.

## Pattern Separation, Healing, and Dance

What if the capability to heal is inherent to humans? What if it resides inside us? What if our brain, the same way it undergoes modifications leading to mental alterations, could alter its structure and function to sustain a healthy and free state of mind? Brain circuitry not only allows for pattern completion. A specific connection in the hippocampus called the mossy fiber synapses, the connection between the granule cells of the dentate gyrus and the CA3 pyramidal neurons, enables another memory-related process called pattern separation. This brain mechanism transforms similar or overlapping inputs into different outputs. Research by Jie Zheng, of the University of California at Irvine, has demonstrated that in the case of emotional events, such as traumatic experiences, pattern separation of emotional stimuli is associated with hippocampal theta oscillations and its interaction with the amygdala, the brain region that provides emotional weight to memories.[19]

As explained above, pattern completion during memory recollection is based on associations encoded during memory consolidation, associations between external stimuli and an individual's internal emotional state. In order to release our mind from the emotional trauma and erase the consequent brain network footprint, these initial distressing associations need to be unlinked—external and internal components must be uncorrelated. In other words, harmful memories need to be re-encoded into positive ones through new neural associations. This may be considered the first step of a healing process. The hippocampal-mediated pattern separation may be the physiological neural mechanism by which we rewire our brain and "shift consciousness," as Nance intends, freeing us from past experiences, traumatic or not. The natural malleability and flexibility of the brain allows for new memory encoding and

---

[19] Jie Zheng, Rebecca F. Stevenson, Bryce A. Mander, Lilit Mnatsakanyan, Frank P.K. Hsu, Sumeet Vadera, Robert T. Knight, Michael A. Yassa, and Jack J. Lin, "Multiplexing of Theta and Alpha Rhythms in the Amygdala-hippocampal Circuit Supports Pattern Separation of Emotional Information," *Neuron* 102, no. 4 (2019): 887–98.

Neuroplasticity 397

**Figure 22.3** *Random Acts of Flyness* directed by Terence Nance (2018-present). Season 1 episode 2. Dance as a means for freedom. Through a captivating musical scene, a world where "the veil," or in other words, the need for healing is nonexistent.

consolidation, facilitating the healing process from which an original and self-guided identity will emerge, or, in the words of Frantz Fanon, "In the world I am heading for, I am endlessly creating myself." In *Black Skin, White Masks*, Fanon defends the idea of stripping off all past experiences and inscribed trauma and making way for a new free self.[20]

Could art facilitate a free state of mind? A flexible mind can change and adapt, the same way a flexible body bends while dancing. The freedom gained from movement-experience while dancing in our earthly body may extend to freedom for our ethereal mind. At the end of the second episode of *RAoF*, a mesmerizing musical retelling of the J. M. Barrie's Neverland, *Nuncaland*, represents a world of permanent childhood created to free the ineludible conversion to a stereotyped adult Black man (figure 22.3). With this dance performance, Nance stages the need for liberating ourselves from a socially expected identity, as the representation of Wendy (Le'Asha Julius) encourages Peter Pan (Kevin Alexis Rivera) to do. The creators of the skit use dance to express emotions and feelings that cannot otherwise be released. Dance, with its different components, physical, cognitive, emotional, and social, may facilitate flexibility, and therefore freedom and healing of bodies, minds, identities, and societies.

[20] Fanon, Frantz. *Black Skin, White Masks* (New York: Grove Press, 1952).

Scientific evidence suggests structural brain modifications related to dance. A larger volume of the dentate gyrus was observed in dancers compared to non-dancers.[21] This hippocampal area is one of the few brain regions where adult neurogenesis takes place, and as mentioned above, the dentate gyrus mediates memory-related processes such as pattern separation.[22] Furthermore, Hanna Poikonen, dancer and neuroscientist at the Professorship for Learning Sciences and Higher Education at ETH Zurich, finds that cortical communication is enhanced in dancers compared to musicians and laymen when observing a dance piece. In her study, she claims that this increased theta synchrony may be induced via the hippocampal-cortical pathway which is associated with cognitive processes such as memory and emotional processing.[23] Although it is tempting to hypothesize about the relationship between dance, the hippocampus, memory, pattern separation, and the possibility to heal, research on the neuroscience of dance is still rare and inconsistent. In the *Nuncaland* musical number, a young boy escapes homophobic taunts to chase his shadow and then dance with other boys, possibly like himself. The dance number could be seen as a re-enactment of the original taunts, and perhaps a moment of healing. Dance facilitates a framework in which measuring up to stereotypes of masculinity is no longer needed. Through a dance style far from a strict technique and where freedom of movement seems the rule, dancers release frustration and express their true emotions and identities. As Terence Nance explained during an interview: "whatever shame and guilt I had internalized as a boy growing into a male, probably the only way to exorcise it is to do some sort of body language, movement, and dance." Within this context of physical and mental freedom, a possibility for healing appears.

In spite of the need for investigation, dance as a liberating and transforming art form is difficult to refute. Dancing goes beyond tangibility, not just as it moves our body, but also as it shifts our identity and our sense of self, which, through brain and mind flexibility, brings these to a freer state. Identities free of burdens can be the foundations for new cycles of experiences, memories, states of mind, self-reflections, and free behaviors.

---

[21] Kathrin Rehfeld, Patrick Müller, Norman Aye, Marlen Schmicker, Milos Dordevic, Jörn Kaufmann, Anita Hökelmann, and Notger G. Müller, "Dancing or Fitness Sport? the Effects of Two Training Programs on Hippocampal Plasticity and Balance Abilities in Healthy Seniors," *Frontiers in Human Neuroscience* 11 (2017): 305.
[22] Neurogenesis: process of generating functional neurons from precursors.
[23] Hanna Poikonen, Petri Toiviainen, and Mari Tervaniemi, "Dance on Cortex: Enhanced Theta Synchrony in Experts When Watching a Dance Piece," *European Journal of Neuroscience* 47, no. 5 (2018): 433–45.

# Bibliography

Bliss, T. V. P., and G. L. Collingridge. "A Synaptic Model of Memory: Long-Term Potentiation in the Hippocampus." *Nature* 361, no. 6407 (January 1993): 31–9.

Dias, Brian G., and Kerry J. Ressler. "Parental Olfactory Experience Influences Behavior and Neural Structure in Subsequent Generations." *Nature Neuroscience* 17, no. 1 (January 2014): 89–96.

Esmail, Sam. Director. "Mr. Robot." *HBO*. 2015. Amazon Prime. https://www.amazon.com/gp/video/detail/B00YBX664Q/ref=atv_dp_season_select_s1. (accessed February 2, 2021).

Fanon, Frantz. *Black Skin, White Masks*. New York: Grove Press, 1952.

Frankland, Paul W., Sheena A. Josselyn, and Stefan Köhler. "The Neurobiological Foundation of Memory Retrieval." *Nature Neuroscience* 22, no. 10 (October 2019): 1576–85.

Griffiths, Benjamin J., George Parish, Frederic Roux, Sebastian Michelmann, Mircea Van Der Plas, Luca D. Kolibius, Ramesh Chelvarajah, et al. "Directional Coupling of Slow and Fast Hippocampal Gamma with Neocortical Alpha/Beta Oscillations in Human Episodic Memory." *Proceedings of the National Academy of Sciences* 116, no. 43 (October 2019): 21834–42.

Kheirbek, Mazen A, Kristen C Klemenhagen, Amar Sahay, and René Hen. "Neurogenesis and Generalization: A New Approach to Stratify and Treat Anxiety Disorders." *Nature Neuroscience* 15, no. 12 (December 2012): 1613–20.

Leutgeb, Stefan, and Jill K. Leutgeb. "Pattern Separation, Pattern Completion, and New Neuronal Codes within a Continuous Ca3 Map." *Learning and Memory* 14, no. 11 (2007): 745–57.

Penfield, Wilder, and Brenda Milner. "Memory Deficit Produced by Bilateral Lesions in the Hippocampal Zone." *Archives of Neurology And Psychiatry* 79, no. 5 (May 1958): 475.

Poikonen, Hanna, Petri Toiviainen, and Mari Tervaniemi. "Dance on Cortex: Enhanced Theta Synchrony in Experts When Watching a Dance Piece." *European Journal of Neuroscience* 47, no. 5 (March 2018): 433–45.

Rechavi, Oded, Gregory Minevich, and Oliver Hobert. "Transgenerational Inheritance of an Acquired Small Rna-based Antiviral Response in C. Elegans." *Cell* 147, no. 6 (November 2011): 1248–56.

Rehfeld, Kathrin, Patrick Müller, Norman Aye, Marlen Schmicker, Milos Dordevic, Jörn Kaufmann, Anita Hökelmann, and Notger G. Müller. "Dancing or Fitness Sport? The Effects of Two Training Programs on Hippocampal Plasticity and Balance Abilities in Healthy Seniors." *Frontiers in Human Neuroscience* 11 (June 2017): 305.

Rolls, Edmund T. "The Mechanisms for Pattern Completion and Pattern Separation in the Hippocampus." *Frontiers in Systems Neuroscience* 7 (January 2013): 74.

Ross, David A., Patrick Sadil, D. Merika Wilson, and Rosemary A. Cowell. "Hippocampal Engagement During Recall Depends on Memory Content." *Cerebral Cortex* 28, no. 8 (August 2018): 2685–98.

Semon, Richard. "Die Mneme Als Erhaltendes Prinzip Im Wechsel Des Organischen Geschehens." *Nature* 73, no. 1893 (Febuary 1906): 338–38.

Staresina, Bernhard P, Sebastian Michelmann, Mathilde Bonnefond, Ole Jensen, Nikolai Axmacher, and Juergen Fell. "Hippocampal Pattern Completion Is Linked to Gamma Power Increases and Alpha Power Decreases During Recollection." *eLife* 5 (August 2016): e17397.

Tulving, Endel, and Donald M. Thomson. "Retrieval Processes in Recognition Memory: Effects of Associative Context." *Journal of Experimental Psychology* 87, no. 1 (1971): 116–24.

Wheeler, M. "Episodic and Autobiographical Memory: Psychological and Neural Aspects." *International Encyclopedia of the Social & Behavioral Sciences* (2001): 4714–17.

Zheng, Jie, Rebecca F. Stevenson, Bryce A. Mander, Lilit Mnatsakanyan, Frank P.K. Hsu, Sumeet Vadera, Robert T. Knight, Michael A. Yassa, and Jack J. Lin. "Multiplexing of Theta and Alpha Rhythms in the Amygdala-Hippocampal Circuit Supports Pattern Separation of Emotional Information." *Neuron* 102, no. 4 (May 2019): 887–98.e5.

23

# Where is my Mind?

## *Mr. Robot* and the Digital Neuropolis

Patricia Pisters

### Introduction: Elliot in Neuropolis

In their article "Living well in the Neuropolis," Des Fitzgerald, Nicholas Rose, and Ilina Singh map the connections between cities and brains, relating the frantic, stressful lives in urban environments to psychological and neurobiological insights about the hectic living conditions of city life. They propose the term "Neuropolis" to describe the figure of the city "embedded in neuropsychological concepts and histories" (as in the idea of the brain as a city); the term also describes "an embodied set of (sometimes pathological) relations and effects that take place between cities and the people who live in them."[1] Drawing on neuropsychological insights and sociological claims, the authors propose future possibilities for residents who might potentially flourish in newly-sustainable cities.

In this chapter, I adopt this concept of the Neuropolis. But as I do so, I shift its focus toward something darker: the overstimulated brain in contemporary urban space. Specifically, I attend to the media ecology of screens and information, as well as to the data of urban global capitalism that has entangled itself with contemporary neuropsychology. Interpreting the television series *Mr. Robot* as a contemporary "neuro-image,"[2] I investigate brain-culture interfaces through the imaginary realms of representation and narrative.[3] The takeaways of my analysis

---

[1] Des Fitzgerald, Nicholas Rose and Ilina Singh, "Living Well in the Neuropolis," *The Sociological Review Monographs* 64, no.1 (April 2016): 221.
[2] Patricia Pisters, *The Neuro-Image: A Deleuzian Film-Philosophy for Digital Screen Cultures* (Stanford: Stanford University Press, 2012).
[3] Alberto Garcia, "A Storytelling Machine: The Complexity and Revolution of Narrative Television," *Between* 6, no.11 (2016): 1–25.

are new, layered insights into the experience of life in the Neuropolis, the "Brain City", what philosopher Gilles Deleuze previsions as contemporary hig-tech screen culture.[4]

Mr. Robot, a television series created by Sam Esmail for USA Network, tells the story of Elliot Alderson (Rami Malek), a millennial hacker vigilante in present-day New York who suffers from depression and delusions.[5] One of Mr. Robot's taglines is, "Life is like a computer."[6] Since audiences experience most of the narrative from within Elliot's mindscape—we gain regular access to his inner life through film techniques including voiceovers—the connections between neuropsychology, the contemporary city, and computer technology feel immediate. These connections evoke links between technological machines, the brain, and the city; as the series unfolds, Mr. Robot highlights how these elements—brain, digital technologies, and global capitalist city—are now completely enmeshed, with fiction, reality, and scientific insight bound tightly together.[7] Here, I argue for a more even appreciation of the show's intricate layers: first, of the television show's story and style of fiction; second, of the digital Neuropolis's realities (its cityscape, global capitalism, cyber war, corporate surveillance and hacking, and mental experiences); and third, of the scientific insights that entwine our experiences of fiction and reality (e.g., media studies, philosophy, neurobiology, and psychiatry). My attention to these layers shows how these fields of knowledge and experience co-evolve. More specifically: I argue that in Mr. Robot, fiction, reality, scientific discourse, and the contemporary digital world, all separated only by porously resonating membranes, co-mingle.

## Isolated in the Screen's Lower Quadrant

Let's look first at the level of the story and style of Mr. Robot. In the first episode, "eps.1.0_hellofriend.mov"— all episodes are presented as file names—we see

---

[4] Gilles Deleuze, *Cinema 2: The Time-Image* (London: The Athlone Press, 1989), 265.
[5] Anthony Smith, "Pursuing 'Generation Snowflake': Mr. Robot and the USA Network's Mission for Millennials," *Television & New Media* 20, no.5 (2018): 1–17.
[6] See *Mr. Robot*, created by Sam Esmail, Prime Video Aanbod, https://primevideoaanbod.nl/amazon-prime-video-aanbod/mr-robot/ (accessed December 27, 2020); Jefferson Grubbs, "This 'Mr. Robot' Theory Suggests Elliot Is Basically A Living Computer," *Bustle*, December 8, 2019, https://www.bustle.com/p/this-mr-robot-season-4-theory-suggests-elliot-is-basically-a-living-computer-19428420 (accessed December 27, 2020).
[7] Douwe Draaisma, *Metaphors of Memory: A History of Ideas about the Mind* (Cambridge: Cambridge University Press, 2000); Uta and Thilo Von Debschitz, *Fritz Kahn: Infographics Pioneer* (London and New York: Routledge, 2017).

Elliot as a smart yet socially awkward black hat (or perhaps gray hat)[8] hacker who exposes a consumer of child pornography, tipping the police off to his probing web searches. All the while, Elliot talks to his imaginary friend in voiceover. In the audience, we hear him warn us, "You're only in my head, we have to remember that," followed by his revelation that there's a "top secret conspiracy bigger than all of us," one organized by a small, powerful cartel secretly running the world. Soon we discover that these powerful people—("the invisible guys that play God without permission," as Elliot tells his imaginary friend)—are a global, hyper-capitalist conglomerate called Evil Corp (or E-Corp). We also discover that Elliot works as a tech engineer at cyber security firm All Safe, whose biggest client is Evil Corp.

We're then quickly introduced to the series' central characters. Underground, Elliot meets a mysterious man (Christian Slater), dressed in a coat sporting a "Mr. Robot" badge (for a computer repair shop), who invites him to become part of a collective, "Fsociety," aiming to hack E-Corp and erase all global debt. In a later episode, viewers discover Mr. Robot is actually Elliot's deceased father, who, as an E-Corp employee, died from exposure to environmental toxins, courtesy of the corporation's nuclear power plant. Fsociety's hacker vigilantes operate from inside an abandoned arcade at Coney Island. At this site, Mr. Robot shares with Elliot Fsociety's plan to change the world. The collective collaborates with a Chinese hacker group, the Dark Army, led by Whiterose (BD Wong), a woman who is sometimes also a man—Minister Zhang, the Chinese Minister of State Security. Another recurrent figure is Tyrell Wellick (Martin Wallström), an ambiguous antagonist who half-way through becomes Elliot's associate. Women in Elliot's life include Darlene Alderson (Carly Chaikin), his sister and fellow Fsociety hacker; Angela Moss (Portia Doubleday), Elliot's trusted childhood friend who similarly lost her mother to leukemia; and Krista Gordon (Gloria Reuben), Elliot's psychotherapist. FBI agent Dominique DiPierro (Grace Gummer) is Elliot's driven, incisive investigator.

Drawn into Elliot's mind, the viewer is interpolated into the unfolding of what I've called "the neuro-image," a form of extended cinema in the digital age that allows us to experience the world directly from within characters' mindscapes and

---

[8] According to *Mr.Robot_DECOD3D.doc* (a special episode), a white hat hacker collaborates with the government. A black hat hacker works toward malicious ends, and a gray hat hacker resides in between, sometimes employing illegal means for worthy ends. See Lev Gartman, *Mr. Robot Dec0d3d. doc*, https://vimeo.com/176633054 (accessed December 27, 2020).

brain worlds.[9] In *Mr. Robot*, we don't only hear Elliot speak to himself through internal monologue; we also *experience* the world through his mind, his neuropsychology: images, sounds, and complete scenes eventually reveal themselves as delusions and hallucinations—vagaries produced by drug abuse, environmental stress, and unaddressed trauma. In effect, Elliot becomes an unreliable narrator; the events we witness on screen are ambiguous and disorienting, highlighting overlaps between reality and delusion, memory, and dream.

Throughout the series, Elliot comments on his own thought processes with varying degrees of clarity and intensity, creating self-reflexive moments about the strangeness of certain situations. Eventually, even Elliot's hallucinations form a *mise-en-abyme*, further scrambling perception and reality. "Error Code 404: Not Found," for instance, is a computer code that appears at different times, most noticeably in an episode called "eps1.3_da3mOns.mp4" (S1E4), in which Elliot plots to destroy E-Corp's back-up tapes.[10] Before he can carry out his plan—reprogram the climate control system of Steel Mountain, a highly-secured data storage facility—Elliot has to detox from morphine. While suffering from withdrawal symptoms, he experiences all kinds of physical and mental torment; his shaking and aching body becomes the physical manifestation of his fears, and, for the audience, the television screen opens a portal into his inner hell.

During one of his nightmarish delusions, Elliot revisits his old family house and finds a note stuck to a utility pole: "Error Code 404: Not Found." A small girl on a kid's scooter asks, "Can you tell me what's your monster?" Metaphorically, the scene helps us understand that it is Elliot himself who suffers an error code 404: he cannot find the connections to his past nor to the people in the present. This disconnection is a symptom of a syndrome that Johann Hari, in *Lost Connections*, has described as a key sign of clinical depression.[11] In the last

---

[9] The neuro-image is considered in Deleuze's *Cinema and Philosophy*. According to Deleuze, aesthetics have permeated the brain. He emphasizes global capitalism's psychopathological effects: capitalism's temporal dimensions are dominated by an obsession with a future from which the present and past can be accessed; truth and certainty of the present becomes ungrounded; the neuro-image's affect is primary. After the movement-image and the time-image, the neuro-image deals with the specific psychopathologies and complex entanglements of contemporary global capitalist media culture. See Pisters, *The Neuro-Image*.

[10] The error summary of the http 404 code reads: "The resource you are looking for has been removed, changed its name or is temporarily unavailable," https://amoghnatu.net/2013/09/16/question-please-help-iis-throwing-http-404-not-found-but-requested-resource-actually-exists-requested-url-also-changing-automatically (accessed December 18, 2020). In the last season all episodes have an error code as title: 401 Unauthorized, 402 Payment Required, 403 Forbidden, 404 Not Found, 405 Not Allowed, etc.

[11] Johann Hari, *Lost Connections: Uncovering the Real Causes of Depression and the Unexpected Solutions* (New York: Bloomsbury, 2018).

section, I will return to this idea, focusing on scientific insights concerning mental conditions brought on by the Neuropolis.

For now, though, I'd like to emphasize how, in another self-reflexive moment in this episode, Elliot quite literally presents the brain as a computer. While withdrawing from his medication in the back of a car on the way to Steel Mountain, Elliot wonders about the devil: "The devil is at its strongest while we are looking the other way, like a program running in the background silently while we are busy doing other shit." "[Daemons] perform action without user interaction: monitoring, logging, notifications, primal urges, repressed memories, unconscious habits. They are always there, always active. Intensions are not relevant; they don't drive us. Daemons drive us" (S1E4). This is a clear example of the conceptual slippage between computational and neurological processing that characterizes *Mr. Robot*'s narrative style.

Moreover, it seems a technological media machine (especially the tv screen, camera, and computer) always provokes and manifests Elliot's psychopathological states of mind. Jeffrey Sconce has called this "the technical delusion," that converts electronic media's messages into insanity and psychosis. In the Information Age "electricity has become the nervous fluid of the entire planet," Sconce argues.[12] In one of the episodes, Elliot experiences a warped childhood memory in the form of a cheaply produced 1980s television sitcom (with canned laughter and grainy, haptic video images), in which the Alderson family is on a road trip with a 1980s-version of Tyrell locked in the trunk of the family car, and Mr. Robot, as his father, behind the wheel. The constant mediation of delusional content and form indicates not just metaphoric connections between brains and machines, but also deeper, material connections between human and non-human assemblages.

I can only nod here to the great cinematographic forms that *Mr. Robot* blends together to express life in the contemporary Neuropolis—from its gritty *mise-en-scène* and remarkable camerawork, which often consists of long takes and strange angles, to its dark, pulsating soundtrack, composed by Mac Quayle. But I can emphasize that the series' filmmakers employ a subtle yet affective technique to express each of the characters' alienation and off-centeredness: anxious framing. Throughout the show, characters are often isolated in the lower quadrant of the screen, framed at a low angle that visually expresses the characters' paranoia, desperation, anxiety, and loneliness. Characters squeezed into the

---

[12] Jeffrey Sconce, *The Technical Delusion: Electronics, Power, Insanity* (Durham and London: Duke University Press, 2019), 6.

screen's bottom seem overwhelmed—as if the city's weight is carried on their shoulders or balanced on their heads. In terms of the fictional level of story and style, as a neuro-image, *Mr. Robot* thus presents the city, the brain, and media machines as tied together, connected by their mutual expression.

## What if Error404 is a Message from the Real World?[13]

Part of *Mr. Robot*'s attraction as a fictional world derives from how it resonates with and connects to the contemporary. Today, all digital information has become vulnerable. Corporate conglomerates, as well as governments, also have unprecedented access to private citizens' data, giving these entities "daemon-like" power. And as I'm working to show here, the show's realistic depictions of hacking culture not only foreground the power structures of global capitalism, but also the psychology (and psychopathologies) of the human brain in the Neuropolis. For instance: if we consider Error Code 404 not only as a metaphor for Elliot's brain and the lost connections to his own past and to the world, but also as a reference to the broken or changed links within contemporary reality itself, then *Mr. Robot* has a lot to say about the experiences of the post-Wikileak world we inhabit.

Elliot's hacking practices have received critical acclaim from hacking communities worldwide. And this realism is, of course, not an accident. Kor Adana, one of *Mr. Robot*'s writers and technology producers, ensured the technological plausibility of all the hacks.[14] Elliot breaches systems by using technologies and skills from his realistic toolkit, including credit card–sized Raspberry Pi computers, DeepSound discs, Slackware Linux, rootkits, and DDoS attacks; he also swaps sim cards, swipes cell phones, and picks locks. Onscreen IP addresses, URLs, and QR codes lead viewers to real destinations. Importantly, all the software and codes are not just Matrix-like number streams, but rather actual tools and programs.[15] In the Steel Mountain data hack, for instance, Elliot uses a

---

[13] An Error 404 meme: See "What if Error 404 is a Message from the Real World," *Quick Meme*, http://www.quickmeme.com/meme/35o13a (accessed January 11, 2020).

[14] See Julia Franz, "How Realistic are the Hacks in 'Mr. Robot'?" *The World*, https://www.pri.org/stories/how-realistic-are-hacks-mr-robot (accessed January 11, 2020); Corey Nachreiner, "'Mr. Robot' Rewind: Rewinding the '5/9' Hack in a Stunning Season Finale," *Geek Wire*, December 18, 2017, https://www.geekwire.com/2017/mr-robot-rewind-rewinding-5-9-hack-stunning-season-finale/ (accessed January 11, 2020; https://www.sentryo.net/mr-robot-8-steps-cyberattack-decoded-sentryo/ (accessed January 11, 2020).

[15] See also Lizzie Plaugic, "A QR Code in Last Night's Mr. Robot Leads to a Mysterious Website," *The Verge*, July 14, 2016, https://www.theverge.com/2016/7/14/12187768/qr-code-mr-robot-confictura-industries-usa (accessed January 11, 2020).

Figure 23.1 A video message by Fsociety using a Guy Fawkes mask in *Mr. Robot* (Sam Esmail, 2015– ).

Raspberry Pi computer to infiltrate the storage facility's temperature control system.[16] Everything can be hacked, from climate control systems to smart phones, even entire buildings. In season 2, Fsociety attacks the FBI via the employer's android cell phones. Further realistic technical details include the use of discarded USBs and CDs as potential baits for victims, and the phishing and hacking of social media accounts to find individuals' pressure points for future blackmail.[17] Fsociety also links to the real world through their video messages (and, later, in their demonstrations). The collective covers their face with Guy Fawkes masks, just like those used by the real-world hacktivists of Anonymous (figure 23.1). As Kriss Ravetto has demonstrated, the use of this mask, "the imagined effigy of the 17th century Gunpowder Plot conspirator [...] tortured and executed for his attempt to blow up the English Parliament in 1605" has often been used as a symbol of resistance.[18] *V for Vendetta* (2006), famously, also employed the mask, specifically as a shield that guaranteed anonymity in the face of corporate security and a totalitarian state. Like many hacktivist groups, *Mr. Robot*'s Fsociety fights for a noble cause, though sets unintended precedents.

[16] In *Mr.Robot_DECOD3D.doc*, Kor Adana and other cyber specialists compare the Steel Mountain hack to the Stuxnet attack of an Iranian nuclear facility, allegedly ordered by US and Israeli governments in 2010. Gartman, *Mr.Robot_DECOD3D.doc*.
[17] Danny Hernandez, "10 Things Mr. Robot Gets Rights About Hacking," *Screen Rant*, February 1, 2019, https://screenrant.com/things-mr-robot-gets-right-hacking/ (accessed January 11, 2020).
[18] Kriss Ravetto-Biagioli, "Anonymous: Social as Political," *Leonardo Electronic Almanac* 19, no. 4 (2013): 186.

In season one, when Fsociety finally hacks E-Corp, that is, new seeds for future problems are planted.

I'll now consider several scenes from the show. Season 3's remarkable episode (S3E5), "eps3.4_runtime-error.r00," brings together a number of the elements I discussed earlier. By this point in the narrative, we know Mr. Robot is actually part of Elliot's disassociated personality—a part that takes possession of him at certain critical moments. Mr. Robot is more radicalized than Elliot—he with Tyrell and Angela have executed the downfall of E-Corp's New York facility. To avenge the death of her mother—a death caused by E-Corp—Angela sedates Elliot in an attempt to keep Mr. Robot in the game. She and Elliot's alternate personality, Mr. Robot, then collaborate with the Dark Army to bring down E-Corp's primary building with a new hack. Elliot, who's concerned about casualties, tries to prevent this, but fails. Angela then accesses the Hardware Security Modus (HSM) on E-Corp's data storage floor through a hard drive and the Dark Army's instructions.

Notably, the episode was filmed in real-time as a single take: a tracking camera accompanies Elliot's traversals. First, Elliot and Angela get evicted from the building; then, Angela returns to provoke the hack. Simultaneously, a group of masked demonstrators—a decoy by the Dark Army—distract attention from the real cyberattack. For a viewer, watching this sequence might feel like actually walking *with* these characters, lulled into a trance by the hypnotic images and throbbing soundtrack.

Now at the episode's beginning, Elliot doesn't know how he got to the E-Corp facility. His inner voice observes that while he is running his routine on autopilot, he feels as if his daily program has crashed. "Is there a runtime error happening to me right now?" he wonders. Whether due to a runtime error or a memory leak, the system crashes.

These programming concepts also serve as metaphors for understanding Elliot's own brain processes and character arc. He has no memory of the past three days during which his alter ego, Mr. Robot, has taken control; he also doesn't share Fsociety's belief that taking down an E-Corps building with the Dark Army will lead to "the right outcome." (Saving the world, perhaps.) Yet again, the brain, technology, and the global city form a tight knot.

This episode also portrays corporate culture, transnational capitalism, and global power relations realistically. Evil Corp is directly modeled after the Trump administration; and season 3's plotline includes footage of a Trump speech, as well as content that intimates that China set Trump up as president, and that Minister

Zhang participated in planting an Fsociety sapling with roots in Iran (S3E3); too, the toxic waste disaster that killed both Elliot's father and Angela's mother resonate with the chemical pollution and criminal negligence that has caused the deaths of innocent people in Flint, Michigan's water crisis.[19] Platform consumerism and everlasting debt, too, are today's building blocks of hyper-capitalism.

The most salient reference to contemporary politics, however, is China's economic and political force. As leader of the Dark Army, Whiterose planned the Stage 2 sabotage on the day he, in his appearance as Minister Zhang, through the UN vote, attempted to annex the Congo to China. Through the "eps3.4_runtime-error.r00" episode, real television news from the UN's controversial resolution is broadcast. While in reality China has so far not yet annexed Congo officially (and certainly not via a UN resolution), Mr. Robot addresses China's stake in the globe's rich materials, including the African continent's copper and coltan (crucial for cell phones). The unsigned resolution illumines shifts in power among US, China, and Iran. When in the following episode "eps.3.5_kill-process.inc" (S3E6) Elliot manages to save E-Corp's New York building and thinks he's turned the tide, he soon realizes that there was a runtime error: news reports now show that 71 other E-Corp buildings have gone up in flames, causing thousands of casualties. It is hard not to see *Mr. Robot*'s parallels to the shifting power dynamics and intricate political games of visible and invisible actors participating in the networks of world politics.

## Psychic Realities

While computer technology, hacktivism, and contemporary global power dynamics seamlessly blend into the fictional world of *Mr. Robot*, another level of reality seeps through the screen. Besides technological accuracy and political truths, the Neuropolis' more subjective mental psychologies are explicitly addressed in the show: Elliot suffers from social anxiety disorder, clinical depression, and dissociative identity disorder. Because we experience *Mr. Robot*'s world through Elliot's mindscape, it's worthwhile to highlight this aspect of the series.

From the opening, it's clear that Elliot is extremely isolated. He declines a birthday party invitation, preferring instead to spend time on his computer, and

---

[19] The Flint water pollution scandals are also addressed by Michael Moore in his documentary *Fahrenheit 11/9* (2018).

he takes care not to share with his hosts his own "source code"; a metaphor for his innermost self (S1E3; S1E7). Elliot's only mode of connecting to others is through hacking them. He represses his pain and sadness with morphine and Suboxone, which his next-door neighbor Shayla (Frankie Shaw) and soon-to-be-murdered girlfriend supplies. Elliot's depression appears clinical. Though his psychiatrist Krista tries to intervene, she too fails at a connection. Instead, Elliot hacks her, uncovering her own loneliness. Yet for Elliot, their shared sense of disconnection allows for some trust. Throughout the show, Krista remains an anchor point for Elliot, eventually becoming the only one who can unlock his "source code."

As the show progresses, we eventually discover Mr. Robot isn't Elliot's imaginary friend, but rather a symptom of his disassociated, multiple personality. At this point, *Mr. Robot* references David Fincher's *Fight Club* (1999), in which Edward Norton and Brad Pitt comprise a split personality; too, a Pixies song, "Where is my Mind" (*Surfer Rosa*, 1988; S1E9), also plays.[20] Like *Mr. Robot*, *Fight Club* also takes us into its characters' experience of the distorted brain-world-city of global capitalism, a world of intense and doubt-inducing perceptions and experiences.[21] Both films draw us into a neurological and psychopathological space.

In *Mr. Robot*, Elliot's consciousness slips constantly—his memory, especially of childhood, is blurred. He fails to recognize his sister Darlene or Mr. Robot as imaginary friend, or father. Throughout the first season, Elliot attempts to account for this slippage. After jumping from a railing in a state of delusion, Elliot finds himself in a hospital, connected to health monitors while a fly crawls over the life support machines. We hear Elliot's inner voice connecting the bug to the computers: "Most coders think debugging software is about fixing a mistake, but that's bullshit. It is actually all about finding the bug, about understanding why the bug was there to begin with; about knowing that its existence was no accident. It came to you to deliver a message, like an unconscious bubble floating to the surface, popping with a revelation you've secretly known all along." (S1E3) Elliot speaks here not just about hacking and coding, but

---

[20] Other films referenced in Mr. Robot include *Back to the Future* (1985), *Pulp Fiction* (1994), *The Shawshank Redemption* (1995), *The Shining* (1975) and *American Psycho* (2000). For a longer list see Emily Manuel, "Every Retro Movie Reference in Mr. Robot," *Screen Rant*, September 21, 2016, https://screenrant.com/every-movie-reference-in-mr-robot-amc/ (accessed January 13, 2020).
[21] Pisters, *The Neuro-Image*, 14–16.

also about the mind and the brain. And while there are many different layers to unpack about what *Mr. Robot* has to say about urban brains and psychopathologies of high-tech global capitalism, I find that one trope, which culminating in *Mr. Robot*'s final episode, ("407 Proxy Authentication Required" (S4E7)), is especially significant.

Contrary to other episodes, the form of episode 407 follows a classic five-act structure derived from Greek and, later, Shakespearian drama: exposition, rising action, climax, falling action, and *dénouement*. The action takes place entirely in Krista's apartment and is set up like a theater play, with the stage divided into two halves: the office and the adjoining kitchen/living space (the camera reveals these adjoining spaces sometimes from a perspective above the set's open ceilings). Krista is held hostage in her office by Fernando Vera (Elliot Villar), a drug dealer and gang leader who supplied Shayla, Elliot's neighbor, and girlfriend, with drugs for Elliot. In season one, Elliot exposed Vera's criminal business by decoding his disguised Twitter transactions. Against her ethics, while bound and threatened by Vera, Krista shares that Mr. Robot is the key to understanding Elliot. And yet Vera, a hardened criminal, reveals himself, surprisingly, as capable of deep psychological insights.

There's much backstory here, but most important is that when Vera threatens to kill Krista, Elliot confesses that he needs her. Vera then forces Krista to start a therapy session in his presence, which is performed in Act Four. Under the pressure of Vera's sharp, incisive observations, within this incredibly subtle session, we enter the darkest corners of Elliot's mind, in which he has buried past memories and traumas too painful to recall. Mr. Robot is an invention of Elliot to protect him from his real father, who, as Krista's sensitive questioning reveals, sexually molested him as a child. This revelation makes Elliot break down. In Act Five, Vera consoles Elliot by telling him that he recognizes his pain (as a child he has been "passed round" by his mum so she could get high) and tells Elliot that now that he knows the truth, he can use it: self-hatred is empowering, it fuels you with power. "You are the storm," Vera tells Elliot, who screams out a city window while thunder and rain echo his anguish. Krista, taking advantage of Vera's distraction, thrusts a knife into his back. Vera dies, and Krista leads Elliot out of the building.

Episodes 6 and 7 of the last season are followed by helpline announcements: "If you or someone you know needs help finding crisis resources, visit the National Suicide Prevention Lifeline (E6) or the National Domestic Violence Hotline (E7)". Late capitalism's psychopathologies, also known as cognitive capitalism,[22] do not escape basic human psychology: childhood or family trauma still inflict harm. But personal history also intersects with the media machines; together, they play specific roles in psychic damage and damage control. In the next section, I'll draw on other scientific perspectives to delve deeper into *Mr. Robot*'s realism and to explore how contemporary neuroscience has influenced philosophy, media studies, and psychiatry.

## The New Wounded: Healing the Fragmented Self in Scientific Discourse

Like many other characters in contemporary media culture, Elliot seems to be suffering from a mental condition that, according to new neurological insights, is described as a brain disorder. Philosopher Catherine Malabou has proposed the term "new wounded" to indicate biology's connection to mental disorders: "The new wounded, people with brain lesions, have replaced the possessed or the madman of ancient medicine and the neurotic of psychoanalysis. The specter of such phenomenon hints at a post-traumatic condition that reigns everywhere today and demands to be thought."[23] Malabou develops three hypotheses from her observations: the Freudian repressed libidinal energy model of neurosis needs to be replaced by a cerebral model of neuropsychopathology; a theory of trauma must extend beyond war-inflicted PTSD to include a broader range of the brain's mind's wounds; and a destructive brain plasticity can deconstitute identity.[24] After her grandmother's contracting of Alzheimer's, Malabou came to these new accounts of psychic make-up. Traditional methods of psychoanalysis (or philosophy) couldn't fully account for what was happening, nor offer solutions or consolation. For Malabou, all are the new wounded: "victims of various cerebral lesions or attacks, head trauma, tumors, encephalitis (Parkinson, Alzheimer). Those with schizophrenia, autism, epilepsy, Tourette syndrome, obsessive-compulsive disturbances, hyperactivity syndrome with attention deficit, or those resistant to psychoanalysis."[25] Common to all these diseases is a

---

[22] Arne De Boever and Warren Neidich, eds *The Psychopathologies of Cognitive Capitalism: Part One* (Berlin: Archive Books, 2013).
[23] Catherine Malabou, *The New Wounded: From Neurosis to Brain Damage* (New York: Fordham University Press, 2012), 17.
[24] Ibid., xix.
[25] Ibid., 10.

brain shift instigated by violence—a trauma that impacts neuronal organization and leads to an emotional deficit (disaffection, panic, withdrawal) and altered personality. The Neuropolis' cinematic characters are also often "newly wounded."

Malabou invites philosophers to rethink trauma as a physical, neurological wound and to reconsider the brain's plasticity in response to psychological distress. She hypothesizes that selves change under pressure. Ian Hackling, in his book *Rewriting the Soul*, proposes a contrasting, scientifically-informed philosophical approach to multiple personality as paradigmatic for the ways in which memory and personality are formed.[26] He begins by referencing the DSM-III (1980) which defines multiple personality disorder (now referred to as "dissociated personality") as the existence of two or more distinct personalities within one individual, each of which can at times gain dominance. These distinct personalities determine the individual's behavior patterns and social relationships. Although Hacking's models were proposed two decades earlier than Malabou's, they still resonate with hers on the new wounded. According to Hacking, dissociative personality traits lie in a traumatic event, a wound to the spirit that is often combined with severe depression and suicidal thoughts. Often triggered through child abuse, dissociation functions as a defense mechanism against hidden memories and psychic pain. A literal rewriting of the soul then transpires through the invention of another personality. In *Mr. Robot*, Elliot's self-internalization of his deceased father fits Hacking's description.

Whereas Hacking centers his discussion on multiple personality in the sciences and ideologies of memory that have long traditionally been linked to these pyschopathological conditions, Janina Fisher, in *Healing the Fragmented Selves of Trauma Survivors*, looks at the neurobiological legacy of redefining trauma.[27] Her argument, based on cases in her own clinical practice, gives a neurobiological explanation for dissociative splitting. Fisher claims this process is a normal adaptation to trauma, and she refers to the Structural Disassociation theory for survival responses. Splitting the self is a solution for survival. Like Malabou and Hacking, Fisher defines trauma broadly and applies it to a spectrum of mental disbalances, from PTSD, bipolar disorder, borderline personality, schizophrenia, to ADHD. Though we may not wish to clinically diagnose Elliot—a fictional character—he does appear to pay a price for his split reactions

---

[26] Ian Hacking, *Rewriting the Soul: Multiple Personality and the Sciences of Memory* (Princeton and New Jersey: Princeton University Press, 1995).
[27] Janina Fisher, *Healing the Fragmented Selves of Trauma Survivors: Overcoming Internal Self-Alienation* (Oxfordshire: Routledge, 2017).

to trauma. While self-alienation (and consequently the alienation of others and the world) might be a defense mechanism, it doesn't protect him from suffering. As Fisher explains: "Over time, self-alienation can only be maintained by most individual[s] at [great] cost."[28]

Fisher claims that, drawing on neurobiological research and a split between the left and right hemisphere[29]—trauma can reduce communication between halves of the brain, leaving clients with "two brains" rather than one that's integrated.[30] Deploying the term "encoding," Fisher argues that the traumatized part can "hijack"[31] or "hack" the normal functioning part and disremember; it can "'disown the bad (victim) child' to whom it happened as 'not me.'"[32] The consequence of such disconnection is depression and loneliness. Fisher suggests concrete measures for attachment repair. So, while these therapies may help alleviate a character like Elliot's sense of lost connections, they need to be placed alongside an understanding of the ways urban life in the digital age of global capitalism reinforces illness.

## Feeling Mediated

Let's now briefly return to the surrounding media ecology, where every connection seems to be mediated and therefore also seems to embody a possible cut between the self and the world. Hacking already observed that multiple personalities seemed to explode after the invention of film and intensified with the television remote control.[33] Media, rather than offering simply a representation of reality, can assume the eerie power of messing with the transmission of reality, upsetting the realms of the private and the public and hence influencing how inner and outer worlds fail to or can connect. Jeffrey Sconce and John Durham Peters, for instance, show that many schizophrenic people claim the radio or television broadcast private thoughts, and mass media is addressing them personally.[34]

---

[28] Ibid., 5.
[29] Michael Gazzaniga, *Tales from Both Sides of the Brain: A Life in Neuroscience* (New York: Harper Collins, 2015).
[30] Fisher, *Healing the Fragmented Selves*, 23.
[31] Ibid., 27.
[32] Ibid., 19.
[33] Hacking, *Rewriting the Soul*, 31–2.
[34] Both Peters and Sconce refer to Victor Tausk's seminal study "On the Origin of the 'Influencing Machine' in Schizophrenia," *The Psychoanalytic Quarterly* 2, (1933): 519–56; John Durham Peters, "Broadcasting and Schizophrenia," *Media, Culture and Society* 32, no. 1 (2010): 123–40; Sconce, *The Technical Delusion*.

Employing a media-archeological approach, Brenton Malin, in *Feeling Mediated*, explores how different media entwine with people's complex thoughts and feelings about machines.[35] Since the nineteenth century and the birth of modern transmission technologies, people have often treated machines like real people and places. Such engagement isn't always pathological, of course; photography, film, television, computers, cell phones, and AI have certainly become integrated into our neurological, psychological, socioeconomic and political worlds. Malin discusses this directly. In Stanley Kubrick's *2001 A Space Odyssey* (1968), the computer HAL sings "Daisy Bell" as he's extinguished. Spike Jonze's *Her* (2013) more recently describes our intimate relation to our computer voiceovers. The brain in the digital age, according to Malin, is quite literally hooked to our computer screens. Positively, these can serve as useful prostheses. But as Tiziana Terranova notes, interactions with contemporary information and communication technologies can in fact change neural structures, which in some cases can culminate in attention deficit disorders and anhedonia. Terranova stresses the costs paid by the psyche when it engages with digital capitalism (device-driven distraction and dissatisfaction) leading to something like Facebook chat as a "Bermuda time zone: [...] a dense curtain of fog and oblivion, jeopardizing deadlines, exams, health ...".[36] With the Neuropolis, in sum, the normal and the pathological are not so much opposed as *pre*supposed by one another. While the pain of real, personal trauma is always singular and difficult to express, a show like *Mr. Robot* conveys the psychic and physical difficulties of living in a Neuropolis in ways that may resonate for us all.

## The Freedom Tower in Neuropolis

I'll close now by returning to the Neuropolis in *Mr. Robot*. At season one's end, after Fsociety's first successful hacks on E-Corp, Elliot finds himself among protesting crowds in the streets of New York—a context inspired by the Occupy movement and the Arab Spring revolt. He appears afraid when, in the company

---

[35] Brenton J. Malin, *Feeling Mediated: A History of Media Technology and Emotion in America* (New York: New York University Press, 2014).
[36] Tiziana Terranova, "Ordinary Psychopathologies of Cognitive Capitalism," in *The Psychopathologies of Cognitive Capitalism: Part One*, eds. Arne de Boever and Warren Neidich (Berlin: Archive Books, 2013), 56; see also Jonathan Crary, *24/7: Late Capitalism and the Ends of Sleep* (London and New York: Verso, 2013).

of his younger self and mother, Mr. Robot appears. Elliot shouts for them to leave, that they're not real. In response, Mr. Robot, surrounded by Times Square's sky-high screens and masked Fsociety demonstrators, tells Elliot the entire world isn't real: it's built on a fantasy supported by medically induced emotions, by advertisements as psychological warfare, by media-generated brainwashing seminars, and by other technologies of delusion. "You have to dig pretty deep to find anything real, kiddo," Mr. Robot warns Elliot. When Elliot asks to be alone, his family members suddenly appear on an enormous screen and tell him they are inside him and will not leave. Elliot takes their advice to return home. (Alabama Shakes' song, "Sound and Color" [2015], here emphasizes the mediated nature of Elliot's mind, consisting of "sound and color").

The show's last episode closes with a final, emblematic image of the digital Neuropolis. New York's post-9/11 Freedom Tower looms large throughout *Mr. Robot*, appearing in almost all shots where the city envelops its characters, who linger, again, at the frame's bottom edges. After preventing a nuclear disaster that kills White Rose and wounds Elliot, Elliot is in hospital, drifting deliriously in and out of consciousness. Only Darlene's voice calls him back to reality. In these final moments, Elliot meets another double of himself (one who chooses the safe life and marries Angela). A twin version of Krista now explains to him that the personalities he has created to protect him from his childhood traumas are not only Mr. Robot and his vindictive mother, but also Elliot himself as a vigilante hacker. Darlene (who was kept out of the imaginary family) tells him that while she has not been dealing with the real Elliot, Fsociety's hacks and other events have been real. Whiterose, Angela, and Shayla really have died.

We then drift again into Elliot's mind: he stands before a skyscraper window, looking out onto the Freedom tower nestled in the Manhattan skyline, and he sees a younger Elliot, his mother, and Mr. Robot. One by one, they leave, and finally Elliot himself opens another door onto a cinema theater, where he takes a place among his other personalities. Watching the screen, we see only the projection lamp as it transforms into a tunnel along which the camera backtracks. Visually, we exit, finally, through a black hole, revealed to be Elliot's pupil (a reference here to *Fight Club*).[37] Through Elliot's watery eyesight, we see Darlene's surprised face, asking "Hello Elliot?" The city, the screen, the brain—they converge in this moment of revelation and interiority, in this exploration of

---

[37] *Fight Club*'s (1999) opening sequence presents a neuro-image where we move directly into characters' brain spaces.

Elliot's disconnections from the world, other people, and himself. But the film's ending remains ambiguous. We don't know whether Elliot's mind will be able to rewire itself in the context of the digital world. Will the city help him reconstitute reality in a healthy way? In a way that will allow him to reconnect to others and to the facts of the physical world? Or will his alter egos and disassociated selves remain protective responses to the realities of the twenty-first Neuropolis?

# Bibliography

Crary, Jonathan. *24/7: Late Capitalism and the Ends of Sleep*. London and New York: Verso, 2013.

De Boever, Arne, and Warren Neidich, eds. *The Psychopathologies of Cognitive Capitalism: Part One*. Berlin: Archive Books, 2013.

Deleuze, Gilles. *Cinema 2: The Time-Image*. Translated by Hugh Tomlinson and Robert Galeta. London: The Athlone Press, 1989.

Draaisma, Douwe. *Metaphors of the Memory: A History of Ideas about the Mind*. Translated by Paul Vincent. Cambridge: Cambridge University Press, 2000.

Esmail, Sam, dir. *Mr. Robot*. Prime Video Aanbod. https://primevideoaanbod.nl/amazon-prime-video-aanbod/mr-robot/. (accessed December 27, 2020).

Fisher, Janina. *Healing the Fragmented Selves of Trauma Survivors: Overcoming Internal Self-Alienation*. Oxfordshire: Routledge, 2017.

Fitzgerald, Des, Nikolas Rose and Ilina Singh. "Living Well in the Neuropolis." *The Sociological Review Monographs* 64, no.1 (April 2016): 221–37.

Franz, Julia. "How Realistic are the Hacks in 'Mr. Robot'?" *The World*. https://www.pri.org/stories/how-realistic-are-hacks-mr-robot. (accessed January 11, 2020).

García, Alberto. "A Storytelling Machine: The Complexity and Revolution of Narrative Television." *Between* 6, no. 11 (2016): 1–25.

Gartman, Lev. *Mr. Robot Dec0d3d.doc*. https://vimeo.com/176633054. (accessed December 27, 2020).

Gazzaniga, Michael. *Tales from Both Sides of the Brain: A Life of Neuroscience*. New York: Harper Collins, 2015.

Grubbs, Jefferson. "This 'Mr. Robot' Theory Suggests Elliot Is Basically A Living Computer." *Bustle,* December 8, 2019. https://www.bustle.com/p/this-mr-robot-season-4-theory-suggests-elliot-is-basically-a-living-computer-19428425 (accessed December 27, 2020).

Hacking, Ian. *Rewriting the Soul: Multiple Personality and the Sciences of Memory*. Princeton and New Jersey: Princeton University Press, 1995.

Hari, Johann. *Lost Connections: Uncovering the Real Causes of Depression and the Unexpected Solutions*. New York: Bloomsbury, 2018.

Hernandez, Danny. "10 Things Mr. Robot Gets Rights About Hacking." *Screen Rant*, February 1, 2019. https://screenrant.com/things-mr-robot-gets-right-hacking/. (accessed January 11, 2020).

Malabou, Catherine. *The New Wounded: From Neurosis to Brain Damage*. Translated by Steven Miller. New York: Fordham University Press, 2012.

Malin, Brenton J. *Feeling Mediated: A History of Media Technology and Emotion in America*. New York: New York University, 2014.

Manuel, Emily. "Every Retro Movie Reference in Mr. Robot." *Screen Rant*, September 21, 2016. https://screenrant.com/every-movie-reference-in-mr-robot-amc/. (accessed January 13, 2020).

Nachreiner, Corey. "'Mr. Robot' Rewind: Rewinding the '5/9' hack in a Stunning Season Finale." *Geek Wire*, December 18, 2017. https://www.geekwire.com/2017/mr-robot-rewind-rewinding-5-9-hack-stunning-season-finale/. (accessed January 11, 2020).

Neidich, Warren, ed. *The Psychopathologies of Cognitive Capitalism, Part Two*. Berlin: Archive Books, 2013.

Peters, John Durham. "Broadcasting and Schizophrenia." *Media, Culture and Society* 32, no.1 (2010): 123–40.

Pisters, Patricia. *The Neuro-Image: A Deleuzian Film-Philosophy of Digital Screen Culture*. Stanford: Stanford University Press, 2012.

Plaugic, Lizzie. "A QR Code in Last Night's Mr. Robot Leads to a Mysterious Website." *The Verge*, July 14, 2016. https://www.theverge.com/2016/7/14/12187768/qr-code-mr-robot-confictura-industries-usa. (accessed January 11, 2020).

Quick Meme. "What if Error 404 is a Message from the Real World." *Quick Meme.com*, http://www.quickmeme.com/meme/35o13a. (accessed January 11, 2020).

Ravetto-Biagioli, Kriss. "Anonymous: Social as Political." *Leonardo Electronic Almanac* 19, no. 4 (2013): 178–95.

Sconce, Jeffrey. *The Technical Delusion: Electronics, Power, Insanity*. Durham and London: Duke University Press, 2019.

Smith, Anthony. "Pursuing 'Generation Snowflake': Mr. Robot and the USA Network's Mission for Millenials." *Television & New Media* 20, no. 5 (2018): 1–17.

Tausk, Victor. "On the Origins of the Influencing Machine in Schizophrenia." *Psychoanalytic Quarterly* 2 (1919): 519–56.

Terranova, Tiziana "Ordinary Psychopathologies of Cognitive Capitalism." In *The Psychopathologies of Cognitive Capitalism: Part One*, edited by Arne De Boever and Warren Neidich, 45–68. Berlin: Archive Books, 2013.

Von Debschitz, Uta and Thilo Von Debschitz. *Fritz Kahn: Infographics Pioneer*. New York and London: Routledge, 2017.

# The Dopamine Circuits of Wanting, Liking, Habit and Goals

## An Interview about *Mr. Robot* with Neuroscientist Talia Lerner

Jonathan Leal, Carol Vernallis, and Patricia Pisters

### Introduction

At her laboratory at Northwestern University, neuroscientist Talia Lerner studies motivation, reward learning, and decision-making; in particular, she focuses on how dopamine circuits regulate reward learning and contribute to the risk for mental disorders. On October 22, 2020, she Zoomed with media scholars Carol Vernallis, Patricia Pisters, and Jonathan Leal to discuss the cyber-thriller television series *Mr. Robot* (2015–2019). In their conversation, lightly condensed below, the group reads *Mr. Robot*'s engagements with addiction and mental disorders and relates them to contemporary research in neuroscience.

**Talia** I'm excited to meet you all. I'm an assistant professor at Northwestern, and my lab studies dopamine and reward learning. We're interested in what reward learning signals do—how they control behavior and facilitate transitions from normal to compulsive or habitual behavior. Examples include when people repeat actions, like in OCD or through drug addictions. What predisposes people genetically or through life experiences to become addicted or suffer neuropsychiatric disorders?

These circuits drive normal behavior in our everyday lives, as well—how we learn to play sports or play the piano. Our days are programmed with lots of different routines. I'm excited to talk to you guys.

**Patricia Pisters**  I'm in Amsterdam, so I'm in a different time zone. I teach in the Media Studies department at the University of Amsterdam, and I've written a book called *The Neuro-Image* on contemporary cinema and its connections to neuroscience. I'm excited to think with all of you about my chapter on *Mr. Robot*.

**Jonathan Leal**  I'm a Postdoctoral Fellow in the Society of Fellows in the Humanities at USC, and I'll be joining the English department soon. I write about music, race, and narrative across media, and I'm currently working on projects that engage the utopian horizons of underground and avant-garde music scenes.

**Carol Vernallis**  And I work on contemporary audiovisual aesthetics!

**Talia**  I liked the questions you sent me because they make me think about how some of these more abstract concepts in the research relate to everyday life. "How can I tell if I'm in a habit or goal-directed activity" is interesting. I don't think it's a complete split. You have both brain systems devoted to the two different tasks working at the same time, so it's not necessarily black and white, but I do think a lot about why brain systems exist. They're not just there to give us neuropsychiatric disorders: you use habit or even punishment-resistant reward seeking for beneficial purposes in your everyday life. Habits, by helping you automate tasks, save cognitive resources for other more creative activities. Often, athletes who repetitively practice routines until they're perfect, talk about being in the zone. I think that's a good, intuitive way to understand what it feels like to be in a habitual sequence, everything flows without thinking, you know? When athletes choke, suddenly their anxiety provokes them to rely on goal-directed systems, and suddenly they're consciously thinking about every behavior, and then the activity falls apart, because it's really their habit system that knows what to do.

And while punishment-resistant reward-seeking behavior sounds bad—we're often talking about it in the context of drug addiction, where your compulsion for the next hit causes you to lose your family—there's lots of things where persisting in the face of negative consequences is considered virtuous. We talk about grit and resilience all the time. When you're pursuing academic studies, for example, you need to keep going even when your paper's rejected, or you're told your ideas are stupid. There's lots of reasons why you'd want to chase more than your most recent reward. Thinking this way helps us consider why we have these brain systems in the first place.

**Carol**  So, thinking about your publications. It seems to me that habit doesn't have to be tied to just motor activity. Before I go to sleep, I regularly loosely rehearse what I've been writing, and when I wake up, the prose is usually better. And I can't remember an acronym to save my life. So I'm going, "VTA, DML, DFL, PFC!" (brain modules). As I was getting ready to speak with you. That's habit, yes?

**Talia**  Yeah, again, our knowledge comes from animal systems, and therefore is somewhat limited. It's easy to read out motor behaviors in a mouse or rat, but you can't ask them what their internal thoughts are, or even if they have them. But I think in humans, at least we can hypothesize that a lot of these functions do come into play, that there's habit systems for patterns of thought. Disorders like depression, which seem to involve the dopamine system, seem to involve certain habitual patterns of thought that people find hard to break out of. I think these are analogous to learned, habitual motor patterns.

**Carol**  You caught something that interests me. When you're an athlete in the zone, or an average person acting out a habitual loop, an affect goes with it. I think a show like *Mr. Robot* can do a lot for us. The actors and the music are so fine grained and specific, it's easy for us to identify and sympathize with when Elliot's in a goal-driven, focused, present state, or inattentive to his body and surroundings and just drifting. We can learn to identify these, right?

**Talia**  Yeah, I think the montage-y sequences of Elliot typing and coding, where he's hacking someone he knew he just met, puts him in an automatic routine. Here's the first thing: I try to crack their password. It's very rote for him. He's not even thinking if he wants to know this about the person. He does it for everyone as a kind of habit.

**Carol**  Do you have an image of the brain where you can show readers where this might play out in the brain, for example, perhaps in the striatum?

**Talia**  I can bring it up quickly, I guess, if I just Google Image search.
There's a lot of different interconnected brain systems that work to create affect. It's important to note that dopamine does not just equal pleasure. Scientists thought this a while ago, and the notion's persisted in popular culture. But the literature in behavioral neuroscience and psychology distinguishes between

wanting and liking, which sounds like it should be similar, right? I want things I like; I like things that I want, but they're not the same because liking is about hedonic pleasure (e.g., enjoying what you're doing in the moment) and wanting is craving something and being willing to work for it. You can actually see, if you block dopamine receptors or study Parkinson's patients who lack dopamine, that people can enjoy things without it, though the effort you'll expend for the pleasurable experience is lower. Rats that have depleted dopamine will display facial expressions of pleasure if you give them sugar right in their mouth, but they won't work to get that sugar. So, the brain systems mediate different things, there's a disconnect. And we think dopamine is really connected to learning, for example, which things predict reward, but not the feeling of reward you'll get with the desired experience.

**Carol**   Would some of this would take place in the hippocampus, and some in the striatum?

**Talia**   So, the hippocampus is more involved in spatial learning. If you're learning about a space and you're trying to memorize a map with cues that are outside yourself, then you'll use your hippocampus to create a map of where you are. If you're using your striatum, it's more about a self-centered-type of navigation, where you're remembering what you're doing relative to yourself, like do I turn left or do I turn right? A more hippocampal way of navigating draws on external landmarks like do I turn toward the greenhouse? Some people say the hippocampus realizes episodic or semantic memory, where you put into words what you're trying to remember. The striatum is more devoted to implicit associations you're making that you're less consciously aware of. These are things you learn through subtle reinforcement, including probabilistic relationships, intuition, which you won't necessarily be able to articulate.

**Talia**   I can try to show a quick image. On Google, there are millions of them. This one's an example of the human brain on the top and the mouse brain on the bottom. They're sliced coronally. [like slices from a loaf of bread]. The striatum appears more prominent in the mouse because the human has a huge expansion of the cortex. In the mouse, the cortex is smooth, just a strip on top of the striatum. In the human, all these folds make up the expanded cortex that still fits in our head. But still, the cells' anatomy, the way they're connected and other stuff, is basically pretty similar. The dopamine inputs are pretty analogous as well. This image shows the dopamine cells going to the striatum. They're located farther back in the mid-brain, and then they project up to the forebrain through a similar path.

**Carol**   Wow, cool. Your articles seem to suggest that regular, repetitive rewards, like when the mice get sugar every twenty seconds—leads to habitual punishment-resistant behavior. There's a "ding, ding, ding," and then you start shocking them, and they won't stop, right?

**Talia**   So, the shock is tied to how we scientists talk about compulsion. Compulsion is partly defined by an insensitivity to punishment. Neuroscientists argue about why it happens and whether it's tied to habit or not. There are two opposing possibilities. The reason you persist even while you're punished is due to habit or goal-directed behavior. The habit possibility would be, you're just so into a habit, you're not thinking about what you're doing anymore, so the punishment doesn't really register, you're having trouble connecting the pain to your actions. Or instead, you so want to get the pleasurable thing so much you're willing to tolerate the pain because you're so focused on the future reward. We think we can actually see, in our experiments, different processes happening in different mice when we record their neural circuits. Depending on whether we stimulate the mouse's DMS (the dorsomedial striatum that helps control goal-directed actions) or DLS (the dorsolateral striatum that is involved in habit among other things), we can identify punishment-resistant behaviors for these different reasons. I think this will help us understand drug addiction, because we may discover that drug addicts are driven because of different causes. If you can subtype these different underlying neural circuits, you might be able to offer more targeted treatment.

**Carol**   So, we could screen *Mr. Robot*'s scene where Elliot's trying to score a hit, but he doesn't know the environment very well. He hasn't seen this apartment building before, but still, he must have sort of a mental map of how it's laid out. I'm curious how that plays out in the hippocampus and the striatum. I really like this scene, because the music describes the lurch back but still being drawn toward the drugs. It's almost as if he's a rat and his little foot is being shocked. It feels painful and awkward. The awkwardness is underscored by the seller's careening back and forth. Thoughts?

**Talia**  Yeah, it's an interesting example of goal-directed and habit systems. I would argue this is a more goal-directed pursuit, because Elliot's not following a rote pattern for drugs. It's the opposite of that. I think it's a dream sequence anyway, but it's going to a new place, taking a new form of opioid, but it shows you the power of that craving, that wanting, to drive actually at goal-directed behavior.

**Carol**  But it also feels like habit, yes? Leaning up against Mr. Robot, Elliot has his head down, which suggests, a kind of a tunnel vision, as if he's not taking in any external stimulus. And when he's going up the steps, he's dragging his foot as if it's been hurt. He stands still for a minute, and the music keeps going something like "wurh, wurh," as if he's heading up those steps not by choice, but as if someone were pushing him that way.

**Talia**  Yeah, it's more of an inner plan. And I think that's something in particular in drug addiction that makes that goal so salient that you can't really do anything else, you know? So the music helps play into that kind of feeling of it's this tunnel that he has to go down, kind of.

**Patricia**  Can I ask about something more general? As a neurobiologist, how do you see a series like *Mr. Robot*? Is it realistic, useful, or beyond the point? To me, it seems like all of these aesthetic techniques bring you completely into his mind—it's not just the music or his focus. For the whole series, we're in Elliot's voice, in his mind. I take this as an aesthetic expression of how it must feel to be in a mind that has been traumatized, that has been marked by so many stimuli, including screens and media machines. I find it very powerful.

**Talia**  Yeah, I think the show's done a better job than anything else I've seen of putting you in the perspective and mindset of a person who's having serious problems. You're involved even as the narrator, as one of Elliot's personalities. This is realistically disorienting. It doesn't need the "Oh, it was a dream," or some psychedelic background. Instead, it gives you the feeling of how a really crazy behavior is justified in the mind of the person going through those experiences. Though honestly, why did none of the real people around him try to institutionalize him?

**Patricia**  Most of them are crazy too!

**Talia**   Yeah. How could his sister and his friend know that he has this dissociative personality disorder and drug addiction problem and not do something, other than be like "I'd like to talk to the other Elliot now"?

**Carol**   So, I found another addiction moment, and it's nice, because the series' aesthetic techniques build over time. We talked about that tunnel, but we see the tunnel more clearly here, because the strip of drugs is already set up as a tunnel.

**Talia**   Yeah, it has that same very visual tunnel when it comes to him going for the drug.

**Carol**   But aren't many processes going on simultaneously? The dopamine pathway must be operating because he seems to have want and drive. But then he's got depression, so he's got anhedonia. He can't act. This seems simultaneous, right?

**Talia**   Yeah, people who have addiction tend to have lower expressions of dopamine receptors, which maybe also can be attributed to a depression-type state. You try to compensate for the low dopamine by stimulating the system in some other way. Elliot's going through withdrawal symptoms, so there's a big drive to get the reward of relieving withdrawal symptoms. That, too, drives you toward that behavior. What the dopamine trains you to do though is—it trains you about, for example, your past actions tell you about what will lead you to certain rewards. That's the distinction between liking and wanting. Your past experience with this drug, the dopamine had solidified into learning, that taking this drug will do something to relieve your withdrawal symptoms, will do something to relieve your depressed mood, and so you want it. But the pleasurable feeling of actually taking the drug, those are separate opioid and serotonin systems. And you can tell he has some apprehension about doing the drugs, right? He tried to make rules for when he is and isn't allowed to have them, but he's breaking those rules again because of that overwhelming feeling of craving.

**Jonathan**   I have a question that picks up on Patricia's. As a neuroscientist, how would you think about the relationship between filmic representations of addiction and the experiences of real people suffering? It's so interesting, for instance, that we're able to converse in such detail about Elliot's drives and habit formations within the film discourse.

**Talia**  Yeah, I guess with film depictions, you're always trying to translate what people are going through in their daily lives into something that creates a dramatic narrative sure to grip viewers' attention. We don't really have these soundtracks playing in our everyday lives, but I think *Mr. Robot* does do a good job through techniques like music and lighting to convey, maybe more coherently, what a person might go through. A person on their last drug line might justify it so eloquently, though they're unconscious processes may be similar. "I won't do it again—I just have to do it this one time," kind of feeling. So, the show's doing a nice job of creating that feeling in a viewer who maybe hasn't been in that exact situation before.

**Jonathan**  That's super interesting. It reminds me of Carol's point about Wittgenstein—about this idea of private languages and the extent to which it's even possible for us to communicate certain types of singular, embodied experiences. Because of the multisensory tools that filmmakers have at their disposal, perhaps films offer more detailed ways to communicate affects than spoken language?

**Talia**  Yeah, I think it's really hard even for humans who do have language to access every part of themselves. The entire nature of habit is not really accessible to our conscious minds. It's easy to not realize when what you're doing is habit. Humans do seem to do a lot of post-hoc rationalization of their actions, but in reality, what drives them is a lot more unconscious processes, and it's interesting for a film to try to express that. Animal researchers struggle with this all the time. Humans can be unreliable reporters of their own inner processes, but animals just can't be reporters at all. We need them to give us some sort of behavioral output to measure, and they can't speak, they can't write, they can make facial expressions, they can show us they're willing to do something to get a reward. I try to make sure the behavior I'm studying in an animal is relevant to human behavior, but I can't be sure. People give themselves away with facial expressions, and mice and rats do make facial expressions. Can we line those up somehow? Can we translate emotion across these different media? It's important for communicating in terms of the arts, and it's an important scientific question too in terms of how we translate biomedical studies into human therapies.

**Carol**  I'm pleased to see that *Mr. Robot*'s large-scale form for the first season plays out in one pass Elliot's progress through learning about and becoming tuned to habits. I think these match the arcs of your experiments with rats. In season 1's first episodes, Elliot hovers trying to decide whether or not to continue with a pattern. He's at the door. Should he or should he not take the drugs and have sex with his dealer? When he's hacking the Dulles Data Farm, he hovers. Do I type F society and close them down, or not? Mid-season, there's running down the rat's T-maze, if I may say so—the long hallways of Steel Mountain, or spinning on Ferris wheels. Finally, near the season's close, he discovers the reward or punishment. He springs the pusher from jail to then discover his lover's, Shana's, dead body in the trunk of his car. He totally freaks out.

**Talia**  Yes, rats will make mistakes as they try to pursue rewards that seem human. This is anthropomorphizing, but they seem to not want to regret a poor choice. You can give them tasks where they can make bets, and where they go into a little chamber and find out, "I'll get a reward but with this delay," and they have to decide if the delay is worth the reward. You'd think they just go in and start their delay countdown timer and then make a decision, because then you've saved yourself some seconds, but they don't do that. They wait outside the door and try to decide if they'll go in at all at first. This seems to be analogous to human behavior—I don't want to be caught making the wrong decision, commit to it, and then give up. I want to decide and then stick with it. It's not the optimal way to solve the task.

**Carol**  Talia, it'd be nice if I can change topics for a second. I think the optogenetics is so exciting and it's so new, and most people don't know about it. I mean we're not just doing lever presses and putting electrodes in the brain. And what's CLARITY?

**Talia**  Yeah, optogenetics and CLARITY are distinct. Optogenetics are a nice way to control specific subsets of brain cells. Rather than just observing neural activity and correlating it with behavior, we can do these very careful, causal experiments to say, was it really this cell that was making you act this way? With optogenetics, we induce neurons to express a light sensitive protein, and then when we shine a light into the brain, we can turn neurons on or off. Light is a nice way to control brain cells, because it's fast—as we switch the light on and off, we have this very fine temporal control over brain activity. And we can use genetic tricks to get these light-sensitive proteins only in very specific cells that we want to study. Electricity affects all brain cells in a certain area, but with optogenetics, we can pick out a subset that are releasing dopamine and simultaneously not stimulate other neurotransmitters.

**Carol**   And you can do it in vivo? (Or in other words, as it's happening?) That's so great.

**Talia**   Yes. That's really nice, because then we can do it as the animals are awake, they're walking around, they're engaging in decision-making.

**Carol**   That seems like a new thing for these experiments. In some of your papers, you talk about the animal walking around before the experiment proper begins.

**Talia**   Yeah, and we can see, again, in terms of dopamine really being about learning patterns, we stimulate dopamine, for example, in these different striatal subregions. We can get different effects on behavior. We actually do the dopamine stimulation during training periods, but we actually test them without stimulation. So, the effects that we see are not due to the immediate turning on and off of dopamine. What they're due to is the effects of dopamine during the training that then helped the animal learn something affecting its current behavior. And I do think that's important, because dopamine is more about learning, and learning from past experience, than it necessarily is about controlling your immediate behavior.

**Carol**   You've done a lot of work on the striatum than most. I did look around at other people, that's special for you, Talia, right, to some extent?

**Talia**   I mean, there's always many people studying stuff, but yeah, all my career has been focused on the striatum and dopamine and looking at it from somewhat different perspectives, using different types of tools.

**Carol**   Can you talk about how CLARITY works?

**Talia**   Yeah, CLARITY is a technique. It's not done on living animals, because it requires fixing the tissue which kills the cells. It allows us to make the tissue optically clear, as if you can see through it. Normally if you take the brain out, it's opaque and tan colored, mostly because of the lipids. The CLARITY method fixes our desired proteins in a hydrogel matrix and gets rid of the surrounding lipids. We can then see all the way through the 3-D structure, without having to thinly slice the brain.

We can trace connections between cells that make really long paths through the brain, like even dopamine neurons. They're toward the back and bottom of your brain, and the striatum's toward the front top of your brain, so those axons have a really long way to go to get to their target. But if we have the 3-D structure, we can trace the axon all the way, including seeing where it's branching, where it's going, and where it's been. How many places did it go to communicate its signal at the same time?

**Jonathan**   That's super neat. That process of extracting lipids—I'm sure it's not the same thing at all, but it reminds me of the way technicians make aerogels out of silica.

**Talia**   Yeah, the chemist I worked for as a postdoc had worked for a company that made hair gels. He knew all about these hydrogel polymers, and then we were like, we could use this for brains! I think this is the value of research basically. You're never going to know how things end up connecting. Scientists may be more like artists than we imagine. We'll both inventively make use of what we can.

**Patricia, Jonathan, and Carol**   Thanks so much! It's been a wonderful conversation.

25

# The Taste of Cybermedia

## An Interview with Hojoon Lee, The Lee Lab at Northwestern University

Julia Peres Guimarães, Selmin Kara, and Carol Vernallis

### Introduction: Julia Peres Guimarães

The recent surge in the production of science fiction films and television series during the last few decades has signaled an increased interest in the unstable boundaries between humans and artificially intelligent creatures and aliens. Whereas considerable attention has been given to the human senses of vision, sound, and touch as they are transformed into data, fewer works have focused on the faculties of taste and smell. In this context, in the films *Ex Machina* and *Under the Skin*, there are critical scenes where a robot and an alien, respectively, experience reactions to food which elicit affective responses that complicate our understanding of non-human agents. In this light, we might ask: how could knowledge about the neurobiological processes involved in tasting help us rethink the differences between humans and bots/aliens? How might the research on the sensory experience of taste unsettle our pre-assumptions about cinematic representations of food? To consider these questions, among others, on September 14, 2020 we interviewed neurobiologist Dr. Hojoon Lee, the principal investigator of the Lee Lab at Northwestern University, via Zoom. We learned that the study of taste reveals that humans have hardwired and prototypical behavioral responses to basic taste qualities (attraction or aversion) which evoke innate affect (pleasure or disgust). Due to the complexity of this system, which is essential to the survival of humans and animals, scientists have investigated how taste information received by receptor cells in the tongue is transmitted through neuronal projections in the brain. Although Dr. Lee observed that this research area is still in relatively early stages, the conversation's insights might push the limits of how we conceive of the role of taste in science fiction film and television.

**Carol Vernallis**   Compared to vision, there is so little research and scholarship on taste. Why do you think that is?

**Hojoon Lee**   Even in the scientific field, taste is vastly understudied. Research on the neurobiology of taste is still largely centered around taste "sensation"—how the basic taste qualities are detected from the tongue, and we have little understanding of how taste "perception" is formed in the brain. Taste in itself is very basic: If you taste something, the brain registers the signal as one of five basic tastes: sweet, umami/savory (the taste of protein), sour, salty and bitter. The responses that we show to pure, basic taste are prototypical: we either like it or we reject it. For example, when you taste something bitter, you will spit it out; if you taste something sweet, you'll like it. These affects are innate and hardwired. There's no ambiguity to how we react. This should be distinguished from "flavor," which is the combination of olfaction and taste. When we consider solely the pure sense of taste, we can't really make associations with food.

**Carol**   But once we smell things, our experience becomes more complex, right?

**Hojoon**   Right. It's often said that smelling is a learned behavior. There's nothing innately bad or good with smell. Take strong cheese: depending on how you are conditioned to associate its smells, you might like it very much or not at all. This is why, when we talk about our memories of food, we are often conjuring olfaction in combination with basic tastes.

**Carol**   I've been trying to conjure up memories of tastes and smells, and I'm surprised at how poorly I do this. Sometimes I get a quick flash, but I can't sustain the memories.

**Hojoon**   Proust aside, try to imagine the smell of madeleines. There are reports regarding how remembering or conjuring up smells is very difficult. Some people say that it is not even possible.

**Carol**   I was surprised when I started reading on artificial intelligence that vision and sound can be transferred into something that is digital, but that is much harder to do with smells. There's a scene in *Ex Machina* (2014) where the robot is chopping raw fish. I think the audience experiences a visceral reaction here, but otherwise, sensory experiences linked to food seem distant.

**Hojoon** There are about 800 different types of olfactory receptors in the human nose. But we have yet to figure out what the vast majority of these receptors are detecting. Finding the exact chemical(s) that are sensed by each receptor, a scientific process called "de-orphaning," would be necessary to digitize chemical information for robots to detect smells from the environment. (In contrast, receptors that detect chemical for each basic taste quality have been discovered, although thirty bitter human taste receptors remain to be de-orphaned). However, for us humans in the audience, eating nutritious and fresh food is essential for survival, therefore we respond with very strong, hardwired behaviors to what we taste. A good example of this is the stereotypical facial expressions we show to taste. If we see someone biting into a lemon, and we see the person's facial expression, but not the fruit, we can detect that that person is experiencing a sour or bitter taste.

**Carol** Is there a different expression for bitter than for sour?

**Hojoon** I'm sure we can easily recognize someone's facial expression from eating something very sour, like a slice of lemon. This has been shown even in animals. I use mice in my research, and researchers used machine learning to analyze their facial expressions as they taste different things. These scientists say they can tell if the mice are tasting bitter sour, sweet, or sweet foods just by "looking" at their facial expressions.

**Carol** That is interesting because in "Nosedive" (*Black Mirror* season 3, episode 1, 2016), Lacie takes a bite of a cookie and drinks some cappuccino. She's supposed to be having a beautiful experience as she takes a picture of the items to share on Instagram, but they both taste bad. We could watch the scene right now and see if she's feigning her response. Conversely, the alien in *Under the Skin* 2013), played by Scarlett Johansson, gags on cake. Yet it is not clear whether or not one can guess, by looking at the facial expressions of an alien, if the cake's bitter, sweet, or sour.

**Hojoon** Of course, the experiences we have with specific food items can affect how we react to them. Researchers are taking small, but significant, steps toward understanding how taste is perceived in the brain. Because of the taste system's complexities, I've had to take a reductionist approach, and look at very defined questions. In my lab, we're currently focused on understanding that first step, on how taste is represented in the brain stem, because this is where the innate reflexive behaviors are encoded.

The sense of taste is organized around taste buds on our tongues. In each taste bud, you have about fifty cells and each one detects one taste quality. In other words, you have a cell that detects bitter and another that detects sweet. (These are distributed across the tongue. It was a mistranslation from German that our tongue has a "taste map," where we're more sensitive to sweet and salty tastes in the front of the tongue and bitter in its back). Cells have to make connections to the brain, and to get to the brain, taste information is passed onto neurons in our middle ear. By the time that input gets from the periphery of taste buds and neurons in the middle ear, onto the next station, which is the brain stem in the central nervous system, it's already too complicated for us to understand.

**Carol**   But much like vision, there must be a stage where you're turning the sensory input into a food you recognize or can categorize, and that involves a lot of pre-assumptions about previous experiences with food, right?

**Hojoon**   Yes, and probably by that point, there is already integration with the olfactory information. There must exist a "flavor center" in the brain that combines the two sensory pieces of information, and that makes associations and forms memories. Yet there is so little known about how taste and olfactory information is processed in the brain. We don't even know where the initial integration of smell and taste happens in the brain yet. There is still a lot of work to be done.

**Julia Guimarães**   There are ways in which sonic signals, and audiovisual signals produce certain illusions. Do we experience illusions for taste?

**Hojoon**   Besides olfaction, other sensory qualities, such as vision, sound and touch (texture) influence our thoughts on taste. This multimodal information is stored in our brains and illusions for taste could be conjured up by stimulating any one of the senses.

**Carol**  How about when we see representations of eating? If we watch someone eating on screen in cinema, how do we respond? I felt highly attuned to Lacie's eating a bad-tasting cookie in "Nosedive" and puzzling out, through her expression, what her experience was. I'm curious about the scene in *Under the Skin* where the alien, Scarlett Johansson, eats cake, because I'm human—I have no idea what it means for her.

But with Lacie's cooking in a kitchen in "Nosedive," I noticed she touched her hair. And I touched my hair, too (we talked about this with Jeff Zacks in another *Cybermedia* interview). In *Under the Skin*, when Johansson gags on the cake, I gagged, too. Perhaps I'm learning some aversive associations to cake, as well?

**Hojoon**  When you hear the film's soundtrack, which in this case might be dystopian or upsetting, it all gets encoded together with signals linked to smell and taste. I think that details such as *Under the Skin*'s sick-looking cherries on top of the cake and the alien's missing a stomach are important. Even though I said that taste is hardwired, perception can change our experience. There are two main layers to the responses to taste. One has to do with the reflexive behaviors, related to sensations like aversion versus attraction, appetitiveness versus disgust, which are inherently hardwired. These are hard to change. Anencephalic babies born without a big chunk of the forebrain or cortex, for example, show stereotypical behaviors to various tastes. When you give these babies sweet or sugar water, they will continue licking. If you give them something sour or bitter, they will start salivating or rejecting it. Most of these primitive responses to taste are performed in the absence of the cortical input, or the hippocampus, which process learning and memory. But then when these primitive responses gets fed in with other sensory information, the visual, the auditory and the olfactory input, then I think you can override reflexive reactions. That's why even though lemonade or coffee are inherently bitter or sour and aversive, we are able to override our initial aversion because our memories and associations help us interpret coffee and lemonade as good.

We can do this in mice, too. We can feed them sugar water and simultaneously inject drugs that gives them a stomachache. They'll stop drinking the sugar water. Like when we get food poisoning, they'll develop a dislike of a food if it makes them sick. We, too, have long memories for when food made us sick.

A lot of this is remembered below the level of consciousness. Our taste preferences also shift based on whether we're deprived of a certain nutrient. At low concentrations, people love saltiness—but no one's going to drink seawater, because over a certain threshold, we'll have an aversive response to the salination. However, if we were salt-deprived and our bodies drive us to seek out salt, we'd gladly drink that seawater.

**Julia**  If I can pivot to more pleasurable experiences: in cybermedia films like *Ex Machina,* we see sensual scenes in which robots are making sushi or spilling wine, and these seem charged with sexual tension. How much is food a part of the process of feeling or expressing sensuality and sexuality?

**Hojoon**  There's, again, not much known about it. In neuroscience, we don't even know the connection between how taste affects eating or feeding behaviors. The interesting aspect I think is that sex, reproduction, and eating are all central to our being. They are the most basic needs for our survival, yet, when you think about robots, these are the things they lack completely.

**Carol**  Has your relationship to food changed with your research?

**Hojoon**  I definitely got to appreciate it a lot more and to be more conscientious of "tastes." I also started to enjoy cooking, although my wife often complains that I try to be too precise when following recipes and cook like a scientist performing experiments.

**Julia**  There's a way in which the West privileges vision as our main sense, so we don't always appreciate our relations to other senses. Perhaps we'll have a deeper relationship to taste and smell now. Thank you very much for sharing your research and insights for our collection.

# Index

Page numbers in **bold** refer to figures, page numbers in *italic* refer to tables.

*2 Dope Queens* (TV show), 316, 317
*2001 A Space Odyssey* (film), 415

Absolute Infinity, 65
acoustic resonance, 157–8
acoustical information, 259
Adana, Kor, 406
Adorno, Theodor, 385–6
affect, 421–2
　economy of, 154–6
affect system, the, 181
affective economies
　in *Black Mirror* "Nosedive", 151–68, **152**, **162**
　gendered assumptions, 160–1
　neoliberalism, 153
　pursuit of likes, 151
　and racial difference, 161–4
　rating systems, 151–2, 153
　social regulation, 156–8
affective labor, 153, 155
Afghanistan, 277
African American identity, 285–6
Afrofuturism, 288–9, 321–2, 337
Afropessimist theory, 294, 303, 337
agency, 27, 83, 92
　automated, 79–81
Aldama, Fredrick, 6
*AlphaGo*, 15–6, 20, 27
alternate worlds, 33
*American Crime* (TV show), 313, 315
analogy problems, 56
android experience
　Cartesian dualism, 219–21
　Cartesian view, 209–10
　duplicate androids, 214–9
　the knowledge argument, 210–1, 219, 220
　learning, 208–10
　mind-body interaction, 219–21
　swamp-Dolores, 214–9
　*Westworld*, 207–21
　zombie-Dolores, 211–4
animacy, 87–8
animals, intrinsic value of, 123–4
animation, 83, 86–7, 89–91
*Annihilation* (film), 36, 40, 93, 139, 140, 141
Anthropocene, the, 9, 11
anthropomorphism, 84–5, 427
*Antichrist* (film), 377
Apple Inc., 312, 319
apps, 311–3, **314**, 315, **316**
artificial environments, fascination with, 93
artificial general intelligence, 145, 146, 147–8
artificial intelligence, 3–4, 4–5, 27, 50–2
　comparison with human intelligence, 15–29
　consciousness, 125
　containment, 146
　depictions of supremacy, 6
　film depictions of supremacy, 4–5
　goal-directed thinking, 26–8
　and human intelligence, 4
　language translation systems, 24
　limits, 68
　metacognition, 20, 22
　mutual simultaneous constraint satisfaction, 19–20
　potential, 146
　reality, 15–6
　risks of, 143
　self-directed, 28
　symbol-based, 50
　tools for thinking, 23, 24, 26
　transcendent, 5, 65–72
artificial life, 83, 88
associative poetics, 5

atemporality, 233
attention economy, 6
Austin, J.L., 161
automated agency, 79–81
automatic filmmaking, 36, 37–8
automation, 89, 90–1
autonomous behavior, 76

backpropagation, 109, **109**
back-propagation algorithm, 51
Bandela, Nelson, 325–6
Barad, Karen, 5
Barrett, Lisa Feldman, 181
Barthes, Roland, 334
Bayesian inference, 332–3, **333**
Bayesian networks, 51
Bazin, André, 379
Bechtel, William, 122
Beckman, Karen, 86
Beethoven, Ludwig van, 25
behavior, perception and, 7–8
behaviorism, 53–4
Benjamin, Walter, 84, 285
Bennett, Jane, 9, 92, 234
Berlant, Lauren, 373
Biederman, Irving, 351
big data, 5, 157–8
biopolitics, 157, 289
Black art, 341
Black bodies, 287
Black consciousness, 295, 306
Black death, 283
Black embodiment, 289, 293, 296
Black essentialism, 305–6
Black false consciousness, 290–1
Black genetic essentialism, 296
Black hapticality, 288–90
Black Interface, the, 294–6
Black lifeworlds, 285
Black Lives Matter, 283
*Black Mirror* "Nosedive" (TV show),
    151–2, 151–68, **152**, **162**, 433, 435
  aesthetic choices, 159–61
  closing image, 168
  color palate, 159–60, 175
  conceit of, 155–6
  denouement, 165–7
  dystopia, 159

economy of affect, 154–6
event perception, 173–8
gendered assumptions, 160–1
genre expectations, 178–80
opening scene, 151
plot, 152, 154
push notifications, 153
and racial difference, 161–4, **162**,
    181–2
rental condo scene, 174–80, 183
resolution, 167
score, 166, 178
social regulation, 156–8
speech acts, 160–1
worldmaking, 157–8
Black power movement, 321
Black programming, 316–7
Black protest, 283
Black representation, 283–4
Black sociality, 297
Black sonics, 283–306
  deployment, 284–5
  *Get Out*, 294–6, **295**, 299, 305
  interface, 285–90
  *Random Acts of Flyness*, 296–305, **298**,
    **301**, **303**, 305
  *Sorry to Bother You*, 290–4, **291**, **293**,
    296, 299, 304
  technics, 285–90
  visual register, 284
Black technology, 287–8
Black theory, 7
Black theory films, 10
Black trauma, 287
Blackness, 277, 284, 286, 289–90, 292,
    293–4, **293**, 299, 306, 321
black-tailed prairie dog, 70–1
*Blade Runner* (film), 45, 61
blaxploitation, 321
Bodomo, Nuotama Frances, 358
body
  and mind, 209–10
  in *Nymphomaniac*, 377
  in *Westworld*, 228–32, 234
body skills, 107–8
body-knowing, 97, 98, 99, 124–5
  dynamical systems approach, 109–10
  modelling, 104–10, **105**, **106**, **107**, **109**

Boehr theory, 33–4, 42
Bohm, David, 134, **135**
Bollmer, Grant, 88
Boltz, Marilyn, 241–2
Borges, Jorge Luis, 65, 66
Bostrom, Nick, 143
brain, the, 8–9
    as a computer, 405
    and dance, 397–8
    interpretive skills, 238–40
    media ecology, 415
    and music, 238–40
    neuroimaging research, 238–40
    overstimulated, 401–17
    re-wiring, 392–3
    structure, 422–3
    and taste, 434
Brain Initiative, 187–201
Branigan, Edward, 84
*Breaking Bad* (TV show), 323
Brooker, Charlie, 151
Bukatman, Scott, 86
Buñuel, Luis, 226

Cantor, Georg, 65
capital, 233–4
capitalism, power structures, 406–9
CAPTCHA, 48
Carpenter, Julie, 49
Cartesian dualism, 219–21
Cartesian intuition, 211
*Casablanca* Effect, 242
causal chains, 115–6
causes and effects, 130–1
Cavell, Stanley, 379, 380
cerebral model of neuropsychopathology, 412
*Changeling* (film), 86
chaos theory, 131
Chapman, Dale, 6
chatbots, 54
China, People's Republic of, Social Credit System, 155
Chion, Michel, 84, 90
Cholodenko, Alan, 83
Chomky, Noam, 53–4
Christiansen, Steen Ledet, 5
CLARITY, 427–9

classical inference, 332, **333**
classical sculptures, color, 348–9
Coates, Ta-Nehisi, 279, 312–3
code-switching, 338, 342
cognitive capitalism, 411–2
cognitive life, source of, 32
cognitive mapping, 225
cognitive processes, 171–84
    event perception, 171–8
    multisensory processes, 176–7
    now print function, 172–4
cognitive revolution, the, 53–4
cognitive robotics, 5
cognitive-affective approach, comic books, 187–201, **196, 197, 198, 199**
    analysis, 191, *192*, 193
    emotion triggers, 200, 201
    and emotions, 188, 191, 193, 193–5, **194, 195, 196,** 197–200, **197, 198, 199**
    materials, 190–1
    methods, 188–93
    narrative units, *192*, 193
    participants, 188–90, *190*
    Picture-Text Synchrony, 195
    procedure, 191
    question, 188
    questionnaire, 191
    reading frequency, 189
    results summary, 193–200, **194**
    the splash-page, 188, 195, **196,** 197–8, 200
Cohen, Annabel, 6, 10, 330
Collective Tacit Knowledge, 98–9
Collins, Harry, 97, 98–9, 112–4, 114, 116, 116–7, 120, 124, 125
Colomer, Sara Ferrando, 7–8
colonization, 276, 312
color, 347–61
    classical sculptures, 348–9
    and colorblindness, 350–1
    conventions, 354
    debates, 349–50
    in film, 357–61, **359**
    and health, 347, 355–6
    neuroscience of, 352–3
    problem of, 348–50
    in *Random Acts of Flyness*, 358–61, **359**

role of, 350–4
skin, 347, 352, 354–7
color consilience, 3
color conventions, 354
comic books, cognitive-affective approach, 187–201, **197**
 analysis, 191, *192*, 193
 emotion triggers, 200, 201
 and emotions, 188, 191, 193, 193–4, **194, 195, 196, 197, 198, 199**
 materials, 190–1
 methods, 188–93
 narrative units, *192*, 193
 participants, 188–90, *190*
 Picture-Text Synchrony, 195
 procedure, 191
 question, 188
 questionnaire, 191
 reading frequency, 189
 results summary, 193–200, **194**
 the splash-page, 188, 195, **196**, 197–8, 200
commodity culture, 233
complementary learning systems theory, 330, 331, 333–4, **334**
compulsion, 423
computational methods, 97
Congruence Associationist Model, 330
Congruence Associationist Model with Working Narrative, 258–64, **259**, 265
connectionist framework, 4
connectionist networks, 124–5
connectionist systems, 104–5, 108
 architecture of, 105, 105–8, **107**
 digital, 116–7
 dynamical systems approach, 109–10
 irreducibility, 110–1
consciousness, 6, 46, 59–61, 255
 definition, 46
 in *Ex Machina*', 5, 143–8
 future of, 124–5
 problem of, 46–7, 60
convergent evolution, 70
Conway, Bevil, 3, 7
Copjec, Joan, 230
COVID-19 pandemic, 2
Crawley, Ashon, 297, 299, 304

creativity, 97, 98–100, 109–10, 115, 258, 324
cruel optimism, 373
Cutting, James, 242
*Cybermedia*, backstory, 8–12, **10**
cybermedia, definition, 1

dance, 175–6, 278, 397–8, **397**
*Dancer in the Dark* (film), 376
De Assis, Leonardo, 5, 10
death drive, 181–2
Deep Blue, 67–8
Deep Learning, 51–2, 53, 67–8
DeepMind, 15, 144
Deleuze, Gilles, 156–7, 368, 402
Denning, Peter J., 58
depression, 404–5
Derrida, Jacques, 76
Descartes, Rene, 47, 50, 209
determinism, 5, 6, 11, 33–4, 98, 115, 130–2, 134, 135–6, 255
Deutsch, David, 34
*Devs*, 33, 34, 36, 40, 98
 determinism, 130–2, 134, 135–6
 music, 138–40
 quantum computers, 129–32, 132, 135–6, 141–2
 quantum theory, 132–5
 simulation of Christ on the cross, 132
 temporality, 140
diegetic awareness, 262–3
digital connectionist systems, 116–7
digital technologies, 2
digital vitalism, 367–86
 light and shadow, 381–4, **382**
 *Nymphomaniac*, 367–8, 369–70, 372–8, 381, 381–6
 political level, 367–8
 quasi-naturalism, 378–81, **380**
 *Under the Skin*, 367–8, 369, 370–6, 378–81, **380**, 381–6, **382**
 stylistic turn, 374–8, **376**
disciplinary society, 156–7
dissociative identity disorder, **394**
distributed representations, 110
Djawadi, Ramin, 237, 240, 251, 253, 254, **256–7**
Dolce, Lodovico, 348
dopamine, 180, 419–29

*D'oú Viens-tu ... Johnny?* (film), 224
double-consciousness, 275–6, 285–6, 291, 303, 390–1
Dretske, Fred, 216–8
Dreyfus, Hubert, 98, 99, 121–2, 124
drug addiction, 420, 425–6
Du Bois, W. E. B., 285, 285–8, 289, 304, 390–1
dual control systems, 113–4, 118–9
Dunning, Stefanie, 337
dynamical systems, 115–6
dynamical systems approach, 109–10
dystopian visions, 2

effortlessness, aesthetics of, 90
Einstein, Albert, 26
Eisenstein, Sergei, 89
elaboration, 112
Eliasmith, Chris, 56
ELIZA, 47, 48, 67–8
Ellison, Ralph, 284
Elman, Jeffrey, 110–1
embodied experiences, 1, 31
embodied knowing, 98
emergence, 123
emergent being, 99–100, 117–8
emergent intentionality, 122
emergent systems, 117–8, 120
emergent technologies, 93
Emerson, Jim, 384
emotional intelligence, 125
emotions, and music, 240, 252–3
empathy, 49, 144
Eno, Brian, 324
epiphylogenesis, 287, 293
Eshun, Kodwo, 284
ethics, and quantum physics, 5
event perception, 171–8
  and genre expectations, 178–80
event segmentation theory, 171–2, 184
event structure, 182–3
Everett, Hugh, 134
everyday life, financialization of, 153
evolution, 122, 122–3
*Ex Machina* (film), 1, 3–4, 15, 45–6, 125, 431, 433, 436
  attraction of, 61
  competing interpretations, 145–8

and consciousness, 5, 143–8
ending, 41, 68
on gender, 40
interior life question, 39
and learning, 52, 59–60
moral challenge, 146
music, 137–8, 139, 140–1
narrative, 36
opening scene, 143
qualia, 42–3
and self-directed thinking, 28
status, 143
structural forms, 38–9
the Turing Test, 37, 46, 47, 49, **50**, 144–8
writing, 33–4
expectations, building and assessing, 331–2
experiences, traumatic, 389, 389–91, **390**, 393–6, **394**, 396
explicability
  elaboration, 112
  string transformation, 112–3
explicit knowledge, 112, 120

Fanon, Frantz, 285–6, 295, 397
fatalism, 132
femaleness, 300–3, **301**
femininity, 367–8
feminization of labor, 159
Feynman, Richard, 43
*Fight Club* (film), 281, 322–3, 410
Figlerowicz, Marta, 7
film, 11, 37
  AI supremacy depictions, 4–5
  Black theory, 10
  color in, 357–61, **359**
  reading, 2–3
financial resources, science, 12
financialization, of everyday life, 153
Fincher, David, 281
Fisher, Janina, 413–4
Fitzgerald, Des, 401
*flâneur*, the, 234
Fodor, Jerry, 19, 23–4, 35
Foerde, Karin, 331
food, 433–4
  representation of, 431, 435–6

formal thinking ability, 23
Forthun, Eric, 341
Foucault, Michel, 156
Fram, Noah, 7, 10
Frankenstein, 66
free will, 6, 8, 11, 34, 255
freedom, 6, 97, 98–100, 109–10, 115, 122, 258, 299
Freud, Sigmund, 226
Friedberg, Anne, 232, 233–4
from-to intentional structure, 103
future states, 134

gamification, 180
Gao, Chuanji, 260
Garland, Alex, 1, 4, 5, 31–44
  automatic filmmaking, 36, 37–8
  determinism, 33–4
  on film, 37
  on gender, 40
  on narrative, 36
  structural forms, 38–9
  on thinking, 32
  understanding of science, 34–7
  use of music, 137–42
Gates, Henry Louis, Jr., 300
Gates, Rachel, 341
gaze, 234
Gegenfurtner, Karl R., 351
gender, 40, 42, 160–1
general intelligence, 145, 147
genetic essentialism, Black, 296
genocide, 312
genre, 329–30, 336
  determining, 330–1
  *Get Out*, 338–40, 341
  and *Random Acts of Flyness*, 340–1
  and *Sorry to Bother You*, 336–8
genre bridges, 329
genre expectations, 178–80
genre perception, and prior knowledge, 341–3
*Get Out* (film), 294–6, **295**, 299, 305, 329, 338–40, 341
Gibson, William, 65–6
Gillespie, Michael, 341
Glazer, Jonathan, 367–8, 378
goal-directed thinking, 26–8

God, 218
Gordon, Peter, 25
GPT-3, 22, 27
Granade, Andrew, 243–4, 252
Greenwood, Jonny, 79, 80, 81
Grusin, Richard, 86–7
Guattari, Felix, 157
Guimaraes, Julia Peres, 7–8
Gunning, Tom, 85, 88
Guy Fawkes masks, 407, **407**

habits, 420–1, 423–4
Hacking, Ian, 413, 414
hacking culture, 406–8, **407**
Hamm, Jon, 319, 360
Hardt, Michael, 153
Hardy, G. H., 71
Hari, Johann, 404–5
Hasantash, Maryam, 355
Hawking, Stephen, 143
healing, 396, 396–8, **397**, 412–4
Hegelianism, 295
Herzog, Amy, 370
hierarchical expectation, 331–2
Hinton, Geoff, 34–5
historical traumas, 7
Ho, Karen, 153
Hoeckner, Berthold, 242
*Home of the Brave* (film), 295
*homo economicus*, 152
horror films, 338–40
housing discrimination, 279
Human Brain Project, 187
human intelligence
  and artificial intelligence, 4
  augmenting role, 29
  comparison with artificial intelligence, 15–29
  extension of, 29
  goal-directed thinking, 26–8
  metacognition, 20–2, **21**
  mutual simultaneous constraint satisfaction, 16–20, **17**
  self-directed, 28
  tools for thinking, 22–6

Ifá, 280
imperialism, 295

independent thinkers, 28
indeterminism, 5
induction, 56
inexplicability, 110–1
inference, 332–3, **333**
infinity, 65
Information Age, 405
information sources, multiple simultaneous, 16–20, **17**
*Inglourious Basterds* (film), 383
innovation, 28, 99–100
*Insecure* (TV show), 316, 317
inspiration, 12
intelligence, interactionist approach, 49
intentionality, 97, 109–10, 115, 122, 122–3, 124
interconnected lives, 2
interdisciplinary studies, 8–10
interior life, 39–40
intuition, 26, 28, 108
irreducibility, 97, 99–100, 117–9
irreducibility, connectionist, 110–1

Jackson, Frank, 210–1
James, Robin, 157–8
Jameson, Fredric, 224, 225, 232, 233
joint comprehension, 100, **101**, **102**, 103–4, **104**, 105–6
*Joker* (film), 318
Jones, Quincy, 275
Joy, Lisa, 258, 265
joy, sense of, 8
Ju, Ginny, 351
Jungian perspective, 322–4
*Jurassic Park* (film), 384

Kaller, Christopher, 241
Kanerva, Pentti, 56
Kant, Immanuel, 82
Kara, Selim, 33
Karlin, Fred, 237, 241, 243–4, 244–50, 245–7, **248**, **250**
kinesthetic index, 88
kinesthetics, 259
knowledge, fully explicable, 111–20
knowledge argument, the, 210–1, 219, 220
Kotsko, Adam, 323–4
Kuhn, Thomas, 35

labor, feminization of, 159
Lacan, Jacques, 232
Lafer-Sousa, Rosa, 355
Lakoff, Robin, 160–1
Landsberg, Alison, 339
Langton, Christopher G., 83
language, 9, 54–9, 98, 426
  animal, 70–1
  connectionist systems, 105–6
  private, 55–6
  sensation words, 57–9
language acquisition, 53–4
language games, 48, 60
language translation systems, 24
Laplace, Pierre-Simon, 130–1
Laplace's demon, 131
Lash, Scott, 88, 90
late capitalism, 224, 232–4
Latour, Bruno, 92
Leal, Jonathan, 6, 33
learning, 208–10, 428
  layers in, 107–8, **107**
  and machine intelligence, 52–4
  reinforcement, 53
  skills, 108–9
  supervised, 53
  through experiences, 59–60
  tools for, 22–6
Lee, Benjamin, 158
Lee, Hojoon, 7–8, 11, 431–6
Lee, Spike, 280
Lem, Stanislaw, 66–7
Lerdahl, Fred, 330
Lerner, Talia, 7–8, 11, 419–29
Levy, Simon, 4–5, 32
life
  artificial, 83, 88
  definition, 82–3
  erasure of, 89–90
  imitation of, 91–3
  and movement, 81–5
  simulating, 78–9
likes, pursuit of, 151
linear regression plots, 9, **10**
LiPuma, Edward, 158
liveness, 86–8
Lockhart, Annalise, 321
Loebner Prize, the, 48

*The Lord of the Rings: Two Towers* (film), 78–9
Lowney, Charles, 4–5, 5
Lyon, Eric, 7

McClelland, James, 330, 331
McClelland, Jay, 3–4, 4, 32
McGarvey, Seamus, 159
McGlotten, Shaka, 299
machine learning, 5, 52–4
McKittrick, Katherine, 7
Malabou, Catherine, 412–3
Malcolm, Norman, 60
Malick, Terrence, 383–4
Malin, Brenton, 415
Manovich, Lev, 86
mantras, 274
many-worlds interpretation, 133–5, **134**, **135**
Marshall, Kingsley, 251, 254
Marx, Karl, 224
masculinity, destructive, 278
Mason, Zachary, 5
Massive technology, 75, 76, 77–9, 82, 88, 89–90, 91, 92–3
materialism, 213
mechanical reproduction, 84
mechanization, 231
media studies, 11
media-studies, 3
memory, 18–9, 287, 293, 330, 389
   for music, 241–2
   now print function, 172–4
   and prior knowledge, 334–6, **335**, **336**
   processing, 392–3
   retrieval, 394–6, **394**, 396–7
   superpositional, 27
   suppressed, 395
   transgenerational, 391–2
memory disorders, 173
memory formation, 6
memory palace, the, 3
mental disorders, 7–8
   healing, 412–3
mental events, 219–20
mental lives, 6
mental states, 47, 60, 390
Merleau-Ponty, Maurice, 98, 122

metacognition, 20–2, **21**
Metz, Christian, 259
Michelangelo, 25
mind, 9, 211, 221
mind-body interaction, 209, 219–21
mindscapes, 403–4
Minz, Christopher, 6, 11
miraculous duplicates, 216–9
mirror-neuron system, 239
mobilized virtual gaze, 234
Moore's Law, 62
moral challenge, 146
*Morgan* (film), 46, 61
Morricone, Ennio, 243
Moten, Fred, 284, 285, 288–90
motivic structures, 182–3
movement, 76, 81–5, 86, 88
*Mr. Robot* (TV show), 7, 12, 54, 389–90, 393–6, **394**
   attraction, 406
   audience experience, 402
   central characters, 403
   depictions of hacking culture, 406–8, **407**
   eps.1.0_hellofriend.mov, 402–4
   eps1.3_da3mOns.mp4, 404–6
   eps3.4_runtime-error./nlr00, 408–9
   experience of, 424
   hallucinations, 404
   and healing, 412–4
   internal monologue, 403, 404
   life in the Neuropolis, 402–6
   media ecology, 414–5
   mindscapes, 403–4
   the Neuropolis, 401–17
   neuroscience of, 419–29
   psychic realities, 409–12
   reference to contemporary politics, 408–9
   season one last episode, 415–7
   soundtrack, 405
multiple memory systems (MMS) theory, 331
multiple realizability, 125
multiple selves, 12
multisensory modalities, 8
multisensory processes, 176–7
Mulvey, Laura, 11, 181–2

Münsterberg, Hugo, 242
music,
music and musical associations, 10, 25–6, 80, 242
  *Black Mirror* "Nosedive", 166
  and brain activity, 238–40
  configural structure, 260–1
  Congruence Associationist Model with Working Narrative, 258–64, **259**, 265
  consonance and dissonance, 239–40
  cues, 244, *245–7*, 247
  *Devs*, 138–40, 141–2
  diegetic awareness, 262–3
  and emotions, 240, 252–3
  engagement with, 238
  *Ex Machina*, 137–8, 139, 140–1
  film resources, 240–1
  film techniques, 240–3
  Garland's use of, 137–42
  individual knowledge, 251–2
  *leitmotivs*, 241–2
  loudness pattern, 249
  memory for, 241–2
  neuroimaging research, 238–40
  non-diegetic music, 253–5
  organic versus electronic, 140–1
  palette, 139–40
  pitch direction, 238–9
  power of, 253
  primary goals, 237–8
  *Random Acts of Flyness*, 278–9, 325–6
  role, 249
  segmented, 175–6
  the sonic episteme, 157–8
  unconscious inference, 261–2
  and viewer experience, 258–64, **259**, 265
  visual structure, 260–1
  *Westworld*, 231–2, 237–65
  *Westworld* (film), 243, 244–50, *245–7*, **248**, **250**
musical embodiment, 239
Musk, Elon, 143
mutual simultaneous constraint satisfaction, 16–20, **17**
mythmaking, 224

Nagel, Thomas, 123
Nance, Terence, 1, 3, 7, 11, 273–81, 296, 305, 311, 313, 315, 317–8, 318, 320–1, 322, 324, 326–7, 341, 358–9, 393
nature, and nurture, 50–2
Negri, Antonio, 153
neoliberalism, 153, 155, 164
neural network theory, 104–5
neural networks, 4, 97, 113, 145
  artificial, 19
  perceptual systems as, 17–8
  usefulness of, 24
neuro-image, the, 403–4, 406
neuropathologies, 7–8
neuroplasticity, 389–99
  definition, 392
  and healing, 396, 396–8, **397**
  and self-identity, 391–3
  and traumatic experiences, 389, 389–91, **390**, 393–6, **394**, 396
Neuropolis, the, 401–17
  and depression, 404–5
  and healing, 412–4
  life in, 402–6
  media ecology, 414–5
  mindscapes, 403–4
  power structures, 406–9
  psychic realities, 409–12
neuroscience, 7–8
new wounded, the, 412–3
*The New York Times* (newspaper), 3
Newark, 164
Newell, Allan, 23
Ngai, Sianne, 86
*Night of the Living Dead* (film), 212, 213
Nolan, Jonathan, 241, 252, 258, 265
non-diegetic music, 253–5
non-physical knowledge, 209–10, 210–1
Nousiopoulos, Elias, 358
now print function, 172–4
number systems, 25
nurture, and nature, 50–2
Nusbaum, Howard, 242
*Nymphomaniac* (film)
  anxiety about reproduction, 373–5
  and the body, 377
  cruel optimism, 373
  feminist tropes, 374

light and shadow, 381–4
narrative arc, 369–70, 372
ontological undercurrent, 368
political level, 367–8
quasi-naturalism, 381
sexual practices, 369–70
stylistic turn, 374–8, **376**

octopuses, 69–70
Oliva, Aude, 351
*Once Upon a Time in the West* (film), 224, 227
operant conditioning, 115–6
optogenetics, 427–8
Osborn, Brad, 75–6

Paley, William, 218
Paley Syndrome, the, 218
Parallel Distributed Processing, 18, 51
*Passengers* (film), 93
pattern recognition, 104–5
PDP Books, the, 51
Pearl, Judea, 47–8
pedagogy, 276
Peeple, 155
perception, 16–7, **17**
  behavior and, 7–8
  constraints influencing, 18
perceptual skills, 103
perceptual systems, as neural networks, 17–8
Peter Pan, 278
Peters, John Durham, 414
Peters, Shawn, 276
Peterson, Christopher, 373
Pettman, Dominic, 76–7, 82, 83, 91
Phillips, Todd, 318
philosophical zombies, 212–3
philosophy, 5
pilot wave interpretation, 134
Pinocchio, 66
Pisters, Patricia, 7
Plato, 50, 382
Poikonen, Hanna, 398
Poincaré, Henri, 26, 131
Polanyi, Michael, 97, 99, 100, 103, 104–5, 105–6, 108, 111, 115, 119, 120, 120–1, 121–2, 122–3
Poldrack, Russell, 331

police, militarized, 283
Pompey, Damani, 278
popular culture, representation in, 316–7
popular press outlets, 3
Portishead, 140
posthuman, the, 91
postmodern society, 233
Postmodernism, 232
post-traumatic stress disorder, 173, 412, 413–4
predictability, 131
prediction, 183–4
prediction error, 172
prior knowledge, 329–43
  contextual information, 342–3
  feedback loops, 335
  genre bridges, 329
  and genre perception, 341–3
  and *Get Out*, 338–40, 341
  hierarchical expectation, 331–2
  and inference, 332–3, **333**
  and memory, 334–6, **335**, **336**
  and *Random Acts of Flyness*, 340–1
  and *Sorry to Bother You*, 336–8, 341, 342
  theoretical background, 330–1
propositional knowing, 98
propositional logic, 23–4
proto-structures, 121–2
psychoanalysis, resistance to, 412–3
public engagement, with science, 1
Pylyshyn, Zenon W, 23–4

qualia, 42–3, 221
quantum computers, 129–36
  *Devs*, 129–32, 132, 135–6, 141–2
quantum mechanics, 34, 43
quantum physics, 5, 10, 34, 40
quantum theory, 132–5, **134**, **135**

race and racism, 7, 10, 161–4, **162**, 181–2
race relations, 313, 315
racial capitalism, 293
racial difference, 161–4, **162**, 181–2
Radiohead, 5, 75–6, 76–7, 231–2
  "Go to Sleep", 75, 76, **77**, **78**, 79–81, 82, 83–4, 85, **87**, 88, 89, 90–1, 91–3
Rae, Issa, 316–7
Ramanujam, Srinivasa, 71

Ramos-Chapman, Naima, 278
Ramos-Zayas, Ana Y., 156, 164
Rancière, Jacques, 380–1
*Random Acts of Flyness* (TV show), 1, 3, 7, 273–81, 305, 306, 329
 "The Apology", 319
 audience, 318–9
 "Bitch Better Have My Money", 279, 311–3, **314**, 315, **316**, 318
 Black sonics, 296–305, **298**, **301**, **303**
 color choices, 358–61, **359**
 and double-consciousness, 275–6
 episode 2, 300–4, **301**, **303**, 392–3, 397–8, **397**
 episode 5, 322–4, 326
 film editing, 277
 final sketch, 325
 as a form of propaganda, 327
 frame shifts, 321–2
 and genre, 340–1
 interviews, 325
 Jungian perspective, 322–4
 last scenes, 304–5
 lullaby sleep sequence, 197–8, 274, 280, **298**
 music, 278–9, 325–6
 Nuncaland scene, 278, 303–5, **304**, 397–8, **397**
 political agenda, 313
 and prior knowledge, 340–1
 scenius, 324
 season 2, 281
 "The Sexual Proclivities of the Black Community", 321
 sleep in, 273–4
 sound in, 325–6
 structure, 317–8
 technological fluency, 311–27
 technologies, 319–22, 326–7
 thematic areas, 318–9
 and trauma, 390–1, **390**
 use of repetition, 274–5
 "White Be Gone", 319–21, **320**, 360
Rasmussen, Daniel, 56
rating systems
 social media, 151–2, 153
 social regulation, 156–8
Raven's Progressive Matrices task, 56, **57**

Ravetto, Kriss, 407
reason, ability to, 98
recurrent connections, 53
Reich, Elizabeth, 7, 10, 340, 390
reinforcement learning, 53
relational tacit knowledge, 112
religiosity, 138
reparations, 312–3
representation, in popular culture, 316–7
representations, 330
respect, 62
Restivo, Angelo, 227–8, 228
reward-seeking behavior, 420
Rieger, Jochem, 351
Riley, Boots, 292
risk, social abstraction of, 158
robots, 2
Rodowick, D. N., 384
Rolston, Holmes, III, 124
Romney, Jonathan, 379
Rosa, Arash, 355
Rose, Nicholas, 401
rules, 111
 following, 56
Rumelhart, David, 18, 19, 32, 34–5
Russell, Bertrand, 23
Rutterford, Alex, 75, 91
Ryle, Gilbert, 98

Said, Edward, 385, 386
Salisbury, Ben, 11, 137–42
Sameroff, Arnold, 330, 331, 336
sampling, 340
scenius, 324
Schrödinger's equation, 133, 134
Schwartz, Louis-Georges, 81
Schyns, Philippe G., 351
science
 arrogance, 35
 financial resources, 12
 Garland's understanding of, 34–7
 as a narrative container, 36
 popular press outlets, 3
 public engagement with, 1
 and truth, 36
science fiction, 28, 135
scientific explanation, limits of, 119–20
scientific process, the, 35–6

scientism, 62
Sconce, Jeffrey, 405, 414
*Searchers, The* (film), 227
self-care, 273
self-driving cars, 99
self-identity, 391–3
sensory experience, 85
sensuality, 436
Shanahan, Murray, 1, 5, 46
Sharpe, Christina, 285
Shaviro, Steven, 76–7, 321–2
Simon, Herbert, 23
Singh, Ilina, 401
Singularity, the, 62
Sinha, Pawan, 351
skill learning, 97
skin color, 347, 352, 354–7
Skokowski, Paul, 6, 31, 32
slavery, 279, 280, 295, 352
sleep, 273–4
smartphones, 173–4
smell, 432–3
Smith, Grag, 181
Smith, Jeff, 263–4
Smolensky, Paul, 111
*Snowpiercer* (film), 93
Sobchack, Vivian, 89–91, 92
social abstraction, of risk, 158
Social Credit System, People's Republic of China, 155
social media, 2
  rating systems, 151–2, 153
social regulation, 156–8
society of control, the, 156–8, 165
software cinema, 77, 83–4, 92–3
Solaris, 66–7
somatic processes, 107–8
somatic tacit knowledge, 99, 112, 115–6
sonic agencies, 79–81
sonic episteme, the, 157–8
sonic unintelligibility, 300
*Sorry to Bother You* (film), 290–4, **291**, **293**, 296, 299, 304, 329
  and prior knowledge, 336–8, 341, 342
soul, 209, 211
sovereignty, 277
Sparse Distributed Memory, 57–9
spatial attention, 184

spatial learning, 422
speech acts, 160–1
Srnicek, Nick, 385
Standing, Guy, 159
Stiegler, Bernard, 285, 286–7, 293
string transformation, 112–3
suffering, 147–8
Sun Ra, 288–9
superintelligence, 131–2
superpositional memory, 27
supervised learning, 53
Sustained Attention to Response Task, 184
symbol-based, artificial intelligence, 50
symbolic reasoning, 24
synchresis, 84
systematic mental processes, 24–6
systems training, 108–9

tacit knowing, 97, 99–100, 100, **101**, **102**, 103–4, **104**, 110, 120–1
  architecture of, 105, 105–8, **107**
  intentionality, 122–3
  irreducibility, 117–9
  machines, 111–20
  proto-structures, 121–2
tacit knowledge, 100, **101**, **102**, 103–4, **104**, 111, 112–4, 120
Tarantino, Quentin, 383
Tarde, Gabriel, 92, 93
taste, sense of, 11, 431–6
Taylor, Charles, 121–2
technical delusion, the, 405
technical invention, 108
technics, 286
telic fields, 122–3
*Terminator, The* (film), 61
Terranova, Tiziana, 415
Thacker, Eugene, 82
theology, 65
theoretical physics, 5
Therriault, David J., 351–2
thinking, 32
  goal-directed, 26–8
  self-directed, 28
  tools for, 22–6
time, **130**
  experiences of, 178
*Toy Story* (film), 86

transcendent AI, 5, 65–72
trauma, 389–91, **390**, 413–4
*Tree of Life* (film), 383–4
Trigg, Dylan, 226, 229
Trump, Donald, 408–9
truth, and science, 36
Turing, Alan, 47, 48, 49, 144
Turing machines, 71–2
Turing Test, the, 37, 40–1, 46, 47–9, **50**, 59, 116–7, 125, 144–8

unconscious, the, 38–9
unconscious inference, 261–2
*Under the Skin* (film), 3–4, 369, 431, 433, 435–6
   aimlessness, 378
   anxiety about reproduction, 373–5
   cruel optimism, 373
   drowning figures, 379, *380*
   feminist tropes, 374
   light and shadow, 381–4, **382**
   narrative arc, 370–2
   ontological undercurrent, 368
   opening shot, 379
   political level, 367–8
   quasi-naturalism, 378–81, **380**
   soundtrack, 378
   stylistic turn, 374–5
*Us* (film), 305–6
U.S. Department of Defense, 51

Vasari, Giorgio, 348
Vector Symbolic Architectures, 55–6, **57**, 59
Verlaine, Paul, 252
Vernallis, Carol, 6, 8–12, 33, 80, 85, 91
viewer experience, Congruence Associationist Model with Working Narrative, 258–64, **259**, 265
virtual worlds, 76
vision, privileging of, 436
visual theory, 5
Von Neumann-Wigner interpretation, 134
von Trier, Lars, 367–8, **376**

Wagner, Laura, 6
wakefulness, 7, 285
*Washington Post* (newspaper), 3
wave function, 133, 133–5, **135**

wealth gap, 279
Weheliye, Alexander, 284
Weizenbaum, Joseph, 47
Wells, Ida B., 100, **101**, **102**, 103
*Westworld* (film), 237, 242, 258
   dormant worlds music, 249, **250**
   loudness pattern, 249
   music, 243–4, 244–50, *245–7*, **248**, **250**
   music cues, 244, *245–7*, 247
*Westworld* (TV show), 6, 10, 68–9
   aberrant eye motif, 226–7
   android experience, 207–21
   android glitchings, 11
   "Back to Black", 252, 262
   bodies, 228–32, 234
   Cartesian dualism, 219–21
   Cartesian view, 209–10
   Congruence Associationist Model with Working Narrative, 258–64, **259**, 265
   destabilization of historical myths, 223
   diegetic awareness, 262–3
   disembodied voices, 230–1
   Dolores, 207–21
   Dolores's poetic laments, 230
   duplicate androids, 214–9
   first season's final moments, 234
   fly motif, 227–8
   glitches, 227
   guests, 223
   horror, 226–7
   the knowledge argument, 210–1, 219, 220
   and late capitalism, 232–4
   the Man in Black, 229–30, 233
   maze, 230
   mirroring process, 224–5
   music, 231–2, 237–65
   music control of android behavior, 252–3, 255
   music conventions, 243–4
   music sources, 251
   non-diegetic music, 253–5
   opening credits, 255–8, **256–7**
   philosophy of, 207–21
   piano treatment, 251, 252, 253, 254–5, 257–8, 262–3, 264
   premise, 211

rebellion, 234
robot cognition, 228–9
and self, 223–34
surgical scenes, 228–9
swamp-Dolores, 214–9
temp track, 241
temporality, 233
theme music, 240
vision of the West, 224
zombie-Dolores, 211–4
Weta Digital, 77–8
White, Rob, 377
white devils, 323–4
white supremacy, 276
white voice, 292–4, 300

whiteness, 300, 312, 319–21, **320**
Williams, Alex, 385
Winckelmann, Johann Joachim, 348
Wittgenstein, Ludwig, 45, 46–7, 48, 54–5, 55, 56, 57, 60, 62, 426
*Wizard of Oz* (film), 357
woke, 274, 299, 319
Wright, Joe, 159

Yip, Andrew W., 351

Zacks, Jeff, 6, 10, 11, 32, 171–84
Zheng, Jie, 396
Žižek, Slavoj, 230
zombies, 211–4, 221